Functionally Graded Materials

Nonlinear Analysis of Plates and Shells

Functionally Graded Materials

Nonlinear Analysis of Plates and Shells

Hui-Shen Shen

CRC Press
Taylor & Francis Group
Boca Raton London New York

CRC Press is an imprint of the
Taylor & Francis Group, an **informa** business

CRC Press
Taylor & Francis Group
6000 Broken Sound Parkway NW, Suite 300
Boca Raton, FL 33487-2742

First issued in paperback 2019

© 2009 by Taylor & Francis Group, LLC
CRC Press is an imprint of Taylor & Francis Group, an Informa business

No claim to original U.S. Government works

ISBN-13: 978-1-4200-9256-1 (hbk)
ISBN-13: 978-0-367-38601-6 (pbk)

Library of Congress Cataloging-in-Publication Data

Shen, Hui-Shen.
    Functionally graded materials : nonlinear analysis of plates and shells /
Hui-Shen Shen.
        p. cm.
    Includes bibliographical references and index.
    ISBN 978-1-4200-9256-1 (alk. paper)
    1. Functionally gradient materials. 2. Shells (Engineering)--Thermal properties.
3. Plates (Engineering)--Thermal properties. I. Title.

TA418.9.F85S54 2009
624.1'776--dc22   2008036386

Visit the Taylor & Francis Web site at
http://www.taylorandfrancis.com

and the CRC Press Web site at
http://www.crcpress.com

Contents

Preface

With the development of new industries and modern processes, many structures serve in thermal environments, resulting in a new class of composite materials called functionally graded materials (FGMs). FGMs were initially designed as thermal barrier materials for aerospace structural applications and fusion reactors. They are now developed for general use as structural components in extremely high-temperature environments. The ability to predict the response of FGM plates and shells when subjected to thermal and mechanical loads is of prime interest to structural analysis. In fact, many structures are subjected to high levels of load that may result in nonlinear load–deflection relationships due to large deformations. One of the important problems deserving special attention is the study of their nonlinear response to large deflection, postbuckling, and nonlinear vibration.

This book consists of five chapters. The chapter and section titles are significant indicators of the content matter. Each chapter contains adequate introductory material to enable engineering graduates who are familiar with the basic understanding of plates and shells to follow the text. The modeling of FGMs and structures is introduced and the derivation of the governing equations of FGM plates in the von Kármán sense is presented in Chapter 1. In Chapter 2, the geometrically nonlinear bending of FGM plates due to transverse static loads or heat conduction is presented. Chapter 3 furnishes a detailed treatment of the postbuckling problems of FGM plates subjected to thermal, electrical, and mechanical loads. Chapter 4 deals with the nonlinear vibration of FGM plates with or without piezoelectric actuators. Finally, Chapter 5 presents postbuckling solutions for FGM cylindrical shells under various loading conditions. Most of the solutions presented in these chapters are the results of investigations conducted by the author and his collaborators since 2001. The results presented herein may be treated as a benchmark for checking the validity and accuracy of other numerical solutions.

Despite a number of existing texts on the theory and analysis of plates and/or shells, there is not a single book that is devoted entirely to the geometrically nonlinear problems of inhomogeneous isotropic and functionally graded plates and shells. It is hoped that this book will fill the gap to some extent and be used as a valuable reference source for postgraduate students, engineers, scientists, and applied mathematicians in this field.

I wish to record my appreciation to the National Natural Science Foundation of China (grant nos. 59975058 and 50375091) for partially funding this work, and I also wish to thank my wife for her encouragement and forbearance.

Hui-Shen Shen

Author

Hui-Shen Shen is a professor of applied mechanics at Shanghai Jiao Tong University. He graduated from Tsinghua University in 1970, and received his MSc in solid mechanics and his PhD in structural mechanics from Shanghai Jiao Tong University in 1982 and 1986, respectively. In 1991–1992, he was a visiting research fellow at the University of Wales (Cardiff) and the University of Liverpool in the United Kingdom. Dr. Shen became a full professor of applied mechanics at Shanghai Jiao Tong University at the end of 1992. In 1995, he was invited again as a visiting professor at the University of Cardiff and in 1998–1999, as a visiting research fellow at the Hong Kong Polytechnic University, and in 2002–2003 as a visiting professor at the City University of Hong Kong. Also in 2002, he was a Tan Chin Tuan exchange fellow at the Nanyang Technological University in Singapore and in 2004, he was a Japan Society for the Promotion of Science (JSPS) invitation fellow at the Shizuoka University in Japan. In 2007, Dr. Shen was a visiting professor at the University of Western Sydney in Australia.

Dr. Shen's research interests include stability theory and, in general, nonlinear response of plate and shell structures. He has published over 190 journal papers, of which 123 are international journal papers. His research publications have been widely cited in the areas of computational mechanics and structural engineering (more than 1500 times by papers published in 387 international archival journals, and 220 local journals, excluding self-citations). He is the coauthor of the books *Buckling of Structures* (with T.-Y. Chen) and *Postbuckling Behavior of Plates and Shells*. He won the second Science and Technology Progress Awards of Shanghai in 1998 and 2003, respectively. Currently, Dr. Shen serves on the editorial boards of the journal *Applied Mathematics and Mechanics* (ISSN: 0253-4827) and the *International Journal of Structural Stability and Dynamics* (ISSN: 0219-4554). He is a member of the American Society of Civil Engineers.

1

Modeling of Functionally Graded Materials and Structures

1.1 Introduction

The most lightweight composite materials with high strength/weight and stiffness/weight ratios have been used successfully in aircraft industry and other engineering applications. However, the traditional composite material is incapable to employ under the high-temperature environments. In general, the metals have been used in engineering field for many years on account of their excellent strength and toughness. In the high-temperature condition, the strength of the metal is reduced similar to the traditional composite material. The ceramic materials have excellent characteristics in heat resistance. However, the applications of ceramic are usually limited due to their low toughness.

Recently, a new class of composite materials known as functionally graded materials (FGMs) has drawn considerable attention. A typical FGM, with a high bending–stretching coupling effect, is an inhomogeneous composite made from different phases of material constituents (usually ceramic and metal). An example of such material is shown in Figure 1.1 (Yin et al. 2004) where spherical or nearly spherical particles are embedded within an isotropic matrix. Within FGMs the different microstructural phases have different functions, and the overall FGMs attain the multistructural status from their property gradation. By gradually varying the volume fraction of constituent materials, their material properties exhibit a smooth and continuous change from one surface to another, thus eliminating interface problems and mitigating thermal stress concentrations. This is due to the fact that the ceramic constituents of FGMs are able to withstand high-temperature environments due to their better thermal resistance characteristics, while the metal constituents provide stronger mechanical performance and reduce the possibility of catastrophic fracture.

The term FGMs was originated in the mid-1980s by a group of scientists in Japan (Yamanoushi et al. 1990, Koizumi 1993). Since then, an effort to develop high-resistant materials using FGMs had been continued. FGMs were initially designed as thermal barrier materials for aerospace structures and fusion reactors (Hirai and Chen 1999, Chan 2001, Uemura 2003). They

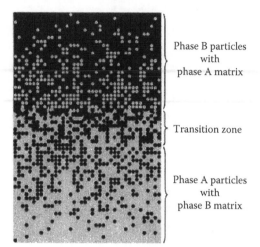

Phase B particles
with
phase A matrix

Transition zone

Phase A particles
with
phase B matrix

FIGURE 1.1
An FGM with the volume fractions of constituent phases graded in one (vertical) direction. (From Yin, H.M., Sun, L.Z., and Paulino, G.H., *Acta Mater.*, 52, 3535, 2004. With permission.)

are now developed for the general use as structural components in high-temperature environments. An example is FGM thin-walled rotating blades as shown in Figure 1.2 (Librescu and Song 2005). Potential applications of FGM are both diverse and numerous. Applications of FGMs have recently been reported in the open literature, e.g., FGM sensors (Müller et al. 2003) and actuators (Qiu et al. 2003), FGM metal/ceramic armor (Liu et al. 2003), FGM photodetectors (Paszkiewicz et al. 2008), and FGM dental implant (Watari et al. 2004, see Figure 1.3). A number of reviews dealing with various

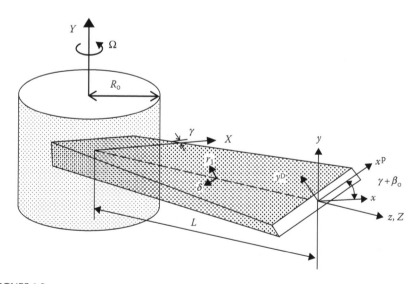

FIGURE 1.2
An FGM thin-walled tapered pretwisted turbine blade. (From Librescu, L. and Song, S.-Y., *J. Therm. Stresses*, 28, 649, 2005. With permission.)

2 mm

FIGURE 1.3
Ti/20HAP FGM dental implant. External appearance (left) and cross-section (right). (From Watari, F., Yokoyama, A., Omori, M., Hirai, T., Kondo, H., Uo, M., and Kawasaki, T., *Compos. Sci. Technol.*, 64, 893, 2004. With permission.)

aspects of FGMs have been published in the past few decades (Fuchiyama and Noda 1995, Markworth et al. 1995, Tanigawa 1995, Noda 1999, Paulino et al. 2003). They show that most of early research studies in FGMs had more focused on thermal stress analysis and fracture mechanics. A comprehensive survey for bending, buckling, and vibration analysis of plate and shell structures made of FGMs was presented by Shen (2004). Recently, Birman and Byrd (2007) presented a review of the principal developments in FGMs that includes heat transfer issues, stress, stability and dynamic analyses, testing, manufacturing and design, applications, and fracture.

1.2 Effective Material Properties of FGMs

Several FGMs are manufactured by two phases of materials with different properties. A detailed description of actual graded microstructures is usually not available, except perhaps for information on volume fraction distribution. Since the volume fraction of each phase gradually varies in the gradation direction, the effective properties of FGMs change along this direction. Therefore, we have two possible approaches to model FGMs. For the first choice, a piecewise variation of the volume fraction of ceramic or metal is assumed, and the FGM is taken to be layered with the same volume fraction in each region, i.e., quasihomogeneous ceramic–metal layers (Figure 1.4a). For the second choice, a continuous variation of the volume fraction of ceramic or metal is assumed (Figure 1.4b), and the metal volume fraction can be represented as the following function of the thickness coordinate Z.

$$V_{\mathrm{m}} = \left(\frac{2Z + h}{2h}\right)^{N} \tag{1.1}$$

where h is the thickness of the structure, and N ($0 \leq N \leq \infty$) is a volume fraction exponent, which dictates the material variation profile through the

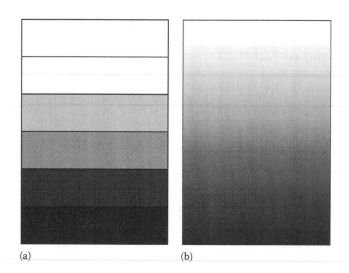

(a)   (b)

FIGURE 1.4
Analytical model for an FGM layer.

FGM layer thickness. As is presented in Figure 1.5, changing the value of N generates an infinite number of composition distributions.

In order to accurately model the material properties of FGMs, the properties must be temperature- and position-dependent. This is achieved by using

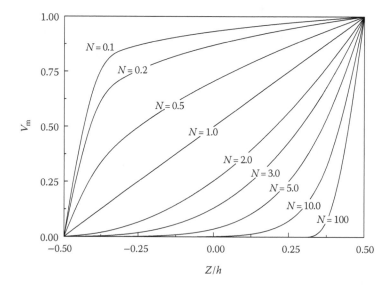

FIGURE 1.5
Volume fraction of metal along the thickness.

a simple rule of mixture of composite materials (Voigt model). The effective material properties P_f of the FGM layer, like Young's modulus E_f, and thermal expansion coefficient α_f, can then be expressed as

$$P_f = \sum_{j=1} P_j V_{f_j} \tag{1.2}$$

where P_j and V_{f_j} are the material properties and volume fraction of the constituent material j, and the sum of the volume fractions of all the constituent materials makes 1, i.e.,

$$\sum_{j=1} V_{f_j} = 1 \tag{1.3}$$

Since functionally graded structures are most commonly used in high-temperature environment where significant changes in mechanical properties of the constituent materials are to be expected (Reddy and Chin 1998), it is essential to take into consideration this temperature-dependency for accurate prediction of the mechanical response. Thus, the effective Young's modulus E_f, Poisson's ratio ν_f, thermal expansion coefficient α_f, and thermal conductivity κ_f are assumed to be temperature dependent and can be expressed as a nonlinear function of temperature (Touloukian 1967):

$$P_j = P_0(P_{-1}T^{-1} + 1 + P_1 T + P_2 T^2 + P_3 T^3) \tag{1.4}$$

where P_0, P_{-1}, P_1, P_2, and P_3 are the coefficients of temperature T (in K) and are unique to the constituent materials. Typical values for Young's modulus E_f (in Pa), Poisson's ratio ν_f, thermal expansion coefficient α_f (in K^{-1}), and the thermal conductivity κ_f (in W mK^{-1}) of ceramics and metals are listed in Tables 1.1 through 1.4 (from Reddy and Chin 1998). From Equations 1.1 through 1.3, one has (Gibson et al. 1995):

$$E_f(Z,T) = [E_m(T) - E_c(T)]\left(\frac{2Z+h}{2h}\right)^N + E_c(T) \tag{1.5a}$$

TABLE 1.1

Modulus of Elasticity of Ceramics and Metals in Pa for E_f

Materials	P_0	P_{-1}	P_1	P_2	P_3
Zirconia	244.27e+9	0	−1.371e−3	1.214e−6	−3.681e−10
Aluminum oxide	349.55e+9	0	−3.853e−4	4.027e−7	−1.673e−10
Silicon nitride	348.43e+9	0	−3.070e−4	2.160e−7	−8.946e−11
Ti-6Al-4V	122.56e+9	0	−4.586e−4	0	0
Stainless steel	201.04e+9	0	3.079e−4	−6.534e−7	0
Nickel	223.95e+9	0	−2.794e−4	−3.998e−9	0

Source: Reddy, J.N. and Chin, C.D., *J. Therm. Stresses*, 21, 593, 1998. With permission.

TABLE 1.2

Coefficient of Thermal Expansion of Ceramics and Metals in K^{-1} for α_f

Materials	P_0	P_{-1}	P_1	P_2	P_3
Zirconia	12.766e − 6	0	−1.491e − 3	1.006e − 5	−6.778e − 11
Aluminum oxide	6.8269e − 6	0	1.838e − 4	0	0
Silicon nitride	5.8723e − 6	0	9.095e − 4	0	0
Ti-6Al-4V	7.5788e − 6	0	6.638e − 4	−3.147e − 6	0
Stainless steel	12.330e − 6	0	8.086e − 4	0	0
Nickel	9.9209e − 6	0	8.705e − 4	0	0

Source: Reddy, J.N. and Chin, C.D., *J. Therm. Stresses*, 21, 593, 1998. With permission.

TABLE 1.3

Thermal Conductivity of Ceramics and Metals in W mK^{-1} for κ_f

Materials	P_0	P_{-1}	P_1	P_2	P_3
Zirconia	1.7000	0	1.276e − 4	6.648e − 8	0
Aluminum oxide	−14.087	−1123.6	−6.227e − 3	0	0
Silicon nitride	13.723	0	−1.032e − 3	5.466e − 7	−7.876e − 11
Ti-6Al-4V	1.0000	0	1.704e − 2	0	0
Stainless steel	15.379	0	−1.264e − 3	2.092e − 6	−7.223e − 10
Nickel[a]	187.66	0	−2.869e − 3	4.005e − 6	−1.983e − 9
Nickel[b]	58.754	0	−4.614e − 4	6.670e − 7	−1.523e − 10

Source: Reddy, J.N. and Chin, C.D., *J. Therm. Stresses*, 21, 593, 1998. With permission.
[a] For 300 K $\leq T \leq$ 635 K.
[b] For 635 K $\leq T$.

TABLE 1.4

Poisson's Ratio of Ceramics and Metals for ν_f

Materials	P_0	P_{-1}	P_1	P_2	P_3
Zirconia	0.2882	0	1.133e − 4	0	0
Aluminum oxide	0.2600	0	0	0	0
Silicon nitride	0.2400	0	0	0	0
Ti-6Al-4V	0.2884	0	1.121e − 4	0	0
Stainless steel	0.3262	0	−2.002e − 4	3.797e − 7	0
Nickel	0.3100	0	0	0	0

Source: Reddy, J.N. and Chin, C.D., *J. Therm. Stresses*, 21, 593, 1998. With permission.

$$\alpha_f(Z, T) = [\alpha_m(T) - \alpha_c(T)]\left(\frac{2Z + h}{2h}\right)^N + \alpha_c(T) \tag{1.5b}$$

$$\kappa_f(Z, T) = [\kappa_m(T) - \kappa_c(T)]\left(\frac{2Z + h}{2h}\right)^N + \kappa_c(T) \tag{1.5c}$$

$$\nu_f(Z, T) = [\nu_m(T) - \nu_c(T)]\left(\frac{2Z + h}{2h}\right)^N + \nu_c(T) \tag{1.5d}$$

It is evident that E_f, ν_f, α_f, and κ_f are both temperature- and position-dependent. This method is simple and convenient to apply for predicting the overall material properties and responses; however, owing to the assumed simplifications the validity is affected by the detailed graded microstructure.

As argued before, precise information about the size, the shape, and the distribution of particles is not available and the effective elastic moduli of the graded microstructures must be evaluated based on the volume fraction distribution and the approximate shape of the dispersed phase. Several micromechanics models have also been developed over the years to infer the effective properties of FGMs. The Mori–Tanaka scheme (Mori and Tanaka 1973, Benveniste 1987) for estimating the effective moduli is applicable to regions of the graded microstructure which have a well-defined continuous matrix and a discontinuous particulate phase as depicted in Figure 1.1. It takes into account the interaction of the elastic fields among neighboring inclusions. It is assumed that the matrix phase, denoted by the subscript 1, is reinforced by spherical particles of a particulate phase, denoted by the subscript 2. In this notation, K_1, G_1, and V_1 denote, respectively, the bulk modulus, the shear modulus, and the volume fraction of the matrix phase; K_2, G_2, and V_2 denote the corresponding material properties and the volume fraction of the particulate phase. It should be noted that $V_1 + V_2 = 1$. The effective local bulk modulus K_f, the shear modulus G_f, thermal expansion coefficient α_f, and thermal conductivity κ_f obtained by the Mori–Tanaka scheme for a random distribution of isotropic particles in an isotropic matrix are given by

$$\frac{K_f - K_1}{K_2 - K_1} = \frac{V_2}{1 + (1 - V_2)(3(K_2 - K_1)/(3K_1 + 4G_1))} \tag{1.6a}$$

$$\frac{G_f - G_1}{G_2 - G_1} = \frac{V_2}{1 + (1 - V_2)((G_2 - G_1)/(G_1 + f_1))} \tag{1.6b}$$

$$\frac{\alpha_f - \alpha_1}{\alpha_2 - \alpha_1} = \frac{(1/K_f) - (1/K_1)}{(1/K_2) - (1/K_1)} \tag{1.6c}$$

$$\frac{\kappa_f - \kappa_1}{\kappa_2 - \kappa_1} = \frac{V_2}{1 + (1 - V_2)((\kappa_2 - \kappa_1)/3\kappa_1)} \tag{1.6d}$$

where

$$f_1 = \frac{G_1(9K_1 + 8G_1)}{6(K_1 + 2G_1)} \tag{1.7}$$

The self-consistent method (Hill 1965) assumes that each reinforcement inclusion is embedded in a continuum material whose effective properties are those of the composite. This method does not distinguish between matrix and reinforcement phases and the same overall moduli are predicted in another composite in which the roles of the phases are interchanged. This makes it particularly suitable for determining the effective moduli in those regions which have an interconnected skeletal microstructure as depicted in Figure 1.6. The locally effective elastic moduli by the self-consistent method are given by

$$\frac{\delta}{K_f} = \frac{V_1}{K_f - K_2} + \frac{V_2}{K_f - K_1} \tag{1.8a}$$

$$\frac{\eta}{G_f} = \frac{V_1}{G_f - G_2} + \frac{V_2}{G_f - G_1} \tag{1.8b}$$

where

$$\delta = 3 - 5\eta = \frac{K_f}{K_f + (4/3)G_f} \tag{1.9}$$

FIGURE 1.6
Skeletal microstructure of FGM material. (From Vel, S.S. and Batra, R.C., *AIAA J.*, 40, 1421, 2002. With permission.)

From Equation 1.8a, one has

$$K_f = \frac{1}{(V_1/(K_1 + (4/3)G_f)) + (V_2/(K_2 + (4/3)G_f))} - \frac{4}{3}G_f \qquad (1.10)$$

and G_f is obtained by solving the following quartic equation:

$$[V_1 K_1/(K_1 + 4G_f/3) + V_2 K_2/(K_2 + 4G_f/3)] \\ + 5[V_1 G_2/(G_f - G_2) + V_2 G_1/(G_f - G_1)] + 2 = 0 \qquad (1.11)$$

Then, the effective Young's modulus E_f and Poisson's ratio ν_f can be found from $E_f = 9K_f G_f/3K_f + G_f$ and $\nu_f = (3K_f - 2G_f)/2(3K_f + G_f)$, respectively.

A comparison between the Mori–Tanaka and self-consistent models and the finite element simulation of FGM was presented in Reuter et al. (1997) and Reuter and Dvorak (1998). The Mori–Tanaka model was shown to yield accurate prediction of the properties with a well-defined continuous matrix and discontinuous inclusions, while the self-consistent model was better in skeletal microstructures characterized by a wide transition zone between the regions with predominance of one of the constituent phases.

1.3 Reddy's Higher Order Shear Deformation Plate Theory

Reddy (1984a,b) developed a simple higher order shear deformation plate theory (HSDPT), in which the transverse shear strains are assumed to be parabolically distributed across the plate thickness. The theory is simple in the sense that it contains the same dependent unknowns as in the first-order shear deformation plate theory (FSDPT), and no shear correction factors are required.

Consider a rectangular plate made of FGMs. The length, width, and total thickness of the plate are a, b, and h. As usual, the coordinate system has its origin at the corner of the plate on the midplane. Let $\overline{U}$, $\overline{V}$, and $\overline{W}$ be the plate displacements parallel to a right-hand set of axes (X, Y, Z), where X is longitudinal and Z is perpendicular to the plate. $\overline{\Psi}_x$ and $\overline{\Psi}_y$ are the midplane rotations of the normal about the Y and X axes, respectively. The displacement components are assumed to be of the following form:

$$U_1 = \overline{U}(X, Y, t) + Z\overline{\Psi}_x(X, Y, t) + Z^2 \xi_x(X, Y, t) + Z^3 \zeta_x(X, Y, t) \qquad (1.12a)$$

$$U_2 = \overline{V}(X, Y, t) + Z\overline{\Psi}_y(X, Y, t) + Z^2 \xi_y(X, Y, t) + Z^3 \zeta_y(X, Y, t) \qquad (1.12b)$$

$$U_3 = \overline{W}(X, Y, t) \qquad (1.12c)$$

where t represents time, $\overline{U}$, $\overline{V}$, $\overline{W}$, $\overline{\Psi}_x$, $\overline{\Psi}_y$, ξ_x, ξ_y, ζ_x, and ζ_y are unknowns.

If the transverse shear stresses σ_4 and σ_5 are to vanish at the bounding planes of the plate (at $Z = \pm h/2$), the transverse shear strains ε_4 and ε_5 should also vanish there. That is

$$\varepsilon_5\left(X, Y, \pm\frac{h}{2}, t\right) = 0, \quad \varepsilon_4\left(X, Y, \pm\frac{h}{2}, t\right) = 0 \tag{1.13}$$

which imply the following conditions

$$\xi_x = 0 \tag{1.14a}$$

$$\xi_y = 0 \tag{1.14b}$$

$$\zeta_x = -\frac{4}{3h^2}\left(\frac{\partial \overline{W}}{\partial X} + \overline{\Psi}_x\right) \tag{1.14c}$$

$$\zeta_y = -\frac{4}{3h^2}\left(\frac{\partial \overline{W}}{\partial Y} + \overline{\Psi}_y\right) \tag{1.14d}$$

Putting the above conditions in Equation 1.12 leads to the following displacement field

$$U_1 = \overline{U} + 2\left[\overline{\Psi}_x - X\frac{4}{3}\left(\frac{2}{h}\right)^2\left(\overline{\Psi}_x + \frac{\partial \overline{W}}{\partial X}\right)\right] \tag{1.15a}$$

$$U_2 = \overline{V} + 2\left[\overline{\Psi}_y - X\frac{4}{3}\left(\frac{2}{h}\right)^2\left(\overline{\Psi}_y + \frac{\partial \overline{W}}{\partial Y}\right)\right] \tag{1.15b}$$

$$U_3 = \overline{W} \tag{1.15c}$$

in which χ is a tracer. If $\chi = 1$, Equation 1.15 is for the case of the HSDPT, which contains the same dependent unknowns ($\overline{U}$, $\overline{V}$, $\overline{W}$, $\overline{\Psi}_x$, and $\overline{\Psi}_y$) as in the FSDPT. If $\chi = 0$, Equation 1.15 is reduced to the case of the FSDPT.

The strains of the plate associated with the displacement field given in Equation 1.15 are

$$\begin{aligned}
\varepsilon_1 &= \varepsilon_1^0 + Z\left(\kappa_1^0 + Z^2\kappa_1^2\right) \\
\varepsilon_2 &= \varepsilon_2^0 + Z\left(\kappa_2^0 + Z^2\kappa_2^2\right) \\
\varepsilon_3 &= 0 \\
\varepsilon_4 &= \varepsilon_4^0 + Z^2\kappa_4^2 \\
\varepsilon_5 &= \varepsilon_5^0 + Z^2\kappa_5^2 \\
\varepsilon_6 &= \varepsilon_6^0 + Z\left(\kappa_6^0 + Z^2\kappa_6^2\right)
\end{aligned} \tag{1.16}$$

where

$$\varepsilon_1^0 = \frac{\partial \overline{U}}{\partial X} + \frac{1}{2}\left(\frac{\partial \overline{W}}{\partial X}\right)^2, \quad \kappa_1^0 = \frac{\partial \overline{\Psi}_x}{\partial X}, \quad \kappa_1^2 = -\chi\frac{4}{3h^2}\left(\frac{\partial \overline{\Psi}_x}{\partial X} + \frac{\partial^2 \overline{W}}{\partial X^2}\right)$$

$$\varepsilon_2^0 = \frac{\partial \overline{V}}{\partial Y} + \frac{1}{2}\left(\frac{\partial \overline{W}}{\partial Y}\right)^2, \quad \kappa_2^0 = \frac{\partial \overline{\Psi}_y}{\partial Y}, \quad \kappa_2^2 = -\chi\frac{4}{3h^2}\left(\frac{\partial \overline{\Psi}_y}{\partial Y} + \frac{\partial^2 \overline{W}}{\partial Y^2}\right)$$

$$\varepsilon_4^0 = \overline{\Psi}_y + \frac{\partial \overline{W}}{\partial Y}, \quad \kappa_4^2 = -\chi\frac{4}{h^2}\left(\overline{\Psi}_y + \frac{\partial \overline{W}}{\partial Y}\right)$$

$$\varepsilon_5^0 = \overline{\Psi}_x + \frac{\partial \overline{W}}{\partial X}, \quad \kappa_5^2 = -\chi\frac{4}{h^2}\left(\overline{\Psi}_x + \frac{\partial \overline{W}}{\partial X}\right) \tag{1.17}$$

$$\varepsilon_6^0 = \frac{\partial \overline{U}}{\partial Y} + \frac{\partial \overline{V}}{\partial X} + \frac{\partial \overline{W}}{\partial X}\frac{\partial \overline{W}}{\partial Y}$$

$$\kappa_6^0 = \frac{\partial \overline{\Psi}_x}{\partial Y} + \frac{\partial \overline{\Psi}_y}{\partial X}$$

$$\kappa_6^2 = -\chi\frac{4}{3h^2}\left(\frac{\partial \overline{\Psi}_x}{\partial Y} + \frac{\partial \overline{\Psi}_y}{\partial X} + 2\frac{\partial^2 \overline{W}}{\partial X \partial Y}\right)$$

The plane stress constitutive equations may then be written in the form:

$$\begin{bmatrix} \sigma_1 \\ \sigma_2 \\ \sigma_6 \end{bmatrix} = \begin{bmatrix} Q_{11} & Q_{12} & 0 \\ Q_{21} & Q_{22} & 0 \\ 0 & 0 & Q_{66} \end{bmatrix} \begin{bmatrix} \varepsilon_1 \\ \varepsilon_2 \\ \varepsilon_6 \end{bmatrix} \tag{1.18a}$$

$$\begin{bmatrix} \sigma_4 \\ \sigma_5 \end{bmatrix} = \begin{bmatrix} Q_{44} & 0 \\ 0 & Q_{55} \end{bmatrix} \begin{bmatrix} \varepsilon_4 \\ \varepsilon_5 \end{bmatrix} \tag{1.18b}$$

where Q_{ij} are the transformed reduced stiffnesses defined by

$$Q_{11} = Q_{22} = \frac{E_f(Z,T)}{1 - v_f^2}, \quad Q_{12} = \frac{v_f E_f(Z,T)}{1 - v_f^2},$$

$$Q_{16} = Q_{26} = 0, \quad Q_{44} = Q_{55} = Q_{66} = \frac{E_f(Z,T)}{2(1 + v_f)} \tag{1.19}$$

As in the classical plate theory, the stress resultants and couples are defined by

$$(\overline{N}_i, \overline{M}_i, \overline{P}_i) = \int_{-h/2}^{h/2} \sigma_i(1, Z, Z^3)dZ, \quad i = 1, 2, 6 \tag{1.20a}$$

$$(\overline{Q}_2, \overline{R}_2) = \int_{-h/2}^{h/2} \sigma_4(1, Z^2)dZ \tag{1.20b}$$

$$(\overline{Q}_1, \overline{R}_1) = \int_{-h}^{h} \sigma_5(1, Z^2) dZ \qquad (1.20c)$$

where
$\overline{N}_i$ and $\overline{Q}_i$ are the membrane and transverse shear forces
$\overline{M}_i$ is the bending moment per unit length
$\overline{P}_i$ and $\overline{R}_i$ are the higher order bending moment and shear force, respectively

Substituting Equation 1.18 into Equation 1.20, and taking Equation 1.16 into account, yields the constitutive relations of the plate

$$\begin{bmatrix} \overline{N} \\ \overline{M} \\ \overline{P} \end{bmatrix} = \begin{bmatrix} A & B & E \\ B & D & F \\ E & F & H \end{bmatrix} \begin{bmatrix} \varepsilon^0 \\ \kappa^0 \\ \kappa^2 \end{bmatrix} \qquad (1.21a)$$

$$\begin{bmatrix} \overline{Q} \\ \overline{R} \end{bmatrix} = \begin{bmatrix} A & D \\ D & F \end{bmatrix} \begin{bmatrix} \varepsilon^0 \\ \kappa^2 \end{bmatrix} \qquad (1.21b)$$

where A_{ij}, B_{ij}, etc. are the plate stiffnesses, defined by

$$(A_{ij}, B_{ij}, D_{ij}, E_{ij}, F_{ij}, H_{ij}) = \int_{-h/2}^{+h/2} (Q_{ij})(1, Z, Z^2, Z^3, Z^4, Z^6) dZ, \quad i, j = 1, 2, 6 \quad (1.22a)$$

$$(A_{ij}, D_{ij}, F_{ij}) = \int_{-h/2}^{+h/2} (Q_{ij})(1, Z^2, Z^4) dZ, \quad i, j = 4, 5 \qquad (1.22b)$$

The Hamilton principle for an elastic body is

$$\int_{t_1}^{t_2} (\delta U + \delta V - \delta K) dt = 0 \qquad (1.23)$$

where
δU is the virtual strain energy
δV is the virtual work done by external forces
δK is the virtual kinetic energy

$$\delta U = \int_{\Omega} \int_{-h/2}^{h/2} (\sigma_i \delta \varepsilon_i) dZ \, dX \, dY$$

$$= \int_{\Omega} (\overline{N}_i \delta \varepsilon_i^0 + \overline{M}_i \delta \kappa_i^0 + \overline{P}_i \delta \kappa_i^2) dZ \, dX \, dY, \quad i = 1, 2, 6 \qquad (1.24a)$$

$$\delta V = - \int_{\Omega} [q(X, Y)\delta U_3] dX \, dY \tag{1.24b}$$

$$\delta K = \int_{\Omega} \int_{-h/2}^{h/2} \rho(\dot{U}_j \delta \dot{U}_j) dZ \, dX \, dY, \quad j = 1, 2, 3 \tag{1.24c}$$

In Equation 1.24c, the superposed dots indicate differentiation with respect to time. Integrating Equation 1.23, and collecting the coefficients of $\delta\overline{U}$, $\delta\overline{V}$, $\delta\overline{W}$, $\delta\overline{\Psi}_x$, and $\delta\overline{\Psi}_y$, we obtain the following equations of motion

$$\delta\overline{U}: \frac{\partial \overline{N}_1}{\partial X} + \frac{\partial \overline{N}_6}{\partial Y} = I_1 \frac{\partial^2 \overline{U}}{\partial t^2} + \bar{I}_2 \frac{\partial^2 \overline{\Psi}_x}{\partial t^2} - c_1 I_4 \frac{\partial^3 \overline{W}}{\partial X \partial t^2}$$

$$\delta\overline{V}: \frac{\partial \overline{N}_6}{\partial X} + \frac{\partial \overline{N}_2}{\partial Y} = I_1 \frac{\partial^2 \overline{V}}{\partial t^2} + \bar{I}_2 \frac{\partial^2 \overline{\Psi}_y}{\partial t^2} - c_1 I_4 \frac{\partial^3 \overline{W}}{\partial Y \partial t^2}$$

$$\delta\overline{W}: \frac{\partial \overline{Q}_1}{\partial X} + \frac{\partial \overline{Q}_2}{\partial Y} + \frac{\partial}{\partial X}\left(\overline{N}_1 \frac{\partial \overline{W}}{\partial X} + \overline{N}_6 \frac{\partial \overline{W}}{\partial Y} \right) + \frac{\partial}{\partial Y}\left(\overline{N}_6 \frac{\partial \overline{W}}{\partial X} + \overline{N}_2 \frac{\partial \overline{W}}{\partial Y} \right)$$

$$+ q - c_2\left(\frac{\partial \overline{R}_1}{\partial X} + \frac{\partial \overline{R}_2}{\partial Y} \right) + c_1 \left(\frac{\partial^2 \overline{P}_1}{\partial X^2} + 2\frac{\partial^2 \overline{P}_6}{\partial X \partial Y} + \frac{\partial^2 \overline{P}_2}{\partial Y^2} \right)$$

$$= I_1 \frac{\partial^2 \overline{W}}{\partial t^2} - c_1^2 I_7 \frac{\partial^2}{\partial t^2}\left(\frac{\partial^2 \overline{W}}{\partial X^2} + \frac{\partial^2 \overline{W}}{\partial Y^2} \right) + c_1 I_4 \frac{\partial^2}{\partial t^2}\left(\frac{\partial \overline{U}}{\partial X} + \frac{\partial \overline{V}}{\partial Y} \right) + c_1 \bar{I}_5 \frac{\partial^2}{\partial t^2}\left(\frac{\partial \overline{\Psi}_x}{\partial X} + \frac{\partial \overline{\Psi}_y}{\partial Y} \right)$$

$$\delta\overline{\Psi}_x: \frac{\partial \overline{M}_1}{\partial X} + \frac{\partial \overline{M}_6}{\partial Y} - \overline{Q}_1 + c_2 \overline{R}_1 - c_1\left(\frac{\partial \overline{P}_1}{\partial X} + \frac{\partial \overline{P}_6}{\partial Y} \right) = \bar{I}_2 \frac{\partial^2 \overline{U}}{\partial t^2} + \bar{I}_3 \frac{\partial^2 \overline{\Psi}_x}{\partial t^2} - c_1 \bar{I}_5 \frac{\partial^3 \overline{W}}{\partial X \partial t^2}$$

$$\delta\overline{\Psi}_y: \frac{\partial \overline{M}_6}{\partial X} + \frac{\partial \overline{M}_2}{\partial Y} - \overline{Q}_2 + c_2 \overline{R}_2 - c_1\left(\frac{\partial \overline{P}_6}{\partial X} + \frac{\partial \overline{P}_2}{\partial Y} \right) = \bar{I}_2 \frac{\partial^2 \overline{V}}{\partial t^2} + \bar{I}_3 \frac{\partial^2 \overline{\Psi}_y}{\partial t^2} - c_1 \bar{I}_5 \frac{\partial^3 \overline{W}}{\partial Y \partial t^2}$$

$$\tag{1.25}$$

where $c_1 = 4/3h^2$, $c_2 = 3c_1$, and

$$\bar{I}_2 = I_2 - c_1 I_4, \quad \bar{I}_5 = I_5 - c_1 I_7, \quad \bar{I}_3 = I_3 - 2c_1 I_5 + c_1^2 I_7, \quad \bar{I}_8 = \bar{I}_3 + \bar{I}_5 \tag{1.26a}$$

and the inertias I_i $(i = 1, 2, 3, 4, 5, 7)$ are defined by

$$(I_1, I_2, I_3, I_4, I_5, I_7) = \int_{-h/2}^{h/2} \rho(Z)(1, Z, Z^2, Z^3, Z^4, Z^6) dZ \tag{1.26b}$$

where ρ is the mass density of the plate, which may also be position dependent.

1.4   Generalized Kármán-Type Nonlinear Equations

Based on Reddy's HSDPT with a von Kármán-type of kinematic nonlinearity (Reddy 1984b) and including thermal effects, Shen (1997) derived a set of general von Kármán-type equations which can be expressed in terms of a stress function $\overline{F}$, two rotations $\overline{\Psi}_x$ and $\overline{\Psi}_y$, and a transverse displacement $\overline{W}$, along with the initial geometric imperfection $\overline{W}^*$. These equations are then extended to the case of shear deformable FGM plates.

Let $\overline{F}\ (X, Y)$ be the stress function for the stress resultants defined by $\overline{N}_x = \overline{F}_{,yy}$, $\overline{N}_y = \overline{F}_{,xx}$, and $\overline{N}_{xy} = -\overline{F}_{,xy}$, where a comma denotes partial differentiation with respect to the corresponding coordinates.

If thermal effect is taken into account, we assume

$$N^* = \overline{N} - \overline{N}^{\mathrm{T}}, \quad M^* = \overline{M} - \overline{M}^{\mathrm{T}}, \quad P^* = \overline{P} - \overline{P}^{\mathrm{T}} \tag{1.27}$$

where $\overline{N}^{\mathrm{T}}$, $\overline{M}^{\mathrm{T}}$, $\overline{S}^{\mathrm{T}}$, and $\overline{P}^{\mathrm{T}}$ are the forces, moments, and higher order moments caused by elevated temperature, and are defined by

$$\begin{bmatrix} \overline{N}_x^{\mathrm{T}} & \overline{M}_x^{\mathrm{T}} & \overline{P}_x^{\mathrm{T}} \\ \overline{N}_y^{\mathrm{T}} & \overline{M}_y^{\mathrm{T}} & \overline{P}_y^{\mathrm{T}} \\ \overline{N}_{xy}^{\mathrm{T}} & \overline{M}_{xy}^{\mathrm{T}} & \overline{P}_{xy}^{\mathrm{T}} \end{bmatrix} = \int_{-h/2}^{+h/2} \begin{bmatrix} A_x \\ A_y \\ A_{xy} \end{bmatrix} (1, Z, Z^3)\Delta T(X, Y, Z)\mathrm{d}Z \tag{1.28a}$$

$$\begin{bmatrix} \overline{S}_x^{\mathrm{T}} \\ \overline{S}_y^{\mathrm{T}} \\ \overline{S}_{xy}^{\mathrm{T}} \end{bmatrix} = \begin{bmatrix} \overline{M}_x^{\mathrm{T}} \\ \overline{M}_y^{\mathrm{T}} \\ \overline{M}_{xy}^{\mathrm{T}} \end{bmatrix} - \frac{4}{3h^2} \begin{bmatrix} \overline{P}_x^{\mathrm{T}} \\ \overline{P}_y^{\mathrm{T}} \\ \overline{P}_{xy}^{\mathrm{T}} \end{bmatrix} \tag{1.28b}$$

where $\Delta T(X, Y, Z) = T(X, Y, Z) - T_0$ is temperature rise from the reference temperature T_0 at which there are no thermal strains, and

$$\begin{bmatrix} A_x \\ A_y \\ A_{xy} \end{bmatrix} = -\begin{bmatrix} Q_{11} & Q_{12} & Q_{16} \\ Q_{12} & Q_{22} & Q_{26} \\ Q_{16} & Q_{26} & Q_{66} \end{bmatrix} \begin{bmatrix} 1 & 0 \\ 0 & 1 \\ 0 & 0 \end{bmatrix} \begin{bmatrix} \alpha_{11} \\ \alpha_{22} \end{bmatrix} \tag{1.29}$$

where α_{11} and α_{22} are the thermal expansion coefficients measured in the longitudinal and transverse directions, respectively.

The partial inverse of Equation 1.21a yields

$$\begin{bmatrix} \varepsilon^0 \\ M^* \\ P^* \end{bmatrix} = \begin{bmatrix} A^* & B^* & E^* \\ -(B^*)^{\mathrm{T}} & D^* & (F^*)^{\mathrm{T}} \\ -(E^*)^{\mathrm{T}} & F^* & H^* \end{bmatrix} \begin{bmatrix} N^* \\ \kappa^0 \\ \kappa^2 \end{bmatrix} \tag{1.30}$$

where the superscript "T" represents the matrix transpose and in which the reduced stiffness matrices $[A_{ij}^*]$, $[B_{ij}^*]$, $[D_{ij}^*]$, $[E_{ij}^*]$, $[F_{ij}^*]$, and $[H_{ij}^*]$ $(i,j = 1, 2, 6)$ are functions of temperature and position, determined through relationships (Shen 1997):

$$A^* = A^{-1}, \quad B^* = -A^{-1}B, \quad D^* = D - BA^{-1}B, \quad E^* = -A^{-1}E,$$
$$F^* = F - EA^{-1}B, \quad H^* = H - EA^{-1}E \tag{1.31}$$

From Equation 1.30, the bending moments, higher order moments, and transverse shear forces can be written in the form:

$$\overline{M}_x = \overline{M}_1 = -B_{11}^*\overline{F}_{,yy} - B_{21}^*\overline{F}_{,xx} + D_{11}^*\overline{\Psi}_{x,x} + D_{12}^*\overline{\Psi}_{y,y}$$
$$- c_1[F_{11}^*(\overline{\Psi}_{x,x} + \overline{W}_{,xx}) + F_{21}^*(\overline{\Psi}_{y,y} + \overline{W}_{,yy})] + \overline{M}_x^T \tag{1.32a}$$

$$\overline{M}_y = \overline{M}_2 = -B_{12}^*\overline{F}_{,yy} - B_{22}^*\overline{F}_{,xx} + D_{12}^*\overline{\Psi}_{x,x} + D_{22}^*\overline{\Psi}_{y,y}$$
$$- c_1[F_{12}^*(\overline{\Psi}_{x,x} + \overline{W}_{,xx}) + F_{22}^*(\overline{\Psi}_{y,y} + \overline{W}_{,yy})] + \overline{M}_y^T \tag{1.32b}$$

$$\overline{M}_{xy} = \overline{M}_6 = B_{66}^*\overline{F}_{,xy} + D_{66}^*(\overline{\Psi}_{x,y} + \overline{\Psi}_{y,x})$$
$$- c_1 F_{66}^*(\overline{\Psi}_{x,y} + \overline{\Psi}_{y,x} + 2\overline{W}_{,xy}) + \overline{M}_{xy}^T \tag{1.32c}$$

$$\overline{P}_x = \overline{P}_1 = -E_{11}^*\overline{F}_{,yy} - E_{21}^*\overline{F}_{,xx} + F_{11}^*\overline{\Psi}_{x,x} + F_{12}^*\overline{\Psi}_{y,y}$$
$$- c_1[H_{11}^*(\overline{\Psi}_{x,x} + \overline{W}_{,xx}) + H_{12}^*(\overline{\Psi}_{y,y} + \overline{W}_{,yy})] + \overline{P}_x^T \tag{1.32d}$$

$$\overline{P}_y = \overline{P}_2 = -E_{12}^*\overline{F}_{,yy} - E_{22}^*\overline{F}_{,xx} + F_{21}^*\overline{\Psi}_{x,x} + D_{22}^*\overline{\Psi}_{y,y}$$
$$- c_1[H_{12}^*(\overline{\Psi}_{x,x} + \overline{W}_{,xx}) + H_{22}^*(\overline{\Psi}_{y,y} + \overline{W}_{,yy})] + \overline{P}_y^T \tag{1.32e}$$

$$Q_1 = (A_{55} - c_2 D_{55})(\overline{\Psi}_x + \overline{W}_{,x}) \tag{1.32f}$$

$$R_1 = (D_{55} - c_2 F_{55})(\overline{\Psi}_x + \overline{W}_{,x}) \tag{1.32g}$$

$$Q_2 = (A_{44} - c_2 D_{44})(\overline{\Psi}_y + \overline{W}_{,y}) \tag{1.32h}$$

$$R_2 = (D_{44} - c_2 F_{44})(\overline{\Psi}_y + \overline{W}_{,y}) \tag{1.32i}$$

Substituting Equation 1.32 into Equation 1.25, and considering the condition of compatibility, which is also expressed in terms of $\overline{F}$, $\overline{\Psi}_x$, $\overline{\Psi}_y$, $\overline{W}$, and $\overline{W}^*$, the equations of motion are obtained in the following

$$\tilde{L}_{11}(\overline{W}) - \tilde{L}_{12}(\overline{\Psi}_x) - \tilde{L}_{13}(\overline{\Psi}_y) + \tilde{L}_{14}(\overline{F}) - \tilde{L}_{15}(\overline{N}^T) - \tilde{L}_{16}(\overline{M}^T)$$
$$= \tilde{L}(\overline{W} + \overline{W}^*, \overline{F}) + \tilde{L}_{17}(\ddot{\overline{W}}) - I_8(\ddot{\overline{\Psi}}_{x,x} + \ddot{\overline{\Psi}}_{y,y}) + q \tag{1.33}$$

$$\tilde{L}_{21}(\overline{F}) + \tilde{L}_{22}(\overline{\Psi}_x) + \tilde{L}_{23}(\overline{\Psi}_y) - \tilde{L}_{24}(\overline{W}) - \tilde{L}_{25}(\overline{N}^T) = -\frac{1}{2}\tilde{L}(\overline{W} + 2\overline{W}^*, \overline{W}) \tag{1.34}$$

$$\tilde{L}_{31}(\overline{W}) + \tilde{L}_{32}(\overline{\Psi}_x) - \tilde{L}_{33}(\overline{\Psi}_y) + \tilde{L}_{34}(\overline{F}) - \tilde{L}_{35}(\overline{N}^{\mathrm{T}}) - \tilde{L}_{36}(\overline{S}^{\mathrm{T}})$$

$$= \bar{I}_5 \ddot{\overline{W}}_{,x} - \bar{I}_3 \ddot{\overline{\Psi}}_x \tag{1.35}$$

$$\tilde{L}_{41}(\overline{W}) - \tilde{L}_{42}(\overline{\Psi}_x) + \tilde{L}_{43}(\overline{\Psi}_y) + \tilde{L}_{44}(\overline{F}) - \tilde{L}_{45}(\overline{N}^{\mathrm{T}}) - \tilde{L}_{46}(\overline{S}^{\mathrm{T}})$$

$$= \bar{I}_5 \ddot{\overline{W}}_{,y} - \bar{I}_3 \ddot{\overline{\Psi}}_y \tag{1.36}$$

where all linear operators $\tilde{L}_{ij}()$ and the nonlinear operator $\tilde{L}()$ are defined by

$$\tilde{L}_{11}() = c_1 \left[F_{11}^* \frac{\partial^4}{\partial X^4} + (F_{12}^* + F_{21}^* + 4F_{66}^*) \frac{\partial^4}{\partial X^2 \partial Y^2} + F_{22}^* \frac{\partial^4}{\partial Y^4} \right]$$

$$\tilde{L}_{12}() = (D_{11}^* - c_1 F_{11}^*) \frac{\partial^3}{\partial X^3} + [(D_{12}^* + 2D_{66}^*) - c_1(F_{12}^* + 2F_{66}^*)] \frac{\partial^3}{\partial X \partial Y^2}$$

$$\tilde{L}_{13}() = [(D_{12}^* + 2D_{66}^*) - c_1(F_{21}^* + 2F_{66}^*)] \frac{\partial^3}{\partial X^2 \partial Y} + (D_{22}^* - c_1 F_{22}^*) \frac{\partial^3}{\partial Y^3}$$

$$\tilde{L}_{14}() = B_{21}^* \frac{\partial^4}{\partial X^4} + (B_{11}^* + B_{22}^* - 2B_{66}^*) \frac{\partial^4}{\partial X^2 \partial Y^2} + B_{12}^* \frac{\partial^4}{\partial Y^4}$$

$$\tilde{L}_{15}(\overline{N}^{\mathrm{T}}) = \frac{\partial^2}{\partial X^2} \left(B_{11}^* \overline{N}_x^{\mathrm{T}} + B_{21}^* \overline{N}_y^{\mathrm{T}} \right) + 2\frac{\partial^2}{\partial X \partial Y} \left(B_{66}^* \overline{N}_{xy}^{\mathrm{T}} \right) + \frac{\partial^2}{\partial Y^2} \left(B_{12}^* \overline{N}_x^{\mathrm{T}} + B_{22}^* \overline{N}_y^{\mathrm{T}} \right)$$

$$\tilde{L}_{16}\left(\overline{M}^{\mathrm{T}}\right) = \frac{\partial^2}{\partial X^2} \left(\overline{M}_x^{\mathrm{T}} \right) + 2\frac{\partial^2}{\partial X \partial Y} \left(\overline{M}_{xy}^{\mathrm{T}} \right) + \frac{\partial^2}{\partial Y^2} \left(\overline{M}_y^{\mathrm{T}} \right)$$

$$\tilde{L}_{17}() = c_1 \left(I_5 - \frac{I_4 I_2}{I_1} \right) \left(\frac{\partial^2}{\partial X^2} + \frac{\partial^2}{\partial Y^2} \right) - I_1$$

$$\tilde{L}_{21}() = A_{22}^* \frac{\partial^4}{\partial X^4} + (2A_{12}^* + A_{66}^*) \frac{\partial^4}{\partial X^2 \partial Y^2} + A_{11}^* \frac{\partial^4}{\partial Y^4}$$

$$\tilde{L}_{22}() = (B_{21}^* - c_1 E_{21}^*) \frac{\partial^3}{\partial X^3} + [(B_{11}^* - B_{66}^*) - c_1(E_{11}^* - E_{66}^*)] \frac{\partial^3}{\partial X \partial Y^2}$$

$$\tilde{L}_{23}() = [(B_{22}^* - B_{66}^*) - c_1(E_{22}^* - E_{66}^*)] \frac{\partial^3}{\partial X^2 \partial Y} + (B_{12}^* - c_1 E_{12}^*) \frac{\partial^3}{\partial Y^3}$$

$$\tilde{L}_{24}() = c_1 \left[E_{21}^* \frac{\partial^4}{\partial X^4} + (E_{11}^* + E_{22}^* - 2E_{66}^*) \frac{\partial^4}{\partial X^2 \partial Y^2} + E_{12}^* \frac{\partial^4}{\partial Y^4} \right]$$

$$\tilde{L}_{25}(\overline{N}^{\mathrm{T}}) = \frac{\partial^2}{\partial X^2} \left(A_{12}^* \overline{N}_x^{\mathrm{T}} + A_{22}^* \overline{N}_y^{\mathrm{T}} \right) - \frac{\partial^2}{\partial X \partial Y} \left(A_{66}^* \overline{N}_{xy}^{\mathrm{T}} \right) + \frac{\partial^2}{\partial Y^2} \left(A_{11}^* \overline{N}_x^{\mathrm{T}} + A_{12}^* \overline{N}_y^{\mathrm{T}} \right)$$

$$\tilde{L}_{31}() = \left(A_{55} - 2c_2 D_{55} + c_2^2 F_{55} \right) \frac{\partial}{\partial X} + c_1(F_{11}^* - c_1 H_{11}^*) \frac{\partial^3}{\partial X^3}$$

$$+ c_1 [(F_{21}^* + 2F_{66}^*) - c_1(H_{12}^* + 2H_{66}^*)] \frac{\partial^3}{\partial X \partial Y^2}$$

$$\tilde{L}_{32}() = \left(A_{55} - 2c_2 D_{55} + c_2^2 F_{55} \right) - \left(D_{11}^* - 2c_1 F_{11}^* + c_1^2 H_{11}^* \right) \frac{\partial^2}{\partial X^2}$$

$$- \left(D_{66}^* - 2c_1 F_{66}^* + c_1^2 H_{66}^* \right) \frac{\partial^2}{\partial Y^2}$$

$$\tilde{L}_{33}() = \left[(D_{12}^* + D_{66}^*) - c_1(F_{12}^* + F_{21}^* + 2F_{66}^*) + c_1^2(H_{12}^* + H_{66}^*)\right]\frac{\partial^2}{\partial X \partial Y}$$

$$\tilde{L}_{34}() = \tilde{L}_{22}()$$

$$\tilde{L}_{35}(\overline{N}^{\mathrm{T}}) = \frac{\partial}{\partial X}\left[(B_{11}^* - c_1 E_{11}^*)\overline{N}_x^{\mathrm{T}} + (B_{21}^* - c_1 E_{21}^*)\overline{N}_y^{\mathrm{T}}\right] + \frac{\partial}{\partial Y}\left[(B_{66}^* - c_1 E_{66}^*)\overline{N}_{xy}^{\mathrm{T}}\right]$$

$$\tilde{L}_{36}(\overline{S}^{\mathrm{T}}) = \frac{\partial}{\partial X}\left(\overline{S}_x^{\mathrm{T}}\right) + \frac{\partial}{\partial Y}\left(\overline{S}_{xy}^{\mathrm{T}}\right)$$

$$\tilde{L}_{41}() = \left(A_{44} - 2c_2 D_{44} + c_2^2 F_{44}\right)\frac{\partial}{\partial Y} + c_1[(F_{12}^* + 2F_{66}^*)$$

$$- c_1(H_{12}^* + 2H_{66}^*)]\frac{\partial^3}{\partial X^2 \partial Y} + c_1(F_{22}^* - c_1 H_{22}^*)\frac{\partial^3}{\partial Y^3}$$

$$\tilde{L}_{42}() = \tilde{L}_{33}()$$

$$\tilde{L}_{43}() = \left(A_{44} - 2c_2 D_{44} + c_2^2 F_{44}\right) - \left(D_{66}^* - 2c_1 F_{66}^* + c_1^2 H_{66}^*\right)\frac{\partial^2}{\partial X^2}$$

$$- \left(D_{22}^* - 2c_1 F_{22}^* + c_1^2 H_{22}^*\right)\frac{\partial^2}{\partial Y^2}$$

$$\tilde{L}_{44}() = \tilde{L}_{23}()$$

$$\tilde{L}_{45}(\overline{N}^{\mathrm{T}}) = \frac{\partial}{\partial X}\left[(B_{66}^* - c_1 E_{66}^*)\overline{N}_{xy}^{\mathrm{T}}\right] + \frac{\partial}{\partial Y}\left[(B_{12}^* - c_1 E_{12}^*)\overline{N}_x^{\mathrm{T}} + (B_{22}^* - c_1 E_{22}^*)\overline{N}_y^{\mathrm{T}}\right]$$

$$\tilde{L}_{46}(\overline{S}^{\mathrm{T}}) = \frac{\partial}{\partial X}(\overline{S}_{xy}^{\mathrm{T}}) + \frac{\partial}{\partial Y}(\overline{S}_y^{\mathrm{T}})$$

$$\tilde{L}() = \frac{\partial^2}{\partial X^2}\frac{\partial^2}{\partial Y^2} - 2\frac{\partial^2}{\partial X \partial Y}\frac{\partial^2}{\partial X \partial Y} + \frac{\partial^2}{\partial Y^2}\frac{\partial^2}{\partial X^2} \tag{1.37}$$

It is worthy to note that the governing differential equations (Equations 1.33 through 1.37) for an FGM plate are identical in form to those of unsymmetric cross-ply laminated plates. These general von Kármán-type equations will be used in solving many nonlinear problems, e.g., nonlinear bending, postbuckling, and nonlinear vibration of shear deformable FGM plates.

References

Benveniste Y. (1987), A new approach to the application of Mori–Tanaka's theory of composite materials, *Mechanics of Materials*, **6**, 147–157.

Birman V. and Byrd L.W. (2007), Modeling and analysis of functionally graded materials and structures, *Applied Mechanics Reviews*, **60**, 195–216.

Chan S.H. (2001), Performance and emissions characteristics of a partially insulated gasoline engine, *International Journal of Thermal Science*, **40**, 255–261.

Fuchiyama T. and Noda N. (1995), Analysis of thermal stress in a plate of functionally gradient material, *JSAE Review*, **16**, 263–268.

Gibson L.J., Ashby M.F., Karam G.N., Wegst U., and Shercliff H.R. (1995), Mechanical properties of natural materials. II. Microstructures for mechanical efficiency, *Proceedings of the Royal Society of London Series A*, **450**, 141–162.

Hill R. (1965), A self-consistent mechanics of composite materials, *Journal of the Mechanics and Physics of Solids*, **13**, 213–222.

Hirai T. and Chen L. (1999), Recent and prospective development of functionally graded materials in Japan, *Materials Science Forum*, **308–311**, 509–514.

Koizumi M. (1993), The concept of FGM, *Ceramic Transactions, Functionally Gradient Materials*, **34**, 3–10.

Librescu L. and Song S.-Y. (2005), Thin-walled beams made of functionally graded materials and operating in a high temperature environment: vibration and stability, *Journal of Thermal Stresses*, **28**, 649–712.

Liu L.-S., Zhang Q.-J., and Zhai P.-C. (2003), The optimization design of metal/ceramic FGM armor with neural net and conjugate gradient method, *Materials Science Forum*, **423–425**, 791–796.

Markworth A.J., Ramesh K.S., and Parks W.P. (1995), Review: modeling studies applied to functionally graded materials, *Journal of Material Sciences*, **30**, 2183–2193.

Mori T. and Tanaka K. (1973), Average stress in matrix and average elastic energy of materials with misfitting inclusions, *Acta Metallurgica*, **2**, 1571–574.

Müller E., Drašar C., Schilz J., and Kaysser W.A. (2003), Functionally graded materials for sensor and energy applications, *Materials Science and Engineering*, **A362**, 17–39.

Noda N. (1999), Thermal stresses in functionally graded material, *Journal of Thermal Stresses*, **22**, 477–512.

Paszkiewicz B., Paszkiewicz R., Wosko M., Radziewicz D., Sciana B., Szyszka A., Macherzynski W., and Tlaczala M. (2008), Functionally graded semiconductor layers for devices application, *Vacuum*, **82**, 389–394.

Paulino G.H., Jin Z.H., and Dodds Jr. R.H. (2003), Failure of functionally graded Materials, in *Comprehensive Structural Integrity*, Vol. 2 (eds. B. Karihallo and W.G. Knauss), Elsevier Science, New York, pp. 607–644.

Qiu J., Tani J., Ueno T., Morita T., Takahashi H., and Du H. (2003), Fabrication and high durability of functionally graded piezoelectric bending actuators, *Smart Materials and Structures*, **12**, 115–121.

Reddy J.N. (1984a), A simple high-order theory for laminated composite plates, *Journal of Applied Mechanics ASME*, **51**, 745–752.

Reddy J.N. (1984b), A refined nonlinear theory of plates with transverse shear deformation, *International Journal of Solids and Structure*, **20**, 881–896.

Reddy J.N. and Chin C.D. (1998), Thermoelastical analysis of functionally graded cylinders and plates, *Journal of Thermal Stresses*, **21**, 593–626.

Reuter T. and Dvorak G.J. (1998), Micromechanical models for graded composite Materials: II. Thermomechanical loading, *Journal of Mechanics and Physics of Solids*, **46**, 1655–1673.

Reuter T., Dvorak G.J., and Tvergaard V. (1997), Micromechanical models for graded composite materials, *Journal of Mechanics and Physics of Solids*, **45**, 1281–1302.

Shen H.-S. (1997), Kármán-type equations for a higher-order shear deformation plate theory and its use in the thermal postbuckling analysis, *Applied Mathematics and Mechanics*, **18**, 1137–1152.

Shen H.-S. (2004), Bending, buckling and vibration of functionally graded plates and shells (in Chinese), *Advances in Mechanics*, **34**, 53–60.

Tanigawa Y. (1995), Some basic thermoelastic problems for nonhomogeneous structural materials, *Applied Mechanics Reviews*, **48**, 287–300.

Touloukian Y.S. (1967), *Thermophysical Properties of High Temperature Solid Materials*, McMillan, New York.

Uemura S. (2003), The activities of FGM on new applications, *Materials Science Forum*, **423–425**, 1–10.

Vel S.S. and Batra R.C. (2002), Exact solution for thermoelastic deformations of functionally graded thick rectangular plates, *AIAA Journal*, **40**, 1421–1433.

Watari F., Yokoyama A., Omori M., Hirai T., Kondo H., Uo M., and Kawasaki T. (2004), Biocompatibility of materials and development to functionally graded implant for bio-medical application, *Composites Science and Technology*, **64**, 893–908.

Yamanoushi M., Koizumi M., Hiraii T., and Shiota I. (eds.) (1990), *Proceedings of the First International Symposium on Functionally Gradient Materials*, Japan.

Yin H.M., Sun L.Z., and Paulino G.H. (2004), Micromechanics-based elastic model for functionally graded materials with particle interactions, *Acta Materialia*, **52**, 3535–3543.

2

Nonlinear Bending of Shear Deformable FGM Plates

2.1 Introduction

The nonlinear bending response of FGM plates subjected to transverse mechanical loads and thermal loading was the subject of recent investigations. Previous works for the linear bending of FGM rectangular, circular, and annular plates can be found in Reddy et al. (1999), Cheng and Batra (2000), Reddy and Cheng (2001), Vel and Batra (2002), Croce and Venini (2004), Kashtalyan (2004), and Chung and Chen (2007). Mizuguchi and Ohnabe (1996) employed the Poincare method to examine the large deflection of heated FGM thin plates with Young's modulus varying symmetrically to the middle plane in thickness direction. Suresh and Mortensen (1997) presented the large deformation problem of graded multilayered composites under thermomechanical loads. When the thermomechanical load reaches a high level, nonlinear strain–displacement relations have to be employed. As a result, a set of nonlinear equations will appear no matter what kind of analysis method is used. Based on the FSDPT, Praveen and Reddy (1998) analyzed nonlinear static and dynamic response of functionally graded ceramic–metal plates subjected to transverse mechanical loads and a one-dimensional (1D) steady heat conduction by using finite element method (FEM). This work was then extended to the case of FGM square plates and shallow shell panels by Woo and Meguid (2001) using Fourier series technique, and to the case of FGM circular plates by Ma and Wang (2003) and Gunes and Reddy (2008), and to the case of FGM rectangular plates by GhannadPour and Alinia (2006) and Ovesy and GhannadPour (2007) using Ritz method and finite strip method, respectively. However, in their studies the formulations were based on the classical plate/shell theory, i.e., the theory based on the Kirchhoff–Love hypothesis and therefore the transverse shear deformations were not accounted for, and the material properties were assumed to be independent of temperature. Reddy (2000) developed theoretical formulations for thick FGM plates according to the HSDPT. In his study, both Navier solutions for linear bending of simply supported rectangular FGM plates and finite element models for nonlinear static and dynamic response were presented. The paper of Cheng (2001) also contains the solution for nonlinear bending of transversely isotropic symmetric shear deformable FGM plates. Moreover, Shen (2002) provided a nonlinear bending analysis of simply supported shear deformable FGM rectangular plates subjected to a transverse uniform or sinusoidal load and in thermal environments. In his study, the material properties were considered to be temperature dependent and the effect of temperature rise on the nonlinear bending response was reported. Subsequently, Yang and Shen (2003a,b) developed a semianalytical-numerical method to examine the large deflection of thin and shear deformable FGM rectangular plates subjected to combined mechanical and thermal loads and under various boundary conditions. This method was then extended to the case of FGM hybrid plates with surface-bonded piezo-electric layers by Yang et al. (2004). In these studies, the material properties were assumed to be temperature independent and temperature dependent, respectively. Recently, Na and Kim (2006) studied nonlinear bending of clamped FGM rectangular plates subjected to a transverse uniform pressure and thermal loads by using a 3D FEM. In their study, the thermal loads were assumed as uniform, linear, and sinusoidal temperature rises across the thickness direction, whereas the material properties were assumed to be temperature independent. On the other hand, ceramics and the metals used in FGMs do store different amounts of heat, and therefore the heat conduction usually occurs (Tanigawa et al. 1996, Kim and Noda 2002). This leads to a nonuniform distribution of temperature through the plate thickness, but it is not accounted for in the above studies. This is because when the material properties are assumed to be functions of temperature and position, and the temperature is also assumed to be a function of position, the problem becomes very complicated. More recently, Shen (2007) provided a nonlinear thermal bending analysis of simply supported shear deformable FGM rectangular plates due to heat conduction. In his study, both heat conduction and temperature-dependent material properties were taken into account.

2.2 Nonlinear Bending of FGM Plates under Mechanical Loads in Thermal Environments

Here, we consider an FGM plate of length a, width b, and thickness h, which is made from a mixture of ceramics and metals. We assume that the composition is varied from the top to the bottom surface, i.e., the top surface $(Z = -h/2)$ of the plate is ceramic-rich whereas the bottom surface $(Z = h/2)$ is metal-rich. The plate is subjected to a transverse uniform load $q = q_0$ or a sinusoidal load $q = q_0 \sin(\pi X/a)\sin(\pi Y/b)$ combined with thermal loads.

It is assumed that E_c, E_m, α_c, and α_m are functions of temperature, but Poisson's ratio ν_f depends weakly on temperature change and is assumed to be a constant. We assume the volume fraction V_m follows a simple power law as expressed by Equation 1.1. According to mixture rules, the effective

Young's modulus E_f and thermal expansion coefficient α_f of an FGM plate can be written as

$$E_f(Z,T) = [E_m(T) - E_c(T)]\left(\frac{2Z+h}{2h}\right)^N + E_c(T) \qquad (2.1a)$$

$$\alpha_f(Z,T) = [\alpha_m(T) - \alpha_c(T)]\left(\frac{2Z+h}{2h}\right)^N + \alpha_c(T) \qquad (2.1b)$$

It is evident that when $Z=-h/2$, $E_f=E_c$ and $\alpha_f=\alpha_c$, and when $Z=h/2$, $E_f=E_m$ and $\alpha_f=\alpha_m$.

In the case of a transverse static load applied at the top surface of an FGM plate, the general von Kármán-type equations (Equations 1.33 through 1.36) can be written in the simple form as

$$\tilde{L}_{11}(\overline{W}) - \tilde{L}_{12}(\overline{\Psi}_x) - \tilde{L}_{13}(\overline{\Psi}_y) + \tilde{L}_{14}(\overline{F}) - \tilde{L}_{15}(\overline{N}^T) - \tilde{L}_{16}(\overline{M}^T) = \tilde{L}(\overline{W},\overline{F}) + q \quad (2.2)$$

$$\tilde{L}_{21}(\overline{F}) + \tilde{L}_{22}(\overline{\Psi}_x) + \tilde{L}_{23}(\overline{\Psi}_y) - \tilde{L}_{24}(\overline{W}) - \tilde{L}_{25}(\overline{N}^T) = -\frac{1}{2}\tilde{L}(\overline{W},\overline{W}) \qquad (2.3)$$

$$\tilde{L}_{31}(\overline{W}) + \tilde{L}_{32}(\overline{\Psi}_x) - \tilde{L}_{33}(\overline{\Psi}_y) + \tilde{L}_{34}(\overline{F}) - \tilde{L}_{35}(\overline{N}^T) - \tilde{L}_{36}(\overline{S}^T) = 0 \qquad (2.4)$$

$$\tilde{L}_{41}(\overline{W}) - \tilde{L}_{42}(\overline{\Psi}_x) + \tilde{L}_{43}(\overline{\Psi}_y) + \tilde{L}_{44}(\overline{F}) - \tilde{L}_{45}(\overline{N}^T) - \tilde{L}_{46}(\overline{S}^T) = 0 \qquad (2.5)$$

Note that the geometric nonlinearity in the von Kármán sense is given in terms of $\tilde{L}()$ in Equations 2.2 and 2.3, and the other linear operators $\tilde{L}_{ij}()$ are defined by Equation 1.37, and the forces, moments, and higher order moments caused by elevated temperature are defined by Equation 1.28.

All the edges are assumed to be simply supported. Depending upon the in-plane behavior at the edges, two cases, case 1 (referred to herein as movable edges) and case 2 (referred to herein as immovable edges), will be considered.

Case 1. The edges are simply supported and freely movable in both the X- and Y-directions, respectively.

Case 2. All four edges are simply supported with no in-plane displacements, i.e., prevented from moving in the X- and Y-directions.

For these two cases the associated boundary conditions are
$X=0, a$:

$$\overline{W} = \overline{\Psi}_y = 0 \qquad (2.6a)$$

$$\int_0^b \overline{N}_x dY = 0 \text{ (movable edges)} \qquad (2.6b)$$

$$\overline{U} = 0 \text{ (immovable edges)} \qquad (2.6c)$$

$Y = 0, b$:

$$\overline{W} = \overline{\Psi}_x = 0 \tag{2.6d}$$

$$\int_0^a \overline{N}_y \mathrm{d}X = 0 \text{ (movable edges)} \tag{2.6e}$$

$$\overline{V} = 0 \text{ (immovable edges)} \tag{2.6f}$$

It is noted that the presence of stretching–bending coupling gives rise to bending curvatures under the action of in-plane loading, no matter how small these loads may be. In this situation, the boundary condition of zero bending moment cannot be incorporated accurately. Because immovable edges are considered in the present analysis, $\overline{M}_x = 0$ (at $X = 0$, a) and $\overline{M}_y = 0$ (at $Y = 0$, b) are not included in Equation 2.6, as previously shown in Kuppusamy and Reddy (1984) and Singh et al. (1994).

The unit end-shortening relationships are

$$\frac{\Delta_x}{a} = -\frac{1}{ab} \int_0^b \int_0^a \frac{\partial \overline{U}}{\partial X} \mathrm{d}X\, \mathrm{d}Y$$

$$= -\frac{1}{ab} \int_0^b \int_0^a \left\{ \left[A_{11}^* \frac{\partial^2 \overline{F}}{\partial Y^2} + A_{12}^* \frac{\partial^2 \overline{F}}{\partial X^2} + \left(B_{11}^* - \frac{4}{3h^2} E_{11}^* \right) \frac{\partial \overline{\Psi}_x}{\partial X} \right.\right.$$

$$+ \left(B_{12}^* - \frac{4}{3h^2} E_{12}^* \right) \frac{\partial \overline{\Psi}_y}{\partial Y} - \frac{4}{3h^2} \left(E_{11}^* \frac{\partial^2 \overline{W}}{\partial X^2} + E_{12}^* \frac{\partial^2 \overline{W}}{\partial Y^2} \right) \right]$$

$$\left. - \frac{1}{2} \left(\frac{\partial \overline{W}}{\partial X} \right)^2 - \left(A_{11}^* \overline{N}_x^T + A_{12}^* \overline{N}_y^T \right) \right\} \mathrm{d}X\, \mathrm{d}Y \tag{2.7a}$$

$$\frac{\Delta_y}{b} = -\frac{1}{ab} \int_0^a \int_0^b \frac{\partial \overline{V}}{\partial Y} \mathrm{d}Y\, \mathrm{d}X$$

$$= -\frac{1}{ab} \int_0^a \int_0^b \left\{ \left[A_{22}^* \frac{\partial^2 \overline{F}}{\partial X^2} + A_{12}^* \frac{\partial^2 \overline{F}}{\partial Y^2} + \left(B_{21}^* - \frac{4}{3h^2} E_{21}^* \right) \frac{\partial \overline{\Psi}_x}{\partial X} \right.\right.$$

$$+ \left(B_{22}^* - \frac{4}{3h^2} E_{22}^* \right) \frac{\partial \overline{\Psi}_y}{\partial Y} - \frac{4}{3h^2} \left(E_{21}^* \frac{\partial^2 \overline{W}}{\partial X^2} + E_{22}^* \frac{\partial^2 \overline{W}}{\partial Y^2} \right) \right]$$

$$\left. - \frac{1}{2} \left(\frac{\partial \overline{W}}{\partial Y} \right)^2 - \left(A_{12}^* \overline{N}_x^T + A_{22}^* \overline{N}_y^T \right) \right\} \mathrm{d}Y\, \mathrm{d}X \tag{2.7b}$$

where Δ_x and Δ_y are plate end-shortening displacements in the X- and Y-directions.

Having developed the theory, we are now in a position to solve Equations 2.2 through 2.5 with boundary condition (Equation 2.6). Before proceeding, it is convenient to first define the following dimensionless quantities for such plates (with γ_{ijk} in Equations 2.14 and 2.16 below are defined as in Appendix A):

$$x = \pi X/a, \quad y = \pi Y/b, \quad \beta = a/b, \quad W = \overline{W}/[D_{11}^* D_{22}^* A_{11}^* A_{22}^*]^{1/4}$$

$$F = \overline{F}/[D_{11}^* D_{22}^*]^{1/2}, \quad (\Psi_x,\Psi_y) = (\overline{\Psi}_x,\overline{\Psi}_y)a/\pi[D_{11}^* D_{22}^* A_{11}^* A_{22}^*]^{1/4}$$

$$\gamma_{14} = [D_{22}^*/D_{11}^*]^{1/2}, \quad \gamma_{24} = [A_{11}^*/A_{22}^*]^{1/2}, \quad \gamma_5 = -A_{12}^*/A_{22}^*$$

$$(\gamma_{T1},\gamma_{T2}) = (A_x^T,A_y^T)a^2/\pi^2[D_{11}^* D_{22}^*]^{1/2},$$

$$(\gamma_{T3},\gamma_{T4},\gamma_{T6},\gamma_{T7}) = (D_x^T,D_y^T,F_x^T,F_y^T)a^2/\pi^2 h^2 D_{11}^*,$$

$$(M_x,M_y,P_x,P_y,M_x^T,M_y^T,P_x^T,P_y^T)$$
$$= (\overline{M}_x,\overline{M}_y,4\overline{P}_x/3h^2,4\overline{P}_y/3h^2,\overline{M}_x^T,\overline{M}_y^T,4\overline{P}_x^T/3h^2,4\overline{P}_y^T/3h^2)a^2/\pi^2 D_{11}^*[D_{11}^* D_{22}^* A_{11}^* A_{22}^*]^{1/4}$$

$$\lambda_q = q_0 a^4/\pi^4 D_{11}^*[D_{11}^* D_{22}^* A_{11}^* A_{22}^*]^{1/4},$$

$$(\delta_x,\delta_y) = (\Delta_x/a,\Delta_y/b)b^2/4\pi^2[D_{11}^* D_{22}^* A_{11}^* A_{22}^*]^{1/2} \tag{2.8}$$

where $A_x^T(=A_y^T)$, $D_x^T(=D_y^T)$, and $F_x^T(=F_y^T)$ are defined by

$$\begin{bmatrix} A_x^T & D_x^T & F_x^T \\ A_y^T & D_y^T & F_y^T \end{bmatrix} = -\int_{-h/2}^{h/2} \begin{bmatrix} A_x \\ A_y \end{bmatrix}(1,Z,Z^3)dZ \tag{2.9}$$

and the details of which can be found in Appendix B.

The nonlinear governing equations (Equations 2.2 through 2.5) can then be written in dimensionless form as

$$L_{11}(W) - L_{12}(\Psi_x) - L_{13}(\Psi_y) + \gamma_{14}L_{14}(F) - L_{16}(M^T) = \gamma_{14}\beta^2 L(W,F) + \lambda_q \tag{2.10}$$

$$L_{21}(F) + \gamma_{24}L_{22}(\Psi_x) + \gamma_{24}L_{23}(\Psi_y) - \gamma_{24}L_{24}(W) = -\frac{1}{2}\gamma_{24}\beta^2 L(W,W) \tag{2.11}$$

$$L_{31}(W) + L_{32}(\Psi_x) - L_{33}(\Psi_y) + \gamma_{14}L_{34}(F) - L_{36}(S^T) = 0 \tag{2.12}$$

$$L_{41}(W) - L_{42}(\Psi_x) + L_{43}(\Psi_y) + \gamma_{14}L_{44}(F) - L_{46}(S^T) = 0 \tag{2.13}$$

where

$$L_{11}() = \gamma_{110}\frac{\partial^4}{\partial x^4} + 2\gamma_{112}\beta^2\frac{\partial^4}{\partial x^2\partial y^2} + \gamma_{114}\beta^4\frac{\partial^4}{\partial y^4}$$

$$L_{12}() = \gamma_{120}\frac{\partial^3}{\partial x^3} + \gamma_{122}\beta^2\frac{\partial^3}{\partial x\partial y^2}$$

$$L_{13}() = \gamma_{131}\beta\frac{\partial^3}{\partial x^2\partial y} + \gamma_{133}\beta^3\frac{\partial^3}{\partial y^3}$$

$$L_{14}() = \gamma_{140}\frac{\partial^4}{\partial x^4} + \gamma_{142}\beta^2\frac{\partial^4}{\partial x^2\partial y^2} + \gamma_{144}\beta^4\frac{\partial^4}{\partial y^4}$$

$$L_{16}(M^T) = \frac{\partial^2}{\partial x^2}(M_x^T) + 2\beta\frac{\partial^2}{\partial x\partial y}(M_{xy}^T) + \beta^2\frac{\partial^2}{\partial y^2}(M_y^T)$$

$$L_{21}() = \frac{\partial^4}{\partial x^4} + 2\gamma_{212}\beta^2\frac{\partial^4}{\partial x^2\partial y^2} + \gamma_{214}\beta^4\frac{\partial^4}{\partial y^4}$$

$$L_{22}() = \gamma_{220}\frac{\partial^3}{\partial x^3} + \gamma_{222}\beta^2\frac{\partial^3}{\partial x\partial y^2}$$

$$L_{23}() = \gamma_{231}\beta\frac{\partial^3}{\partial x^2\partial y} + \gamma_{233}\beta^3\frac{\partial^3}{\partial y^3}$$

$$L_{24}() = \gamma_{240}\frac{\partial^4}{\partial x^4} + \gamma_{242}\beta^2\frac{\partial^4}{\partial x^2\partial y^2} + \gamma_{244}\beta^4\frac{\partial^4}{\partial y^4}$$

$$L_{31}() = \gamma_{31}\frac{\partial}{\partial x} + \gamma_{310}\frac{\partial^3}{\partial x^3} + \gamma_{312}\beta^2\frac{\partial^3}{\partial x\partial y^2}$$

$$L_{32}() = \gamma_{31} - \gamma_{320}\frac{\partial^2}{\partial x^2} - \gamma_{322}\beta^2\frac{\partial^2}{\partial y^2}$$

$$L_{33}() = \gamma_{331}\beta\frac{\partial^2}{\partial x\partial y}$$

$$L_{34}() = L_{22}()$$

$$L_{36}(S^T) = \frac{\partial}{\partial x}(S_x^T) + \beta\frac{\partial}{\partial y}(S_{xy}^T)$$

$$L_{41}() = \gamma_{41}\beta\frac{\partial}{\partial y} + \gamma_{411}\beta\frac{\partial^3}{\partial x^2\partial y} + \gamma_{413}\beta^3\frac{\partial^3}{\partial y^3}$$

$$L_{42}() = L_{33}()$$

$$L_{43}() = \gamma_{41} - \gamma_{430}\frac{\partial^2}{\partial x^2} - \gamma_{432}\beta^2\frac{\partial^2}{\partial y^2}$$

$$L_{44}() = L_{23}()$$

$$L_{46}(S^T) = \frac{\partial}{\partial x}(S_{xy}^T) + \beta\frac{\partial}{\partial y}(S_y^T)$$

$$L() = \frac{\partial^2}{\partial x^2}\frac{\partial^2}{\partial y^2} - 2\frac{\partial^2}{\partial x\partial y}\frac{\partial^2}{\partial x\partial y} + \frac{\partial^2}{\partial y^2}\frac{\partial^2}{\partial x^2}$$

(2.14)

The boundary conditions of Equation 2.6 become

$x = 0, \pi$:

$$W = \Psi_y = 0 \tag{2.15a}$$

$$\int_0^\pi \beta^2 \frac{\partial^2 F}{\partial y^2} \, dy = 0 \text{ (movable edges)} \tag{2.15b}$$

$$\delta_x = 0 \text{ (immovable edges)} \tag{2.15c}$$

$y = 0, \pi$:

$$W = \Psi_x = 0 \tag{2.15d}$$

$$\int_0^\pi \frac{\partial^2 F}{\partial x^2} \, dx = 0 \text{ (movable edges)} \tag{2.15e}$$

$$\delta_y = 0 \text{ (immovable edges)} \tag{2.15f}$$

and the unit end-shortening relationships become

$$\delta_x = -\frac{1}{4\pi^2 \beta^2 \gamma_{24}} \int_0^\pi \int_0^\pi \left\{ \left[\gamma_{24}^2 \beta^2 \frac{\partial^2 F}{\partial y^2} - \gamma_5 \frac{\partial^2 F}{\partial x^2} + \gamma_{24} \left(\gamma_{511} \frac{\partial \Psi_x}{\partial x} + \gamma_{233} \beta \frac{\partial \Psi_y}{\partial y} \right) \right. \right.$$
$$\left. \left. - \gamma_{24} \left(\gamma_{611} \frac{\partial^2 W}{\partial x^2} + \gamma_{244} \beta^2 \frac{\partial^2 W}{\partial y^2} \right) \right] - \frac{1}{2} \gamma_{24} \left(\frac{\partial W}{\partial x} \right)^2 + (\gamma_{24}^2 \gamma_{T1} - \gamma_5 \gamma_{T2}) \Delta T \right\} dx \, dy \tag{2.16a}$$

$$\delta_y = -\frac{1}{4\pi^2 \beta^2 \gamma_{24}} \int_0^\pi \int_0^\pi \left\{ \left[\frac{\partial^2 F}{\partial x^2} - \gamma_5 \beta^2 \frac{\partial^2 F}{\partial y^2} + \gamma_{24} \left(\gamma_{220} \frac{\partial \Psi_x}{\partial x} + \gamma_{522} \beta \frac{\partial \Psi_y}{\partial y} \right) \right. \right.$$
$$\left. \left. - \gamma_{24} \left(\gamma_{240} \frac{\partial^2 W}{\partial x^2} + \gamma_{622} \beta^2 \frac{\partial^2 W}{\partial y^2} \right) \right] - \frac{1}{2} \gamma_{24} \beta^2 \left(\frac{\partial W}{\partial y} \right)^2 + (\gamma_{T2} - \gamma_5 \gamma_{T1}) \Delta T \right\} dy \, dx \tag{2.16b}$$

Applying Equations 2.10 through 2.16, the nonlinear bending response of a simply supported FGM plate subjected to a transverse uniform or sinusoidal load and in thermal environments is now determined by means of a two-step perturbation technique, for which the small perturbation parameter has no physical meaning at the first step, and is then replaced by a dimensionless central deflection at the second step. The essence of this procedure, in the present case, is to assume that

$$W(x, y, \varepsilon) = \sum_{j=1} \varepsilon^j w_j(x, y), \quad F(x, y, \varepsilon) = \sum_{j=0} \varepsilon^j f_j(x, y)$$

$$\Psi_x(x, y, \varepsilon) = \sum_{j=1} \varepsilon^j \psi_{xj}(x, y), \quad \Psi_y(x, y, \varepsilon) = \sum_{j=1} \varepsilon^j \psi_{yj}(x, y), \quad \lambda_q = \sum_{j=1} \varepsilon^j \lambda_j$$

$$\tag{2.17}$$

where ε is a small perturbation parameter and the first term of $w_j(x, y)$ is assumed to have the form:

$$w_1(x, y) = A_{11}^{(1)} \sin mx \sin ny \tag{2.18}$$

Then, we expand the thermal bending moments in the double Fourier sine series as

$$\begin{bmatrix} M_x^T & S_x^T \\ M_y^T & S_y^T \end{bmatrix} = -\varepsilon \begin{bmatrix} M_x^{(1)} & S_x^{(1)} \\ M_y^{(1)} & S_y^{(1)} \end{bmatrix} \sum_{i=1,3,\dots} \sum_{j=1,3,\dots} \frac{1}{ij} \sin ix \sin jy \tag{2.19}$$

where $M_x^{(1)}$, $M_y^{(1)}$, $S_x^{(1)}$, and $S_y^{(1)}$ and all coefficients in Equations 2.32 through 2.34 below are given in detail in Appendix C.

Substituting Equation 2.17 into Equations 2.10 through 2.13, collecting the terms of the same order of ε, we obtain a set of perturbation equations which can be written, for example, as

$O(\varepsilon^0)$:

$$L_{14}(f_0) = 0 \tag{2.20a}$$

$$L_{21}(f_0) = 0 \tag{2.20b}$$

$$L_{34}(f_0) = 0 \tag{2.20c}$$

$$L_{44}(f_0) = 0 \tag{2.20d}$$

$O(\varepsilon^1)$:

$$L_{11}(w_1) - L_{12}(\psi_{x1}) - L_{13}(\psi_{y1}) + \gamma_{14}L_{14}(f_1) = \gamma_{14}\beta^2 L(w_1, f_0) + \lambda_1 \tag{2.21a}$$

$$L_{21}(f_1) + \gamma_{24}L_{22}(\psi_{x1}) + \gamma_{24}L_{23}(\psi_{y1}) - \gamma_{24}L_{24}(w_1) = 0 \tag{2.21b}$$

$$L_{31}(w_1) + L_{32}(\psi_{x1}) - L_{33}(\psi_{y1}) + \gamma_{14}L_{34}(f_1) = 0 \tag{2.21c}$$

$$L_{41}(w_1) - L_{42}(\psi_{x1}) + L_{43}(\psi_{y1}) + \gamma_{14}L_{44}(f_1) = 0 \tag{2.21d}$$

$O(\varepsilon^2)$:

$$L_{11}(w_2) - L_{12}(\psi_{x2}) - L_{13}(\psi_{y2}) + \gamma_{14}L_{14}(f_2) = \gamma_{14}\beta^2 [L(w_2, f_0) + L(w_1, f_1)] + \lambda_2 \tag{2.22a}$$

$$L_{21}(f_2) + \gamma_{24}L_{22}(\psi_{x2}) + \gamma_{24}L_{23}(\psi_{y2}) - \gamma_{24}L_{24}(w_2) = -\frac{1}{2}\gamma_{24}\beta^2 L(w_1, w_1) \tag{2.22b}$$

$$L_{31}(w_2) + L_{32}(\psi_{x2}) - L_{33}(\psi_{y2}) + \gamma_{14}L_{34}(f_2) = 0 \tag{2.22c}$$

$$L_{41}(w_2) - L_{42}(\psi_{x2}) + L_{43}(\psi_{y2}) + \gamma_{14}L_{44}(f_2) = 0 \tag{2.22d}$$

$O(\varepsilon^3)$:

$$L_{11}(w_3) - L_{12}(\psi_{x3}) - L_{13}(\psi_{y3}) + \gamma_{14}L_{14}(f_3)$$
$$= \gamma_{14}\beta^2[L(w_3,f_0) + L(w_2,f_1) + L(w_1,f_2)] + \lambda_3 \qquad (2.23a)$$

$$L_{21}(f_3) + \gamma_{24}L_{22}(\psi_{x3}) + \gamma_{24}L_{23}(\psi_{y3}) - \gamma_{24}L_{24}(w_3) = -\frac{1}{2}\gamma_{24}\beta^2 L(w_1, w_2) \quad (2.23b)$$

$$L_{31}(w_3) + L_{32}(\psi_{x3}) - L_{33}(\psi_{y3}) + \gamma_{14}L_{34}(f_3) = 0 \qquad (2.23c)$$

$$L_{41}(w_3) - L_{42}(\psi_{x3}) + L_{43}(\psi_{y3}) + \gamma_{14}L_{44}(f_3) = 0 \qquad (2.23d)$$

To solve these perturbation equations of each order, the amplitudes of the terms $w_j(x, y)$, $f_j(x, y)$, $\psi_{xj}(x, y)$, and $\psi_{yj}(x, y)$ can be determined step by step, and λ_j can be determined by the Galerkin procedure. As a result, up to third-order asymptotic solutions can be obtained as

$$W = \varepsilon\left[A_{11}^{(1)} \sin mx \sin ny\right] + \varepsilon^3\left[A_{13}^{(3)} \sin mx \sin 3ny + A_{31}^{(3)} \sin 3mx \sin ny\right] + O(\varepsilon^4)$$
$$(2.24)$$

$$\Psi_x = \varepsilon\left[C_{11}^{(1)} \cos mx \sin ny\right] + \varepsilon^2\left[C_{20}^{(2)} \sin 2mx\right]$$
$$+ \varepsilon^3\left[C_{13}^{(3)} \cos mx \sin 3ny + C_{31}^{(3)} \cos 3mx \sin ny\right] + O(\varepsilon^4) \qquad (2.25)$$

$$\Psi_y = \varepsilon\left[D_{11}^{(1)} \sin mx \cos ny\right] + \varepsilon^2\left[D_{02}^{(2)} \sin 2ny\right]$$
$$+ \varepsilon^3\left[D_{13}^{(3)} \sin mx \cos 3ny + D_{31}^{(3)} \sin 3mx \cos ny\right] + O(\varepsilon^4) \qquad (2.26)$$

$$F = -B_{00}^{(0)}\frac{y^2}{2} - b_{00}^{(0)}\frac{x^2}{2} + \varepsilon\left[B_{11}^{(1)} \sin mx \sin ny\right]$$
$$+ \varepsilon^2\left[-B_{00}^{(2)}\frac{y^2}{2} - b_{00}^{(2)}\frac{x^2}{2} + B_{20}^{(2)} \cos 2mx + B_{02}^{(2)} \cos 2ny\right]$$
$$+ \varepsilon^3\left[B_{13}^{(3)} \sin mx \sin 3ny + B_{31}^{(3)} \sin 3mx \sin ny\right] + O(\varepsilon^4) \qquad (2.27)$$

Note that for boundary condition case 1, it is just necessary to take $B_{00}^{(i)} = b_{00}^{(i)} = 0$ ($i = 0, 2$) in Equation 2.27, so that the asymptotic solutions have a similar form, and

$$\lambda_q = \varepsilon\lambda_1 + \varepsilon^2\lambda_2 + \varepsilon^3\lambda_3 + O(\varepsilon^4) \qquad (2.28)$$

All coefficients in Equations 2.24 through 2.27 are related and can be written as functions of $A_{11}^{(1)}$, for example

$$B_{11}^{(1)} = \gamma_{24}\frac{g_{05}}{g_{06}}A_{11}^{(1)}, \quad B_{20}^{(2)} = \frac{\gamma_{24}n^2\beta^2}{32m^2\gamma_6}\left(A_{11}^{(1)}\right)^2, \quad B_{02}^{(2)} = \frac{\gamma_{24}m^2}{32n^2\beta^2\gamma_7}\left(A_{11}^{(1)}\right)^2$$

$$C_{11}^{(1)} = -m\left(\frac{g_{04}}{g_{00}} - \gamma_{14}\gamma_{24}\frac{g_{02}g_{05}}{g_{00}g_{06}}\right)A_{11}^{(1)}, \quad C_{20}^{(2)} = -\gamma_{14}\gamma_{24}\gamma_{220}\frac{mn^2\beta^2}{4(\gamma_{31} + \gamma_{320}4m^2)\gamma_6}\left(A_{11}^{(1)}\right)^2$$

$$D_{11}^{(1)} = -n\beta\left(\frac{g_{03}}{g_{00}} - \gamma_{14}\gamma_{24}\frac{g_{01}g_{05}}{g_{00}g_{06}}\right)A_{11}^{(1)}, \quad D_{02}^{(2)} = -\gamma_{14}\gamma_{24}\gamma_{233}\frac{m^2n\beta}{4(\gamma_{41} + \gamma_{432}4n^2\beta^2)\gamma_7}\left(A_{11}^{(1)}\right)^2$$

$$(2.29)$$

All symbols used in Equation 2.29 are also described in detail in Appendix C. Hence Equations 2.24 and 2.28 can be rewritten as

$$W = W^{(1)}(x,y)\left(A_{11}^{(1)}\varepsilon\right) + W^{(3)}(x,y)\left(A_{11}^{(1)}\varepsilon\right)^3 + \cdots \tag{2.30}$$

and

$$\lambda_q = \lambda_q^{(1)}\left(A_{11}^{(1)}\varepsilon\right) + \lambda_q^{(2)}\left(A_{11}^{(1)}\varepsilon\right)^2 + \lambda_q^{(3)}\left(A_{11}^{(1)}\varepsilon\right)^3 + \cdots \tag{2.31}$$

From Equations 2.30 and 2.31 the load–central deflection relationship can be written as

$$\frac{q_0 a^4}{D_{11}^* h} = A_W^{(0)} + A_W^{(1)}\left(\frac{\overline{W}}{h}\right) + A_W^{(2)}\left(\frac{\overline{W}}{h}\right)^2 + A_W^{(3)}\left(\frac{\overline{W}}{h}\right)^3 + \cdots \tag{2.32}$$

Similarly, the bending moment–central deflection relationships can be written as

$$\frac{\overline{M}_x a^2}{D_{11}^* h} = A_{MX}^{(0)} + A_{MX}^{(1)}\left(\frac{\overline{W}}{h}\right) + A_{MX}^{(2)}\left(\frac{\overline{W}}{h}\right)^2 + A_{MX}^{(3)}\left(\frac{\overline{W}}{h}\right)^3 + \cdots \tag{2.33}$$

$$\frac{\overline{M}_y a^2}{D_{11}^* h} = A_{MY}^{(0)} + A_{MY}^{(1)}\left(\frac{\overline{W}}{h}\right) + A_{MY}^{(2)}\left(\frac{\overline{W}}{h}\right)^2 + A_{MY}^{(3)}\left(\frac{\overline{W}}{h}\right)^3 + \cdots \tag{2.34}$$

Equations 2.32 through 2.34 can be employed to obtain numerical results for the load–deflection and load–bending moment curves of an FGM plate subjected to a transverse uniform or sinusoidal load and in thermal environments. Zirconia and titanium alloys were selected for the two constituent materials of the plate in the present examples, referred to as ZrO_2/Ti-6Al-4V. The material properties of these two constituents are assumed to be nonlinear function of temperature of Equation 1.4 (Touloukian 1967), and typical values for Young's modulus E_f (in Pa) and thermal expansion coefficient α_f

(in K^{-1}) of Zirconia and Ti-6Al-4V can be found in Tables 1.1 and 1.2 (from Reddy and Chin 1998). Poisson's ratio ν_f is assumed to be a constant, and $\nu_f = 0.28$. The results presented herein are for "movable" in-plane boundary conditions, unless it is stated otherwise.

The load–center deflection curves for a zirconia/aluminum square plate with different values of the volume fraction index subjected to a uniform transverse load are compared in Figure 2.1 with numerical results of Praveen and Reddy (1998), using their material properties, i.e., for aluminum $E = 70$ GPa, $\nu = 0.3$, $\alpha = 23.0 \times 10^{-6}$ °C^{-1}, and for zirconia $E = 151$ GPa, $\nu = 0.3$, and $\alpha = 10.0 \times 10^{-6}$ °C^{-1}. Note that in Figure 2.1 the volume fraction index N is defined for V_c, and E_0 is a referenced value of Young's modulus, and $E_0 = 70$ GPa. Clearly, the comparison is reasonably well.

Figure 2.2 gives the load–deflection and load–bending moment curves of ZrO$_2$/Ti-6Al-4V square plate with different values of volume fraction index N ($= 0$, 0.5, 1.0, 2.0, 5.0, and ∞) subjected to a uniform pressure and under thermal environmental condition $\Delta T = 0$ K. The results show that a fully titanium alloy plate ($N = 0$) has highest deflection and lowest bending moment. It can also be seen that the plate has higher deflection and lower bending moment when it has lower volume fraction. This is expected because the metallic plate is the one with the lower stiffness than the ceramic plate.

Figure 2.3 gives the load–deflection and load–bending moment curves of ZrO$_2$/Ti-6Al-4V square plates subjected to a uniform pressure and under three sets of thermal environmental conditions, referred to as I, II, and III. For environmental condition I, $\Delta T = 0$ K; for environmental condition II,

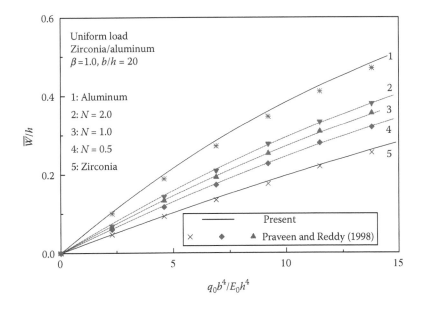

FIGURE 2.1
Comparisons of load–central deflection curves for a zirconia/aluminum square plate.

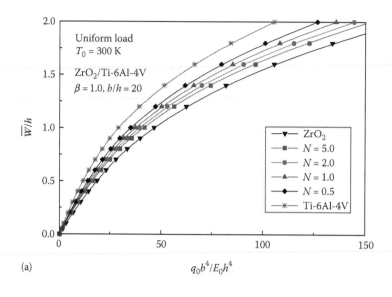

(a)

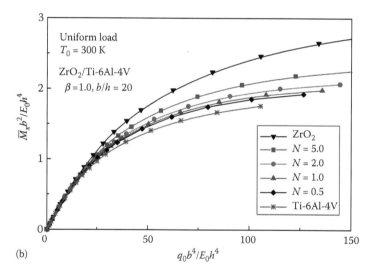

(b)

FIGURE 2.2

Effect of volume fraction index N on the nonlinear bending behavior of ZrO_2/Ti-6Al-4V square plates under uniform pressure: (a) load–central deflection; (b) load–bending moment.

$\Delta T = 200$ K; and for environmental condition III, $\Delta T = 300$ K. Because the thermal expansion at the top surface is higher than that at the bottom surface, this expansion results in an upward deflection. It is seen that the deflections are reduced, but the bending moments are increased with increases in temperature. Note that for environmental conditions II and III the deflections are close to each other when $\overline{W}/h > 1.5$.

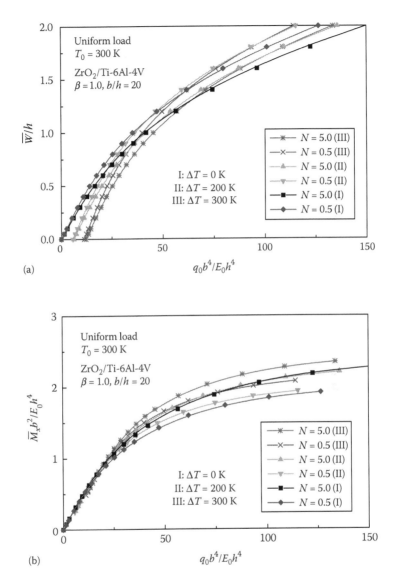

FIGURE 2.3
Effect of temperature rise on the nonlinear bending behavior of ZrO_2/Ti-6Al-4V square plates under uniform pressure: (a) load–central deflection; (b) load–bending moment.

Figure 2.4 shows the effect of in-plane boundary conditions on the nonlinear bending behavior of ZrO_2/Ti-6Al-4V square plates under two environmental conditions. To this end, the load–deflection and load–bending moment curves of ZrO_2/Ti-6Al-4V square plates under "movable" and "immovable" in-plane boundary conditions are displayed. The results

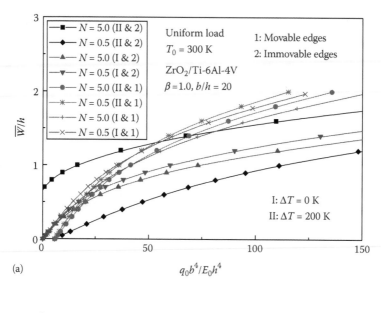

(a)

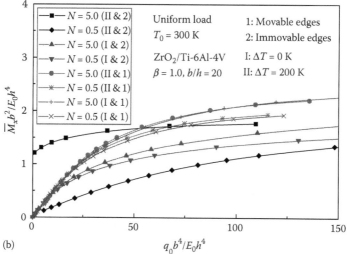

(b)

FIGURE 2.4
Effect of in-plane boundary conditions on the nonlinear bending behavior of ZrO_2/Ti-6Al-4V square plates under uniform pressure: (a) load–central deflection; (b) load–bending moment.

show that the plate with immovable edges will undergo less deflection with smaller bending moments.

Figure 2.5 compares the load–deflection and load–bending moment curves of ZrO_2/Ti-6Al-4V square plates under two cases of transverse loading conditions along with two environmental conditions. It can be seen that both

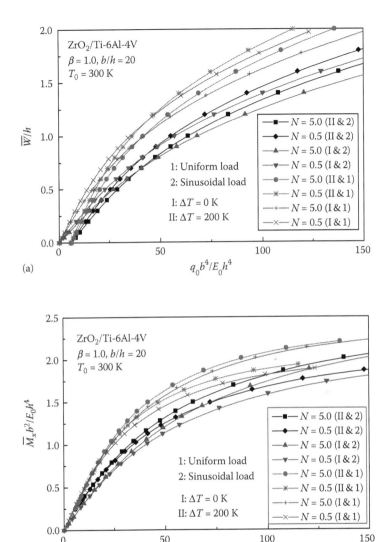

(a)

(b)

FIGURE 2.5
Comparisons of nonlinear responses of ZrO_2/Ti-6Al-4V square plates subjected to a uniform or sinusoidal load and in thermal environments: (a) load–central deflection; (b) load–bending moment.

load–deflection and load–bending moment curves of the plate subjected to a sinusoidal load are lower than those of the plate subjected to a uniform load.

It is appreciated that in Figures 2.2 through 2.5, $\overline{W}/h$, $\overline{M}_x b^2/E_0 h^4$, and $q_0 b^4/E_0 h^4$ denote the dimensionless central deflection of the plate, central bending moment, and lateral pressure, respectively, where $E_0 =$ Young's modulus of Ti-6Al-4V at $T = 300$ K.

2.3 Nonlinear Thermal Bending of FGM Plates due to Heat Conduction

In the case of a temperature field applied at the top and bottom surfaces of an FGM plate, thermal bending usually occurs due to heat conduction. In the present case, we assume that the effective Young's modulus E_f, thermal expansion coefficient α_f, and thermal conductivity κ_f are functions of temperature, so that E_f, α_f, and κ_f are both temperature- and position-dependent. The Poisson ratio ν_f depends weakly on temperature change and is assumed to be a constant. According to mixture rules, we have

$$E_f(Z,T) = [E_m(T) - E_c(T)]\left(\frac{2Z+h}{2h}\right)^N + E_c(T) \tag{2.35a}$$

$$\alpha_f(Z,T) = [\alpha_m(T) - \alpha_c(T)]\left(\frac{2Z+h}{2h}\right)^N + \alpha_c(T) \tag{2.35b}$$

$$\kappa_f(Z,T) = [\kappa_m(T) - \kappa_c(T)]\left(\frac{2Z+h}{2h}\right)^N + \kappa_c(T) \tag{2.35c}$$

We assume that the temperature variation occurs in the thickness direction only and 1D temperature field is assumed to be constant in the XY plane of the plate. In such a case, the temperature distribution along the thickness can be obtained by solving a steady-state heat transfer equation:

$$-\frac{d}{dZ}\left[\kappa\frac{dT}{dZ}\right] = 0 \tag{2.36}$$

Equation 2.36 is solved by imposing the boundary conditions $T = T_U$ at $Z = -h/2$ and $T = T_L$ at $Z = h/2$. The solution of this equation, by means of polynomial series, is (Javaheri and Eslami 2002):

$$T(Z) = T_U + (T_L - T_U)\eta(Z) \tag{2.37}$$

where T_U and T_L are the temperatures at top and bottom surfaces of the plate, and

$$\eta(Z) = \frac{1}{C}\left[\left(\frac{2Z+h}{2h}\right) - \frac{\kappa_{mc}}{(N+1)\kappa_c}\left(\frac{2Z+h}{2h}\right)^{N+1} + \frac{\kappa_{mc}^2}{(2N+1)\kappa_c^2}\left(\frac{2Z+h}{2h}\right)^{2N+1}\right.$$
$$- \frac{\kappa_{mc}^3}{(3N+1)\kappa_c^3}\left(\frac{2Z+h}{2h}\right)^{3N+1} + \frac{\kappa_{mc}^4}{(4N+1)\kappa_c^4}\left(\frac{2Z+h}{2h}\right)^{4N+1}$$
$$\left. - \frac{\kappa_{mc}^5}{(5N+1)\kappa_c^5}\left(\frac{2Z+h}{2h}\right)^{5N+1}\right] \tag{2.38a}$$

$$C = 1 - \frac{\kappa_{mc}}{(N+1)\kappa_c} + \frac{\kappa_{mc}^2}{(2N+1)\kappa_c^2} - \frac{\kappa_{mc}^3}{(3N+1)\kappa_c^3} + \frac{\kappa_{mc}^4}{(4N+1)\kappa_c^4} - \frac{\kappa_{mc}^5}{(5N+1)\kappa_c^5}$$

$$(2.38b)$$

where $\kappa_{mc} = \kappa_m - \kappa_c$. In particular, for an isotropic material, Equation 2.37 may then be expressed as

$$T(Z) = \frac{T_U + T_L}{2} + \frac{T_U - T_L}{h} Z \qquad (2.39)$$

In the present case, since there is no transverse static load applied, the nonlinear governing equations can be written in dimensionless form as

$$L_{11}(W) - L_{12}(\Psi_x) - L_{13}(\Psi_y) + \gamma_{14}L_{14}(F) = \gamma_{14}\beta^2 L(W,F) + L_{16}(M^T) \qquad (2.40)$$

$$L_{21}(F) + \gamma_{24}L_{22}(\Psi_x) + \gamma_{24}L_{23}(\Psi_y) - \gamma_{24}L_{24}(W) = -\frac{1}{2}\gamma_{24}\beta^2 L(W,W) \qquad (2.41)$$

$$L_{31}(W) + L_{32}(\Psi_x) - L_{33}(\Psi_y) + \gamma_{14}L_{34}(F) = L_{36}(S^T) \qquad (2.42)$$

$$L_{41}(W) - L_{42}(\Psi_x) + L_{43}(\Psi_y) + \gamma_{14}L_{44}(F) = L_{46}(S^T) \qquad (2.43)$$

where all nondimensional linear operators $L_{ij}()$ and nonlinear operator $L()$ are defined as expressed by Equation 2.14. Note that Equation 2.9 is now redefined by

$$\begin{bmatrix} A_x^T & D_x^T & F_x^T \\ A_y^T & D_y^T & F_y^T \end{bmatrix} \Delta T = - \int_{-h/2}^{+h/2} \begin{bmatrix} A_x \\ A_y \end{bmatrix} (1, Z, Z^3) \Delta T(X, Y, Z) dZ \qquad (2.44)$$

where ΔT is a constant and is defined by $\Delta T = T_U - T_L$. When $\Delta T = 0$, $A_x^T = (A_y^T)$, $D_x^T = (D_y^T)$, and $F_x^T = (F_y^T)$ can be found in Appendix B, and if $\Delta T \neq 0$, then A_x^T, D_x^T, and F_x^T can be found in Appendix D.

$L_{16}(M^P)$, $L_{36}(S^P)$, and $L_{46}(S^P)$ in Equations 2.40, 2.42, and 2.43 are now treated as "pseudoloads," and we expand the thermal bending moments in the double Fourier sine series as

$$\begin{bmatrix} M_x^T & S_x^T \\ M_y^T & S_y^T \end{bmatrix} = - \begin{bmatrix} M_x^0 & S_x^0 \\ M_y^0 & S_y^0 \end{bmatrix} \sum_{i=1,3,\ldots} \sum_{j=1,3,\ldots} \frac{1}{ij} \sin ix \sin jy \qquad (2.45)$$

where M_x^0, M_y^0, S_x^0, and S_y^0 are given in detail in Appendix E.

By using a two-step perturbation technique, we obtain up to third-order asymptotic solutions in the same form as expressed by Equations 2.24 through 2.27, from which we have

$$\frac{\overline{W}}{t} = A_W^{(1)}\left(A_{11}^{(1)}\varepsilon\right) - A_W^{(3)}\left(A_{11}^{(1)}\varepsilon\right)^3 + \cdots \qquad (2.46)$$

and the bending moments can be written as

$$\frac{\overline{M}_x a^2}{D_{11}^* t} = A_{MX}^{(0)} + A_{MX}^{(1)}\left(A_{11}^{(1)}\varepsilon\right) + A_{MX}^{(2)}\left(A_{11}^{(1)}\varepsilon\right)^2 + A_{MX}^{(3)}\left(A_{11}^{(1)}\varepsilon\right)^3 + \cdots \qquad (2.47)$$

$$\frac{\overline{M}_y a^2}{D_{11}^* t} = A_{MY}^{(0)} + A_{MY}^{(1)}\left(A_{11}^{(1)}\varepsilon\right) + A_{MY}^{(2)}\left(A_{11}^{(1)}\varepsilon\right)^2 + A_{MY}^{(3)}\left(A_{11}^{(1)}\varepsilon\right)^3 + \cdots \qquad (2.48)$$

in Equations 2.46 through 2.48, $\left(A_{11}^{(2)}\varepsilon\right)$ is taken as the second perturbation parameter relating to the temperature variation, i.e.,

$$A_{11}^{(2)}\varepsilon = \lambda + \Theta_2(\lambda)^2 + \Theta_3(\lambda)^3 + \cdots \qquad (2.49)$$

All symbols used in Equations 2.46 through 2.49 are also described in detail in Appendix E.

Equations 2.46 through 2.49 can be employed to obtain numerical results for the thermal load–deflection and thermal load–bending moment curves of an FGM plate under heat conduction. Silicon nitride and stainless steel were selected for the two constituent materials of the substrate FGM layer, referred to as Si$_3$N$_4$/SUS304, in the present examples. The material properties of these two constituents are assumed to be nonlinear function of temperature of Equation 1.4, and typical values for Young's modulus E_f (in Pa), thermal expansion coefficient α_f (in K^{-1}), and the thermal conductivity κ_f (in W mK^{-1}) of silicon nitride and stainless steel can be found in Tables 1.1 through 1.3. Poisson's ratio ν_f is assumed to be a constant, and $\nu_f = 0.28$. For these examples, the lower surface is hold at a prescribed temperature of 300 K, so that $\Delta T_L = 0$ and $\Delta T = \Delta T_U$.

Figure 2.6 presents the thermal load–deflection and thermal load–bending moment curves for square FGM plates with different values of volume fraction index N ($= 0$, 0.2, 0.5, 1.0, 2.0, 5.0, and ∞) under temperature variation ΔT_U at upper surface. It can be seen that the deflection of FGM plates with lower values of volume fraction index N is positive (downward), whereas for the plate with higher values of N the deflection becomes negative. This is due to the fact that the thermal expansion coefficient at the lower surface is larger than that experienced by the upper surface. The results show that the plate has higher bending moment (except for $N = 0$ and $N = 0.2$)

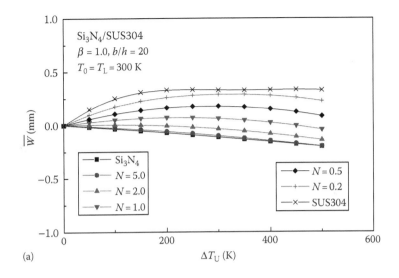

(a)

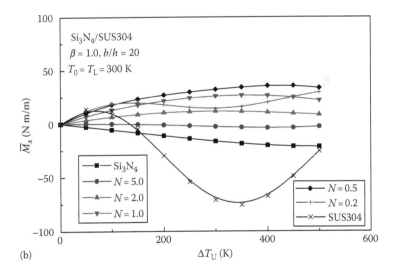

(b)

FIGURE 2.6
Effect of volume fraction index N on the nonlinear bending behavior of $Si_3N_4/SUS304$ square plates due to heat conduction: (a) load–central deflection; (b) load–bending moment.

when it has lower volume fraction. It can also be seen that the bending moment for a fully stainless steel plate ($N=0$) changes from positive to negative when the temperature variation $\Delta T_U > 150$ K. The results reveal that the nonlinear bending responses of an FGM plate due to heat conduction are quite different to those of an FGM plate subjected to transverse mechanical loads.

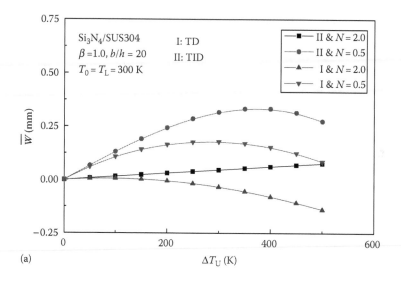

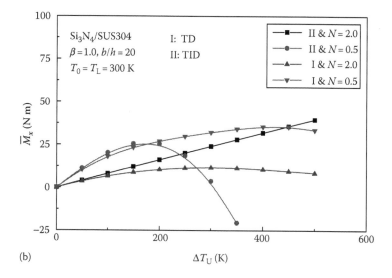

FIGURE 2.7
Effect of temperature-dependency on the nonlinear bending behavior of Si_3N_4/SUS304 square plates due to heat conduction: (a) load–central deflection; (b) load–bending moment.

Figure 2.7 presents the thermal load–deflection and thermal load–bending moment curves for square FGM plates with two values of volume fraction index $N = 0.5$ and 2.0 under two cases of thermoelastic properties TD and TID. Here, TD and TID represent, respectively, the material properties are temperature dependent and temperature independent, i.e., in a fixed temperature $T_0 = 300$ K. Great differences could be seen in these two cases and

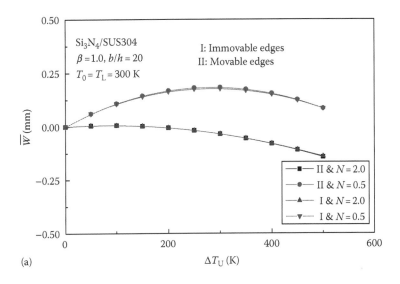

(a)

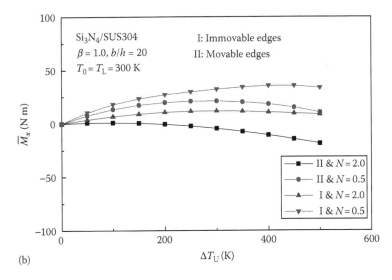

(b)

FIGURE 2.8
Effect of in-plane boundary conditions on the nonlinear bending behavior of Si_3N_4/SUS304 plates due to heat conduction: (a) load–central deflection; (b) load–bending moment.

we believe that the temperature dependency of FGMs could not be neglected in the thermal bending analysis.

Figure 2.8 shows the effect of in-plane boundary conditions on the non-linear thermal bending behavior of FGM plates with $N = 0.5$ and 2.0. The thermal load–deflection and thermal load–bending moment curves of square FGM plates under "movable" and "immovable" in-plane boundary conditions are displayed. The results show that the plate with immovable edges will undergo less deflection with larger bending moments. It can also be seen that for the plate with "movable" edges both deflection and bending moment are decreased by increasing the volume fraction index N.

References

Cheng Z.-Q. (2001), Nonlinear bending of inhomogeneous plates, *Engineering Structures*, **23**, 1359–1363.

Cheng Z.-Q. and Batra R.C. (2000), Three-dimensional thermoelastic deformations of a functionally graded elliptic plate, *Composites Part B*, **31**, 97–106.

Chung Y.-L. and Chen W.-T. (2007), Bending behavior of FGM-coated and FGM-undercoated plates with two simply supported opposite edges and two free edges, *Composite Structures*, **81**, 157–167.

Croce L.D. and Venini P. (2004), Finite elements for functionally graded Reissner-Mindlin plates, *Computer Methods in Applied Mechanics and Engineering*, **193**, 705–725.

GhannadPour S.A.M. and Alinia M.M. (2006), Large deflection behavior of functionally graded plates under pressure loads, *Composite Structures*, **75**, 67–71.

Gunes R. and Reddy J.N. (2008), Nonlinear analysis of functionally graded circular plates under different loads and boundary conditions, *International Journal of Structural Stability and Dynamics*, **8**, 131–159.

Javaheri R. and Eslami M.R. (2002), Thermal buckling of functionally graded plates, *AIAA Journal*, **40**, 162–169.

Kashtalyan M. (2004), Three dimensional elasticity solution for bending of functionally graded rectangular plates, *European Journal of Mechanics A/Solids*, **23**, 853–864.

Kim K.-S. and Noda N. (2002), A Green's function approach to the deflection of a FGM plate under transient thermal loading, *Archive of Applied Mechanics*, **72**, 127–137.

Kuppusamy T. and Reddy J.N. (1984), A three-dimensional nonlinear analysis of cross-ply rectangular composite plates, *Computers and Structures*, **18**, 263–272.

Ma L.S. and Wang T.J. (2003), Nonlinear bending and post-buckling of a functionally graded circular plate under mechanical and thermal loadings, *International Journal of Solids and Structures*, **40**, 3311–3330.

Mizuguchi F. and Ohnabe H. (1996), Large deflections of heated functionally graded simply supported rectangular plates with varying rigidity in thickness direction, in *Proceedings of the 11th Technical Conference of the American Society for Composites*, October 7–9, Atlanta, Georgia, Technomic Publ. Co., Inc., Lancaster, PA, pp. 957–966.

Na K.-S. and Kim J.-H. (2006), Nonlinear bending response of functionally graded plates under thermal loads, *Journal of Thermal Stresses*, **29**, 245–261.

Ovesy H.R. and GhannadPour S.A.M. (2007), Large deflection finite strip analysis of functionally graded plates under pressure loads, *International Journal of Structural Stability and Dynamics*, **7**, 193–211.

Praveen G.N. and Reddy J.N. (1998), Nonlinear transient thermoelastic analysis of functionally graded ceramic-metal plates, *International Journal of Solids and Structures*, **35**, 4457–4476.

Reddy J.N. (2000), Analysis of functionally graded plates, *International Journal for Numerical Methods in Engineering*, **47**, 663–684.

Reddy J.N. and Cheng Z.-Q. (2001), Three-dimensional thermomechanical deformations of functionally graded rectangular plates, *European Journal of Mechanics A/Solids*, **20**, 841–855.

Reddy J.N. and Chin C.D. (1998), Thermoelastical analysis of functionally graded cylinders and plates, *Journal of Thermal Stresses*, **21**, 593–626.

Reddy J.N., Wang C.M., and Kitipornchai S. (1999), Axisymmetric bending of functionally grade circular and annular plates, *European Journal of Mechanics A/Solids*, **18**, 185–199.

Shen H.-S. (2002), Nonlinear bending response of functionally graded plates subjected to transverse loads and in thermal environments, *International Journal of Mechanical Sciences*, **44**, 561–584.

Shen H.-S. (2007), Nonlinear thermal bending response of FGM plates due to heat conduction, *Composites Part B*, **38**, 201–215.

Singh G., Rao G.V., and Iyengar N.G.R. (1994), Geometrically nonlinear flexural response characteristics of shear deformable unsymmetrically laminated plates, *Computers and Structures*, **53**, 69–81.

Suresh S. and Mortensen A. (1997), Functionally graded metals and metal-ceramic composites: Part 2. Thermomechanical behaviour, *International Materials Reviews*, **42**, 85–116.

Tanigawa Y., Akai T., Kawamura R., and Oka N. (1996), Transient heat conduction and thermal stress problems of a nonhomogeneous plate with temperature-dependent material properties, *Journal of Thermal Stresses*, **19**, 77–102.

Touloukian Y.S. (1967), *Thermophysical Properties of High Temperature Solid Materials*, Macmillan, New York.

Vel S.S. and Batra R.C. (2002), Exact solution for thermoelastic deformations of functionally graded thick rectangular plates, *AIAA Journal*, **40**, 1421–1433.

Woo J. and Meguid S.A. (2001), Nonlinear analysis of functionally graded plates and shallow shells, *International Journal of Solids and Structures*, **38**, 7409–7421.

Yang J. and Shen H.-S. (2003a), Nonlinear analysis of functionally graded plates under transverse and in-plane loads, *International Journal of Non-Linear Mechanics*, **38**, 467–482.

Yang J. and Shen H.-S. (2003b), Nonlinear bending analysis of shear deformable functionally graded plates subjected to thermo-mechanical loads under various boundary conditions, *Composites Part B*, **34**, 103–115.

Yang J., Kitipornchai S., and Liew K.M. (2004), Non-linear analysis of the thermo-electro-mechanical behaviour of shear deformable FGM plates with piezoelectric actuators, *International Journal for Numerical Methods in Engineering*, **59**, 1605–1632.

3

Postbuckling of Shear Deformable FGM Plates

3.1 Introduction

When a flat plate is under the action of edge compression in its middle plane, the plate is deformed but remains completely flat when the edge forces are sufficiently small unless there is an initial geometric imperfection. By increasing the load, a state is reached when the plate bends slightly. The in-plane compressive load which is just sufficient to keep the plate in a slightly bent form is called the critical load or buckling load. Once the buckling load is exceeded, the load–deflection relationship exhibits a stable character due to membrane forces which come into play. Actually, the buckling mode will change in the postbuckling range. These changes occur when the energy stored in the plate is sufficient to carry the plate from one buckled form to the other. To obtain an accurate analysis of FGM plates in a wide postbuckling range, the changes in buckling mode must be taken into account. In the usual postbuckling analysis, the buckling mode of the plate is assumed to remain unchanged. This is reasonable assumption in the immediate postbuckling range, e.g., the postbuckling load less than about three times the buckling load.

Many studies have been reported on the buckling and postbuckling analysis of FGM plates subjected to mechanical or thermal loading. Among those, thermal and mechanical buckling of simply supported FGM rectangular plates was studied by Javaheri and Eslami (2002a–c) based on the classical and higher order shear deformation plate theories. Na and Kim (2004, 2006a,b) used solid finite elements to calculate buckling temperature of FGM plates with fully clamped edges. Najafizadeh and Eslami (2002a,b) and Najafizadeh and Heydari (2004a,b, 2008) considered axisymmetric buckling of simply supported and clamped circular FGM plates under a uniform temperature rise or a radial compression based on the first order and higher order shear deformation theory, respectively. Ma and Wang (2003a,b) did the postbuckling analysis of simply supported and clamped FGM circular plates under a radial compression or nonlinear temperature change across the plate thickness based on the classical von Kármán plate theory. Subsequently, they gave the relationships between axisymmetric buckling solutions of FGM circular plates based on third-order plate theory and classical plate theory (Ma and Wang 2004). The effect of initial geometric imperfections on the postbuckling behavior of FGM circular plates subjected to mechanical edge loads and heat conduction was then studied by Li et al. (2007). Naei et al. (2007) calculated buckling loads of radially loaded FGM circular thin plates with variable thickness by using finite element method (FEM). Buckling of FGM plates without or with piezoelectric layers subjected to various nonuniform in-plane loads, along with heat and applied voltage, was considered by Chen and Liew (2004) and Chen et al. (2008) using the first-order shear deformation theory. Ganapathi et al. (2006) and Ganapathi and Prakash (2006) presented the buckling loads for simply supported FGM skew plates subjected to in-plane mechanical loads and heat conduction. In their analysis, the material properties were based on the Mori–Tanaka scheme and rule of mixture, respectively. This work was then extended to the case of thermal postbuckling of FGM skew plates by Prakash et al. (2008). Furthermore, Yang and Shen (2003) studied the postbuckling behavior of FGM thin plates under fully clamped boundary conditions. This work was then extended to the case of shear deformable FGM plates with various boundary conditions and various possible initial geometric imperfections by Yang et al. (2006). Woo et al. (2005) studied the postbuckling behavior of FGM plates and shallow shells under edge compressive loads and a temperature field based on the higher order shear deformation theory. Wu (2004) studied the thermal buckling behavior of simply supported FGM rectangular plates under uniform temperature rise and gradient through the thickness based on the first-order shear deformation plate theory. Shariat and Eslami (2005, 2006) performed the thermal buckling of imperfect FGM rectangular plates under three types of thermal loading as uniform temperature rise, nonlinear temperature rise through the thickness, and axial temperature rise, based on the first-order shear deformation plate theory and the classical thin plate theory, respectively. Wu et al. (2007) studied the postbuckling of FGM rectangular plates under various boundary conditions subjected to uniaxial compression or uniform temperature rise based on the first-order shear deformation plate theory. In the above studies, however, the materials properties were virtually assumed to be temperature-independent (T-ID). Park and Kim (2006) presented thermal postbuckling and vibration of simply supported FGM plates with temperature-dependent (T-D) materials properties by using FEM. Shukla et al. (2007) studied the postbuckling of clamped FGM rectangular plates subjected to thermomechanical loads. In their analysis, the temperature-dependent materials properties were considered and the analytical approach was based on fast-converging Chebyshev polynomials. It has been pointed out by Shen (2002) that the governing differential equations for an FGM plate are identical in form to those of unsymmetric cross-ply laminated plates, and applying in-plane compressive edge loads to such plates will cause bending curvature to appear. Consequently, the bifurcation buckling did not exist due to the stretching/bending coupling effect, as previously proved by Leissa (1986) and Qatu and Leissa (1993), and the solutions are physically incorrect for simply supported FGM rectangular plates subjected to in-plane compressive edge loads and/or temperature variation.

Moreover, Liew et al. (2003, 2004) studied thermal postbuckling behavior of FGM hybrid plates with different kinds of boundary conditions. In their analysis, the material properties were assumed to be temperature-independent and temperature-dependent, respectively, but their results were only for the simple thermal loading case of uniform temperature rise. They confirmed that the FGM plates with all four edges simply supported (SSSS) have no bifurcation buckling temperature, even for the loading case of uniform temperature change. Obviously, when the FGM plate is geometrically midplane symmetric, as reported in Birman (1995) and Feldman and Aboudi (1997), a bifurcation buckling load under in-plane compressive edge loads and/or temperature variation does exist. Recently, Shen (2005) provided a postbuckling analysis for simply supported, midplane symmetric FGM plates with fully covered or embedded piezoelectric actuators subjected to the combined action of mechanical, thermal, and electronic loads. In his study, the material properties were considered to be temperature-dependent and the effect of temperature rise and applied voltage on the postbuckling response was reported. This work was then extended to the case of postbuckling analysis of sandwich plates with FGM face sheets subjected to mechanical and thermal loads (Shen and Li 2008). On the other hand, due to the temperature gradient the plate is subjected to additional moments along with the membrane forces and the problem cannot be posed as an eigenvalue problem, when the four edges of the plate are simply supported. Therefore, the bifurcation solutions for FGM plates subjected to transverse temperature variation, i.e., linear and/or nonlinear gradient through the thickness, may also be physically incorrect. More recently, Shen (2007b) provided a thermal postbuckling analysis for simply supported, midplane symmetric FGM plates under in-plane nonuniform parabolic temperature distribution and heat conduction, and concluded that for the case of heat conduction, the postbuckling path for geometrically perfect plates is no longer of the bifurcation type.

3.2 Postbuckling of FGM Plates with Piezoelectric Actuators under Thermoelectromechanical Loads

Here we consider two types of hybrid laminated plate, referred to as (P/FGM)$_S$ and (FGM/P)$_S$, which consists of four plies and is midplane symmetric, as shown in Figure 3.1. The length, width, and total thickness of the hybrid laminated plate are a, b, and h. The thickness of the FGM layer is h_f, whereas the thickness of the piezoelectric layer is h_p. The substrate FGM

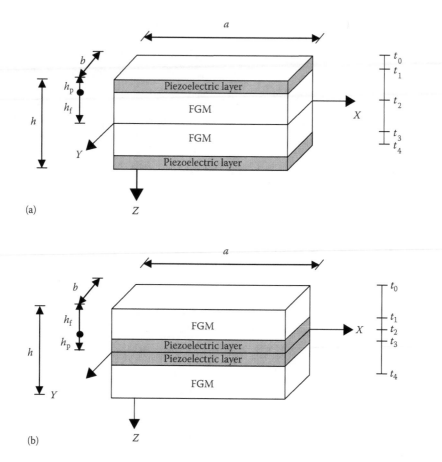

FIGURE 3.1
Configurations of two types of hybrid laminated plates: (a) $(P/FGM)_s$ plate; (b) $(FGM/P)_s$ plate.

layer is made from a mixture of ceramics and metals, the mixing ratio of which is varied continuously and smoothly in the Z-direction. It is assumed that the effective Young's modulus E_f and thermal expansion coefficient α_f of the FGM layer are temperature-dependent, whereas Poisson's ratio ν_f depends weakly on temperature change and is assumed to be a constant. We assume the volume fraction V_m follows a simple power law. According to rule of mixture, we have

$$E_f(Z, T) = [E_m(T) - E_c(T)]\left(\frac{Z - t_1}{t_2 - t_1}\right)^N + E_c(T) \qquad (3.1a)$$

$$\alpha_f(Z, T) = [\alpha_m(T) - \alpha_c(T)]\left(\frac{Z - t_1}{t_2 - t_1}\right)^N + \alpha_c(T) \qquad (3.1b)$$

It is evident that when $Z = t_1$, $E_f = E_c$ and $\alpha_f = \alpha_c$, and when $Z = t_2$, $E_f = E_m$ and $\alpha_f = \alpha_m$.

The plate is assumed to be geometrically imperfect, and is subjected to a compressive edge load in the X-direction combined with thermal and electric loads. Hence the general von Kármán-type equations (Equations 1.33 through 1.36) can be written in the simple form as

$$\tilde{L}_{11}(\overline{W}) - \tilde{L}_{12}(\overline{\Psi}_x) - \tilde{L}_{13}(\overline{\Psi}_y) + \tilde{L}_{14}(\overline{F}) - \tilde{L}_{15}(\overline{N}^P) - \tilde{L}_{16}(\overline{M}^P) = \tilde{L}(\overline{W} + \overline{W}^*, \overline{F}) \quad (3.2)$$

$$\tilde{L}_{21}(\overline{F}) + \tilde{L}_{22}(\overline{\Psi}_x) + \tilde{L}_{23}(\overline{\Psi}_y) - \tilde{L}_{24}(\overline{W}) - \tilde{L}_{25}(\overline{N}^P) = -\frac{1}{2}\tilde{L}(\overline{W} + 2\overline{W}^*, \overline{W}) \quad (3.3)$$

$$\tilde{L}_{31}(\overline{W}) + \tilde{L}_{32}(\overline{\Psi}_x) - \tilde{L}_{33}(\overline{\Psi}_y) + \tilde{L}_{34}(\overline{F}) - \tilde{L}_{35}(\overline{N}^P) - \tilde{L}_{36}(\overline{S}^P) = 0 \quad (3.4)$$

$$\tilde{L}_{41}(\overline{W}) - \tilde{L}_{42}(\overline{\Psi}_x) + \tilde{L}_{43}(\overline{\Psi}_y) + \tilde{L}_{44}(\overline{F}) - \tilde{L}_{45}(\overline{N}^P) - \tilde{L}_{46}(\overline{S}^P) = 0 \quad (3.5)$$

where all linear operators $\tilde{L}_{ij}()$ and nonlinear operator $\tilde{L}()$ are defined by Equation 1.37.

In the above equations, the equivalent thermopiezoelectric loads are defined by

$$\begin{bmatrix} \overline{N}^P \\ \overline{M}^P \\ \overline{P}^P \\ \overline{S}^P \end{bmatrix} = \begin{bmatrix} \overline{N}^T \\ \overline{M}^T \\ \overline{P}^T \\ \overline{S}^T \end{bmatrix} + \begin{bmatrix} \overline{N}^E \\ \overline{M}^E \\ \overline{P}^E \\ \overline{S}^E \end{bmatrix} \quad (3.6)$$

where $\overline{N}^T, \overline{M}^T, \overline{S}^T, \overline{P}^T$ and $\overline{N}^E, \overline{M}^E, \overline{S}^E, \overline{P}^E$ are the forces, moments, and higher order moments caused by the elevated temperature and electric field, respectively.

The temperature field is assumed to be uniformly distributed over the plate surface and through the plate thickness. For the plate-type piezoelectric material, only the transverse direction electric field component E_Z is dominant, and E_Z is defined as $E_Z = -\Phi_{,Z}$, where Φ is the potential field. If the voltage applied to the actuator is in the thickness only, then

$$E_Z = \frac{V_k}{h_p} \quad (3.7)$$

where V_k is the applied voltage across the kth ply.

The forces and moments caused by elevated temperature or electric field are defined by

$$\begin{bmatrix} \overline{N}_x^T & \overline{M}_x^T & \overline{P}_x^T \\ \overline{N}_y^T & \overline{M}_y^T & \overline{P}_y^T \\ \overline{N}_{xy}^T & \overline{M}_{xy}^T & \overline{P}_{xy}^T \end{bmatrix} = \sum_{k=1}^{t_k} \int_{t_{k-1}}^{t_k} \begin{bmatrix} A_x \\ A_y \\ A_{xy} \end{bmatrix}_k (1, Z, Z^3)\Delta T \, dZ \quad (3.8a)$$

$$
\begin{bmatrix} \overline{S}_x^{\mathrm{T}} \\ \overline{S}_y^{\mathrm{T}} \\ \overline{S}_{xy}^{\mathrm{T}} \end{bmatrix} = \begin{bmatrix} \overline{M}_x^{\mathrm{T}} \\ \overline{M}_y^{\mathrm{T}} \\ \overline{M}_{xy}^{\mathrm{T}} \end{bmatrix} - \frac{4}{3h^2} \begin{bmatrix} \overline{P}_x^{\mathrm{T}} \\ \overline{P}_y^{\mathrm{T}} \\ \overline{P}_{xy}^{\mathrm{T}} \end{bmatrix} \tag{3.8b}
$$

and

$$
\begin{bmatrix} \overline{N}_x^{\mathrm{E}} & \overline{M}_x^{\mathrm{E}} & \overline{P}_x^{\mathrm{E}} \\ \overline{N}_y^{\mathrm{E}} & \overline{M}_y^{\mathrm{E}} & \overline{P}_y^{\mathrm{E}} \\ \overline{N}_{xy}^{\mathrm{E}} & \overline{M}_{xy}^{\mathrm{E}} & \overline{P}_{xy}^{\mathrm{E}} \end{bmatrix} = \sum_{k=1}^{t_k} \int_{t_{k-1}}^{t_k} \begin{bmatrix} B_x \\ B_y \\ B_{xy} \end{bmatrix}_k (1, Z, Z^3) \frac{V_k}{h_p} \mathrm{d}Z \tag{3.8c}
$$

$$
\begin{bmatrix} \overline{S}_x^{\mathrm{E}} \\ \overline{S}_y^{\mathrm{E}} \\ \overline{S}_{xy}^{\mathrm{E}} \end{bmatrix} = \begin{bmatrix} \overline{M}_x^{\mathrm{E}} \\ \overline{M}_y^{\mathrm{E}} \\ \overline{M}_{xy}^{\mathrm{E}} \end{bmatrix} - \frac{4}{3h^2} \begin{bmatrix} \overline{P}_x^{\mathrm{E}} \\ \overline{P}_y^{\mathrm{E}} \\ \overline{P}_{xy}^{\mathrm{E}} \end{bmatrix} \tag{3.8d}
$$

in which

$$
\begin{bmatrix} A_x \\ A_y \\ A_{xy} \end{bmatrix} = - \begin{bmatrix} \overline{Q}_{11} & \overline{Q}_{12} & \overline{Q}_{16} \\ \overline{Q}_{12} & \overline{Q}_{22} & \overline{Q}_{26} \\ \overline{Q}_{16} & \overline{Q}_{26} & \overline{Q}_{66} \end{bmatrix} \begin{bmatrix} 1 & 0 \\ 0 & 1 \\ 0 & 0 \end{bmatrix} \begin{bmatrix} \alpha_{11} \\ \alpha_{22} \end{bmatrix} \tag{3.9a}
$$

$$
\begin{bmatrix} B_x \\ B_y \\ B_{xy} \end{bmatrix} = - \begin{bmatrix} \overline{Q}_{11} & \overline{Q}_{12} & \overline{Q}_{16} \\ \overline{Q}_{12} & \overline{Q}_{22} & \overline{Q}_{26} \\ \overline{Q}_{16} & \overline{Q}_{26} & \overline{Q}_{66} \end{bmatrix} \begin{bmatrix} 1 & 0 \\ 0 & 1 \\ 0 & 0 \end{bmatrix} \begin{bmatrix} d_{31} \\ d_{32} \end{bmatrix} \tag{3.9b}
$$

where α_{11} and α_{22} are the thermal expansion coefficients measured in the longitudinal and transverse directions, respectively, d_{31} and d_{32} are the piezoelectric strain constants of a single ply, and $\overline{Q}_{ij}$ are the transformed elastic constants, defined by

$$
\begin{bmatrix} \overline{Q}_{11} \\ \overline{Q}_{12} \\ \overline{Q}_{22} \\ \overline{Q}_{16} \\ \overline{Q}_{26} \\ \overline{Q}_{66} \end{bmatrix} = \begin{bmatrix} c^4 & 2c^2s^2 & s^4 & 4c^2s^2 \\ c^2s^2 & c^4+s^4 & c^2s^2 & -4c^2s^2 \\ s^4 & 2c^2s^2 & c^4 & 4c^2s^2 \\ c^3s & cs^3-c^3s & -cs^3 & -2cs(c^2-s^2) \\ cs^3 & c^3s-cs^3 & -c^3s & 2cs(c^2-s^2) \\ c^2s^2 & -2c^2s^2 & c^2s^2 & (c^2-s^2)^2 \end{bmatrix} \begin{bmatrix} Q_{11} \\ Q_{12} \\ Q_{22} \\ Q_{66} \end{bmatrix} \tag{3.10a}
$$

$$\begin{bmatrix} \overline{Q}_{44} \\ \overline{Q}_{45} \\ \overline{Q}_{55} \end{bmatrix} = \begin{bmatrix} c^2 & s^2 \\ -cs & cs \\ s^2 & c^2 \end{bmatrix} \begin{bmatrix} Q_{44} \\ Q_{55} \end{bmatrix} \tag{3.10b}$$

where

$$Q_{11} = \frac{E_{11}}{(1 - \nu_{12}\nu_{21})}, \quad Q_{22} = \frac{E_{22}}{(1 - \nu_{12}\nu_{21})}, \quad Q_{12} = \frac{\nu_{21}E_{11}}{(1 - \nu_{12}\nu_{21})}, \tag{3.10c}$$

$$Q_{44} = G_{23}, \quad Q_{55} = G_{13}, \quad Q_{66} = G_{12}$$

$E_{11}, E_{22}, G_{12}, G_{13}, G_{23}, \nu_{12}$ and ν_{21} have their usual meanings, and

$$c = \cos\theta, \quad s = \sin\theta \tag{3.10d}$$

where θ is the lamination angle with respect to the plate X-axis. Note that for an FGM layer, $\alpha_{11} = \alpha_{22} = \alpha_f$ is given in detail in Equation 3.1b, and $\overline{Q}_{ij} = Q_{ij}$, which is expressed in Equation 1.19.

All four edges are assumed to be simply supported. Depending upon the in-plane behavior at the edges, two cases, case 1 (referred to herein as movable edges) and case 2 (referred to herein as immovable edges), will be considered. These correspond to the case when the motion of the unloaded edges in the plane tangent to the plate structure's midsurface, normal to the respective edge is either unrestrained or completely restrained, respectively. As a result, we have

Case 1: The edges are simply supported and freely movable in the in-plane directions. In addition the plate is subjected to uniaxial compressive edge loads.

Case 2: All four edges are simply supported. Uniaxial edge loads are acting in the X-direction. The edges $X = 0, a$ are considered freely movable (in the in-plane direction), the remaining two edges being unloaded and immovable (i.e., prevented from moving in the Y-direction).

For both cases, the associated boundary conditions are

$X = 0, a$:

$$\overline{W} = \overline{\Psi}_y = 0 \tag{3.11a}$$

$$\overline{N}_{xy} = 0, \quad \overline{M}_x = \overline{P}_x = 0 \tag{3.11b}$$

$$\int_0^b \overline{N}_x dY + P = 0 \tag{3.11c}$$

$Y = 0, b$:

$$\overline{W} = \overline{\Psi}_x = 0 \tag{3.11d}$$

$$\overline{N}_{xy} = 0, \quad \overline{M}_y = \overline{P}_y = 0 \tag{3.11e}$$

$$\int_0^a \overline{N}_y dX = 0 \quad \text{(movable edges)} \tag{3.11f}$$

$$\overline{V} = 0 \quad \text{(immovable edges)} \tag{3.11g}$$

where P is a compressive edge load in the X-direction, $\overline{M}_x$ and $\overline{M}_y$ are the bending moments, and $\overline{P}_x$ and $\overline{P}_y$ are the higher order moments, defined by

$$\overline{M}_x = -B_{11}^*\overline{F}_{,yy} - B_{21}^*\overline{F}_{,xx} + B_{61}^*\overline{F}_{,xy} + D_{11}^*\overline{\Psi}_{x,x} + D_{12}^*\overline{\Psi}_{y,y} + D_{16}^*(\overline{\Psi}_{x,y} + \overline{\Psi}_{y,x})$$
$$- c_1[F_{11}^*(\overline{\Psi}_{x,x} + \overline{W}_{,xx}) + F_{21}^*(\overline{\Psi}_{y,y} + \overline{W}_{,yy}) + F_{61}^*(\overline{\Psi}_{x,y} + \overline{\Psi}_{y,x} + 2\overline{W}_{,xy})] + \overline{M}_x^T \tag{3.12a}$$

$$\overline{M}_y = -B_{12}^*\overline{F}_{,yy} - B_{22}^*\overline{F}_{,xx} + B_{62}^*\overline{F}_{,xy} + D_{12}^*\overline{\Psi}_{x,x} + D_{22}^*\overline{\Psi}_{y,y} + D_{26}^*(\overline{\Psi}_{x,y} + \overline{\Psi}_{y,x})$$
$$- c_1[F_{12}^*(\overline{\Psi}_{x,x} + \overline{W}_{,xx}) + F_{22}^*(\overline{\Psi}_{y,y} + \overline{W}_{,yy}) + F_{62}^*(\overline{\Psi}_{x,y} + \overline{\Psi}_{y,x} + 2\overline{W}_{,xy})] + \overline{M}_y^T \tag{3.12b}$$

$$\overline{P}_x = -E_{11}^*\overline{F}_{,yy} - E_{21}^*\overline{F}_{,xx} + E_{61}^*F_{,xy} + F_{11}^*\overline{\Psi}_{x,x} + F_{12}^*\overline{\Psi}_{y,y} + F_{16}^*(\overline{\Psi}_{x,y} + \overline{\Psi}_{y,x})$$
$$- c_1[H_{11}^*(\overline{\Psi}_{x,x} + \overline{W}_{,xx}) + H_{12}^*(\overline{\Psi}_{y,y} + \overline{W}_{,yy}) + H_{16}^*(\overline{\Psi}_{x,y} + \overline{\Psi}_{y,x} + 2\overline{W}_{,xy})] + \overline{P}_x^T \tag{3.12c}$$

$$\overline{P}_y = -E_{12}^*\overline{F}_{,yy} - E_{22}^*\overline{F}_{,xx} + E_{62}^*F_{,xy} + F_{21}^*\overline{\Psi}_{x,x} + F_{22}^*\overline{\Psi}_{y,y} + F_{26}^*(\overline{\Psi}_{x,y} + \overline{\Psi}_{y,x})$$
$$- c_1[H_{12}^*(\overline{\Psi}_{x,x} + \overline{W}_{,xx}) + H_{22}^*(\overline{\Psi}_{y,y} + \overline{W}_{,yy}) + H_{26}^*(\overline{\Psi}_{x,y} + \overline{\Psi}_{y,x} + 2\overline{W}_{,xy})] + \overline{P}_y^T \tag{3.12d}$$

The condition expressing the immovability condition $\overline{V} = 0$ (on $Y = 0, b$) is fulfilled on the average sense as

$$\int_0^a \int_0^b \frac{\partial \overline{V}}{\partial Y} dY \, dX = 0 \tag{3.13}$$

This condition in conjunction with Equation 3.14b below provides the compressive stresses acting on the edges $Y = 0, b$.

The average end-shortening relationships are

$$
\frac{\Delta_x}{a} = -\frac{1}{ab} \int_0^b \int_0^a \frac{\partial \overline{U}}{\partial X} \, dX \, dY
$$

$$
= -\frac{1}{ab} \int_0^b \int_0^a \left\{ \left[\left(A_{11}^* \frac{\partial^2 \overline{F}}{\partial Y^2} + A_{12}^* \frac{\partial^2 \overline{F}}{\partial X^2} - A_{16}^* \frac{\partial^2 \overline{F}}{\partial X \partial Y} \right) + \left(B_{11}^* - \frac{4}{3h^2} E_{11}^* \right) \frac{\partial \overline{\Psi}_x}{\partial X} \right. \right.
$$

$$
+ \left(B_{12}^* - \frac{4}{3h^2} E_{12}^* \right) \frac{\partial \overline{\Psi}_y}{\partial Y} + \left(B_{16}^* - \frac{4}{3h^2} E_{16}^* \right) \left(\frac{\partial \overline{\Psi}_x}{\partial Y} + \frac{\partial \overline{\Psi}_y}{\partial X} \right)
$$

$$
- \frac{4}{3h^2} \left(E_{11}^* \frac{\partial^2 \overline{W}}{\partial X^2} + E_{12}^* \frac{\partial^2 \overline{W}}{\partial Y^2} + 2 E_{16}^* \frac{\partial^2 \overline{W}}{\partial X \partial Y} \right) \right]
$$

$$
\left. - \frac{1}{2} \left(\frac{\partial \overline{W}}{\partial X} \right)^2 - \frac{\partial \overline{W}}{\partial X} \frac{\partial \overline{W}^*}{\partial X} - \left(A_{11}^* \overline{N}_x^P + A_{12}^* \overline{N}_y^P + A_{16}^* N_{xy}^P \right) \right\} dX \, dY \qquad (3.14a)
$$

$$
\frac{\Delta_y}{b} = -\frac{1}{ab} \int_0^a \int_0^b \frac{\partial \overline{V}}{\partial Y} \, dY \, dX
$$

$$
= -\frac{1}{ab} \int_0^a \int_0^b \left\{ \left[\left(A_{22}^* \frac{\partial^2 \overline{F}}{\partial X^2} + A_{12}^* \frac{\partial^2 \overline{F}}{\partial Y^2} - A_{26}^* \frac{\partial^2 \overline{F}}{\partial X \partial Y} \right) + \left(B_{21}^* - \frac{4}{3h^2} E_{21}^* \right) \frac{\partial \overline{\Psi}_x}{\partial X} \right. \right.
$$

$$
+ \left(B_{22}^* - \frac{4}{3h^2} E_{22}^* \right) \frac{\partial \overline{\Psi}_y}{\partial Y} + \left(B_{26}^* - \frac{4}{3h^2} E_{26}^* \right) \left(\frac{\partial \overline{\Psi}_x}{\partial Y} + \frac{\partial \overline{\Psi}_y}{\partial X} \right)
$$

$$
- \frac{4}{3h^2} \left(E_{21}^* \frac{\partial^2 \overline{W}}{\partial X^2} + E_{22}^* \frac{\partial^2 \overline{W}}{\partial Y^2} + 2 E_{26}^* \frac{\partial^2 \overline{W}}{\partial X \partial Y} \right) \right]
$$

$$
\left. - \frac{1}{2} \left(\frac{\partial \overline{W}}{\partial Y} \right)^2 - \frac{\partial \overline{W}}{\partial Y} \frac{\partial \overline{W}^*}{\partial Y} - \left(A_{12}^* \overline{N}_x^P + A_{22}^* \overline{N}_y^P + A_{26}^* N_{xy}^P \right) \right\} dY \, dX \qquad (3.14b)
$$

where Δ_x and Δ_y are plate end-shortening displacements in the X- and Y-directions.

It is evident that the above equations involve the stretching/bending coupling, as predicted by B_{ij} and E_{ij}. As argued previously, even for an FGM plate with all four edges simply supported, no bifurcation buckling could occur. For this reason, we consider here geometrically midplane symmetric FGM plates with fully covered or embedded piezoelectric actuators. In such a case, the stretching/bending coupling is zero-valued, i.e., $B_{ij} = E_{ij} = 0$. As a result, $\tilde{L}_{14} = \tilde{L}_{15} = \tilde{L}_{22} = \tilde{L}_{23} = \tilde{L}_{24} = \tilde{L}_{34} = \tilde{L}_{35} = \tilde{L}_{44} = \tilde{L}_{45} = 0$ in Equations 3.2 through 3.5.

Introducing dimensionless quantities of Equation 2.8, and

$$(\gamma_{T1}, \gamma_{T2}, \gamma_{P1}, \gamma_{P2}) = \left(A_x^T, A_y^T, B_x^P, B_y^P\right) a^2 / \pi^2 [D_{11}^* D_{22}^*]^{1/2},$$
$$\lambda_x = Pb/4\pi^2 [D_{11}^* D_{22}^*]^{1/2} \tag{3.15}$$

in which

$$\begin{bmatrix} A_x^T \\ A_y^T \end{bmatrix} \Delta T = -\sum_{k=1}^{t_k} \int_{t_{k-1}}^{t_k} \begin{bmatrix} A_x \\ A_y \end{bmatrix}_k \Delta T \, dZ \tag{3.16a}$$

$$\begin{bmatrix} B_x^P \\ B_y^P \end{bmatrix} \Delta V = -\sum_{k=1}^{t_k} \int_{t_{k-1}}^{t_k} \begin{bmatrix} B_x \\ B_y \end{bmatrix}_k \frac{V_k}{h_P} \, dZ \tag{3.16b}$$

The nonlinear equations (Equations 3.2 through 3.5) may then be written in dimensionless form as

$$L_{11}(W) - L_{12}(\Psi_x) - L_{13}(\Psi_y) = \gamma_{14}\beta^2 L(W + W^*, F) \tag{3.17}$$

$$L_{21}(F) = -\frac{1}{2}\gamma_{24}\beta^2 L(W + 2W^*, W) \tag{3.18}$$

$$L_{31}(W) + L_{32}(\Psi_x) - L_{33}(\Psi_y) = 0 \tag{3.19}$$

$$L_{41}(W) - L_{42}(\Psi_x) + L_{43}(\Psi_y) = 0 \tag{3.20}$$

where all nondimensional linear operators $L_{ij}()$ and nonlinear operator $L()$ are defined by Equation 2.14.

The boundary conditions expressed by Equation 3.11 become

$x = 0, \pi$:

$$W = \Psi_y = 0 \tag{3.21a}$$

$$F_{,xy} = M_x = P_x = 0 \tag{3.21b}$$

$$\frac{1}{\pi} \int_0^{\pi} \beta^2 \frac{\partial^2 F}{\partial y^2} \, dy + 4\lambda_x \beta^2 = 0 \tag{3.21c}$$

$y = 0, \pi$:

$$W = \Psi_x = 0 \tag{3.21d}$$

$$F_{,xy} = M_y = P_y = 0 \tag{3.21e}$$

$$\int_0^\pi \frac{\partial^2 F}{\partial x^2}\, dx = 0 \text{ (movable edges)} \tag{3.21f}$$

$$\delta_y = 0 \text{ (immovable edges)} \tag{3.21g}$$

and the unit end-shortening relationships become

$$\delta_x = -\frac{1}{4\pi^2\beta^2\gamma_{24}} \int_0^\pi \int_0^\pi \left\{ \left[\gamma_{24}^2\beta^2\frac{\partial^2 F}{\partial y^2} - \gamma_5\frac{\partial^2 F}{\partial x^2}\right] - \frac{1}{2}\gamma_{24}\left(\frac{\partial W}{\partial x}\right)^2 - \gamma_{24}\frac{\partial W}{\partial x}\frac{\partial W^*}{\partial x} \right.$$

$$\left. +(\gamma_{24}^2\gamma_{T1} - \gamma_5\gamma_{T2})\Delta T + (\gamma_{24}^2\gamma_{P1} - \gamma_5\gamma_{P2})\Delta V \right\} dx\, dy \tag{3.22a}$$

$$\delta_y = -\frac{1}{4\pi^2\beta^2\gamma_{24}} \int_0^\pi \int_0^\pi \left\{ \left[\frac{\partial^2 F}{\partial x^2} - \gamma_5\beta^2\frac{\partial^2 F}{\partial y^2}\right] - \frac{1}{2}\gamma_{24}\beta^2\left(\frac{\partial W}{\partial y}\right)^2 - \gamma_{24}\beta^2\frac{\partial W}{\partial y}\frac{\partial W^*}{\partial y} \right.$$

$$\left. +(\gamma_{T2} - \gamma_5\gamma_{T1})\Delta T + (\gamma_{P2} - \gamma_5\gamma_{P1})\Delta V \right\} dy\, dx \tag{3.22b}$$

By virtue of the fact that ΔV and ΔT are assumed to be uniform, the thermopiezoelectric coupling in Equations 3.2 through 3.5 vanishes, but terms in ΔV and ΔT intervene in Equation 3.22.

Applying Equations 3.17 through 3.22, the compressive postbuckling behavior of perfect and imperfect, FGM hybrid plates with piezoelectric actuators under thermoelectromechanical loads is now determined by means of a two-step perturbation technique. The essence of this procedure, in the present case, is to assume that

$$W(x,y,\varepsilon) = \sum_{j=1} \varepsilon^j w_j(x,y), \quad F(x,y,\varepsilon) = \sum_{j=0} \varepsilon^j f_j(x,y),$$

$$\Psi_x(x,y,\varepsilon) = \sum_{j=1} \varepsilon^j \psi_{xj}(x,y), \quad \Psi_y(x,y,\varepsilon) = \sum_{j=1} \varepsilon^j \psi_{yj}(x,y) \tag{3.23}$$

where ε is a small perturbation parameter and the first term of $w_j(x, y)$ is assumed to have the form

$$w_1(x,y) = A_{11}^{(1)} \sin mx \sin ny \tag{3.24}$$

and the initial geometric imperfection is assumed to have a similar form

$$W^*(x,y,\varepsilon) = \varepsilon a_{11}^* \sin mx \sin ny = \varepsilon\mu A_{11}^{(1)} \sin mx \sin ny \tag{3.25}$$

where $\mu = a_{11}^*/A_{11}^{(1)}$ is the imperfection parameter.

Substituting Equation 3.23 into Equations 3.15 through 3.20, collecting the terms of the same order of ε, we obtain a set of perturbation equations which can be written, for example, as

$$O(\varepsilon^1): L_{11}(w_1) - L_{12}(\psi_{x1}) - L_{13}(\psi_{y1}) = \gamma_{14}\beta^2 L(w_1 + W^*, f_0) \tag{3.26a}$$

$$L_{21}(f_1) = 0 \tag{3.26b}$$

$$L_{31}(w_1) + L_{32}(\psi_{x1}) - L_{33}(\psi_{y1}) = 0 \tag{3.26c}$$

$$L_{41}(w_1) - L_{42}(\psi_{x1}) + L_{43}(\psi_{y1}) = 0 \tag{3.26d}$$

$$O(\varepsilon^2): L_{11}(w_2) - L_{12}(\psi_{x2}) - L_{13}(\psi_{y2}) = \gamma_{14}\beta^2 [L(w_2, f_0) + L(w_1 + W^*, f_1)] \tag{3.27a}$$

$$L_{21}(f_2) = -\frac{1}{2}\gamma_{24}\beta^2 L(w_1 + 2W^*, w_1) \tag{3.27b}$$

$$L_{31}(w_2) + L_{32}(\psi_{x2}) - L_{33}(\psi_{y2}) = 0 \tag{3.27c}$$

$$L_{41}(w_2) - L_{42}(\psi_{x2}) + L_{43}(\psi_{y2}) = 0 \tag{3.27d}$$

By using Equations 3.24 and 3.25 to solve these perturbation equations of each order, the amplitudes of the terms $w_j(x, y)$, $f_j(x, y)$, $\psi_{xj}(x, y)$, and $\psi_{yj}(x, y)$ are determined step by step. As a result, up to fourth-order asymptotic solutions can be obtained.

$$W = \varepsilon \left[A_{11}^{(1)} \sin mx \sin ny \right]$$
$$+ \varepsilon^3 \left[A_{13}^{(3)} \sin mx \sin 3ny + A_{31}^{(3)} \sin 3mx \sin ny \right] + O(\varepsilon^5) \tag{3.28}$$

$$F = -B_{00}^{(0)}\frac{y^2}{2} - b_{00}^{(0)}\frac{x^2}{2} + \varepsilon^2 \left[-B_{00}^{(2)}\frac{y^2}{2} - b_{00}^{(2)}\frac{x^2}{2} + B_{20}^{(2)}\cos 2mx + B_{02}^{(2)}\cos 2ny \right]$$
$$+ \varepsilon^4 \left[-B_{00}^{(4)}\frac{y^2}{2} - b_{00}^{(4)}\frac{x^2}{2} + B_{20}^{(4)}\cos 2mx + B_{02}^{(4)}\cos 2ny + B_{22}^{(4)}\cos 2mx\cos 2ny \right.$$
$$+ B_{40}^{(4)}\cos 4mx + B_{04}^{(4)}\cos 4ny + B_{24}^{(4)}\cos 2mx \cos 4ny$$
$$\left. + B_{42}^{(4)}\cos 4mx \cos 2ny \right] + O(\varepsilon^5) \tag{3.29}$$

$$\Psi_x = \varepsilon \left[C_{11}^{(1)}\cos mx \sin ny \right]$$
$$+ \varepsilon^3 \left[C_{13}^{(3)}\cos mx \sin 3ny + C_{31}^{(3)}\cos 3mx \sin ny \right] + O(\varepsilon^5) \tag{3.30}$$

$$\Psi_y = \varepsilon \left[D_{11}^{(1)}\sin mx \cos ny \right]$$
$$+ \varepsilon^3 \left[D_{13}^{(3)}\sin mx \cos 3ny + D_{31}^{(3)}\sin 3mx \cos ny \right] + O(\varepsilon^5) \tag{3.31}$$

It is mentioned that all coefficients in Equations 3.28 through 3.31 are related and can be expressed in terms of $A_{11}^{(1)}$, for example

$$C_{11}^{(1)} = -m \frac{g_{04}}{g_{00}} A_{11}^{(1)}, \quad D_{11}^{(1)} = -n\beta \frac{g_{03}}{g_{00}} A_{11}^{(1)}$$

$$B_{20}^{(2)} = \frac{\gamma_{24} n^2 \beta^2}{32 m^2} (1 + 2\mu) \left(A_{11}^{(1)}\right)^2, \quad B_{02}^{(2)} = \frac{m^2}{32 \gamma_{24} n^2 \beta^2} (1 + 2\mu) \left(A_{11}^{(1)}\right)^2 \quad (3.32)$$

Next, upon substitution of Equations 3.28 through 3.31 into the boundary conditions (Equations 3.21c and 3.22a), the postbuckling equilibrium path can be written as

$$\lambda_x = \lambda_x^{(0)} + \lambda_x^{(2)} W_m^2 + \lambda_x^{(4)} W_m^4 + \cdots \quad (3.33)$$

and

$$\delta_x = \delta_x^{(0)} + \delta_x^{(2)} W_m^2 + \delta_x^{(4)} W_m^4 + \cdots \quad (3.34)$$

in which W_m is the dimensionless form of maximum deflection, which is assumed to be at the point $(x, y) = (\pi/2m, \pi/2n)$ and $\lambda_x^{(i)}$ and $\delta_x^{(i)}$ ($i = 0$, $2, 4, \ldots$) are all temperature-dependent and given in detail in Appendix F.

Equations 3.33 and 3.34 can be employed to obtain numerical results for the postbuckling load–deflection or load–end-shortening curves of simply supported shear deformable FGM plates with piezoelectric actuators subjected to uniaxial compression combined with thermal and electric loads. From Appendix F, the buckling load of a perfect plate can readily be obtained numerically, by setting $\mu = 0$ (or $\overline{W}^*/h = 0$), while taking $W_m = 0$ (or $\overline{W}/h = 0$). In such a case, the minimum buckling load is determined by applying Equation 3.33 for various values of the buckling mode (m, n), which determine the number of half-waves in the X- and Y-directions, respectively.

For numerical illustrations, the thickness of the FGM layer $h_f = 1$ mm whereas the thickness of piezoelectric layers $h_p = 0.1$ mm, so that the total thickness of the plate $h = 2.2$ mm. Two sets of material mixture for FGMs are considered. One is silicon nitride and stainless steel, referred to as $Si_3N_4/SUS304$, and the other is zirconium oxide and titanium alloy, referred to as ZrO_2/Ti-6Al-4V. The material properties of these constituents are assumed to be nonlinear function of temperature of Equation 1.4, and typical values for Young's modulus E_f (in Pa) and thermal expansion coefficient α_f (in K^{-1}) of them can be found in Tables 1.1 and 1.2. Poisson's ratio ν_f is assumed to be a constant, and $\nu_f = 0.28$. PZT-5A is selected for the piezoelectric layers. The material properties of which are assumed to be linear functions of temperature change, i.e.,

$$E_{11}(T) = E_{110}(1 + E_{111}\Delta T), \quad E_{22}(T) = E_{220}(1 + E_{221}\Delta T)$$
$$G_{12}(T) = G_{120}(1 + G_{121}\Delta T), \quad G_{13}(T) = G_{130}(1 + G_{131}\Delta T), \quad G_{23}(T) = G_{230}(1 + G_{231}\Delta T)$$
$$\alpha_{11}(T) = \alpha_{110}(1 + \alpha_{111}\Delta T), \quad \alpha_{22}(T) = \alpha_{220}(1 + \alpha_{221}\Delta T) \quad (3.35)$$

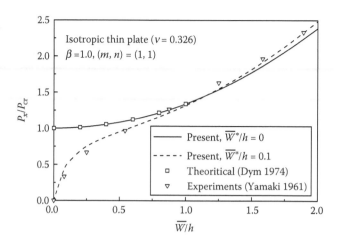

FIGURE 3.2
Comparisons of postbuckling load–deflection curves for isotropic thin plates under uniaxial compression.

where E_{110}, E_{220}, G_{120}, G_{130}, G_{230}, α_{110}, α_{220}, E_{111}, E_{221}, G_{121}, G_{131}, G_{231}, α_{111}, α_{221} are constants. Typical values adopted, as given in Oh et al. (2000), $E_{110} = E_{220} = 61$ GPa, $G_{120} = G_{130} = G_{230} = 24.2$ GPa, $\nu_{12} = 0.3$, $\alpha_{110} = \alpha_{220} = 0.9 \times 10^{-6}$ K^{-1} and $d_{31} = d_{32} = 2.54 \times 10^{-10}$ m V^{-1}; and $E_{111} = -0.0005$, $E_{221} = G_{121} = G_{131} = G_{231} = -0.0002$, $\alpha_{111} = \alpha_{221} = 0.0005$.

The postbuckling load–deflection curves for perfect and imperfect, isotropic thin square plates ($\nu = 0.326$) subjected to uniaxial compression are compared in Figure 3.2 with the analytical solutions of Dym (1974) and the experimental results of Yamaki (1961). These comparisons show that the results from the present method are in good agreement with the existing results.

Tables 3.1 through 3.4 present the buckling loads P_{cr} (in kN) for perfect, moderately thick ($b/h = 20$), (P/FGM)$_S$ and (FGM/P)$_S$ hybrid laminated plates with unloaded edges immovable and with different values of the volume fraction index N ($= 0.0, 0.2, 0.5, 1.0, 2.0$, and 5.0) subjected to uniaxial compression under three sets of temperature rise ($\Delta T = 0$, 100, 200 K). Here, TD represents material properties for both substrate FGM layer and piezoelectric layers are temperature-dependent. TD-F represents material properties of substrate FGM layer are temperature-dependent but material properties of piezoelectric layers are temperature-independent, i.e., $E_{111} = E_{221} = G_{121} = G_{131} = G_{231} = \alpha_{111} = \alpha_{221} = 0$ in Equation 3.35. TID represents material properties for both piezoelectric layers and substrate FGM layer are temperature-independent, i.e., in a fixed temperature $T_0 = 300$ K for FGM layer, as previously used in Yang and Shen (2003). The control voltages with the same sign are also applied to the upper, lower, or middle piezoelectric layers, and are referred to as V_U, V_L, and V_M. Three electrical

TABLE 3.1

Comparisons of Buckling Loads P_{cr} (kN) for Uniaxial Compressed, (P/FGM)$_S$ Plates with a Substrate Made of Si$_3$N$_4$/SUS304 and with Unloaded Edges Immovable under Uniform Temperature Rise and Three Sets of Electrical Loading Conditions ($b/h = 20$, $a/b = 1.0$, $T_0 = 300$ K)

ΔT (in K)	$V_U = V_L =$ (in V)	$N = 0$	$N = 0.2$	$N = 0.5$	$N = 1.0$	$N = 2.0$	$N = 5.0$
(P/FGM)$_S$, TID, $(m, n) = (1, 1)$							
0	−500	115.2026	131.6953	146.1075	157.9360	166.5642	171.3940
	0	114.6376	131.1302	145.5423	157.3708	165.9990	170.8288
	+500	114.0725	130.5651	144.9771	156.8056	165.4337	170.2635
100	−500	93.3042	109.9468	124.9499	137.6636	147.3829	153.4537
	0	92.7391	109.3817	124.3848	137.0984	146.8177	152.8885
	+500	92.1741	108.8166	123.8196	136.5332	146.2524	152.3232
200	−500	71.4058	88.1983	103.7924	117.3912	128.2016	135.5134
	0	70.8407	87.6332	103.2272	116.8260	127.6364	134.9482
	+500	70.2757	87.0681	102.6621	116.2608	127.0711	134.3829
(P/FGM)$_S$, TD-F, $(m, n) = (1, 1)$							
100	−500	90.7119	106.8642	121.4538	133.8422	143.3414	149.3148
	0	90.1468	106.2991	120.8886	133.2770	142.7762	148.7495
	+500	89.5818	105.7340	120.3235	132.7118	142.2110	148.1842
200	−500	63.4408	79.6302	94.7616	108.0310	118.6513	125.9229
	0	62.8758	79.0652	94.1965	107.4658	118.0861	125.3576
	+500	62.3107	78.5001	93.6314	106.9006	117.5209	124.7924
(P/FGM)$_S$, TD, $(m, n) = (1, 1)$							
100	−500	90.4283	106.5790	121.1672	133.5544	143.0529	149.0257
	0	89.8680	106.0187	120.6067	132.9939	142.4923	148.4651
	+500	89.3077	105.4583	120.0463	132.4334	141.9318	147.9045
200	−500	62.8991	79.0828	94.2088	107.4735	118.0899	125.3583
	0	62.3435	78.5270	93.6530	106.9176	117.5339	124.8023
	+500	61.7878	77.9712	93.0971	106.3616	116.9779	124.2462

loading cases are considered. Here $V_U = V_L = 0$ V (or $V_M = 0$ V) implies that the buckling occurs under a grounding condition. Two kinds of substrate FGM layers, i.e., Si$_3$N$_4$/SUS304 and ZrO$_2$/Ti-6Al-4V are considered. It can be found that the buckling load of (P/FGM)$_S$ plate is lower than that of (FGM/P)$_S$ plate. It can be seen that, for the hybrid plates with Si$_3$N$_4$/SUS304 substrate, a fully metallic plate ($N = 0$) has lowest buckling load and that the buckling load increases as the volume fraction index N increases. This is expected because the metallic plate has a lower stiffness than the ceramic plate. It is found that the increase is about +65% for the (P/FGM)$_S$ plate, and about +67% for the (FGM/P)$_S$ one, from $N = 0$ to $N = 5$, under temperature change $\Delta T = 100$ K. It can also be seen that the temperature reduces the

TABLE 3.2

Comparisons of Buckling Loads P_{cr} (kN) for Uniaxial Compressed, $(FGM/P)_S$ Plates with a Substrate Made of $Si_3N_4/SUS304$ and with Unloaded Edges Immovable under Uniform Temperature Rise and Three Sets of Electrical Loading Conditions ($b/h = 20$, $a/b = 1.0$, $T_0 = 300$ K)

ΔT (in K)	V_M (in V)	$N=0$	$N=0.2$	$N=0.5$	$N=1.0$	$N=2.0$	$N=5.0$
$(FGM/P)_S$, TID, $(m, n) = (1, 1)$							
0	−500	138.5191	159.3017	177.6971	193.1219	204.8333	212.0199
	0	137.9540	158.7366	177.1320	192.5568	204.2680	211.4547
	+500	137.3889	158.1714	176.5668	191.9915	203.7028	210.8894
100	−500	116.6206	137.5532	156.5396	172.8495	185.6520	194.0796
	0	116.0556	136.9881	155.9744	172.2843	185.0867	193.5144
	+500	115.4905	136.4229	155.4093	171.7191	184.5215	192.9491
200	−500	94.7222	115.8047	135.3821	152.5771	166.4707	176.1393
	0	94.1571	115.2396	134.8169	152.0119	165.9054	175.5741
	+500	93.5921	114.6745	134.2518	151.4467	165.3402	175.0088
$(FGM/P)_S$, TD-F, $(m, n) = (1, 1)$							
100	−500	113.5414	133.8505	152.2997	168.1724	180.6583	188.9140
	0	112.9764	133.2854	151.7345	167.6072	180.0931	188.3488
	+500	112.4113	132.7203	151.1694	167.0420	179.5278	187.7835
200	−500	85.3580	105.6978	124.6823	141.4299	155.0314	164.5812
	0	84.7929	105.1327	124.1171	140.8647	154.4662	164.0159
	+500	84.2279	104.5676	123.5520	140.2995	153.9010	163.4507
$(FGM/P)_S$, TD, $(m, n) = (1, 1)$							
100	−500	113.5180	133.8241	152.2709	168.1422	180.6273	188.8832
	0	112.9577	133.2637	151.7105	167.5817	180.0668	188.3226
	+500	112.3974	132.7033	151.1500	167.0212	179.5062	187.7620
200	−500	85.3354	105.6667	124.6441	141.3861	154.9840	164.5320
	0	84.7797	105.1109	124.0882	140.8302	154.4280	163.9759
	+500	84.2241	104.5552	123.5323	140.2742	153.8719	163.4198

buckling load when the temperature dependency is put into consideration. The percentage decrease is about −14% for the $(P/FGM)_S$ plate and about −12% for the $(FGM/P)_S$ one from temperature changes from $\Delta T = 0$ K to $\Delta T = 100$ K under the same volume fraction distribution $N = 2$. Also the buckling loads under TD-F and TD cases are very close under the same volume fraction distribution and the same temperature change. In contrast, for the ZrO_2/Ti-6Al-4V hybrid laminated plate, the buckling load is lower than that of the $Si_3N_4/SUS304$ hybrid laminated plate and erratic behavior can be observed in thermal loading conditions $\Delta T = 100$ and 200 K. It can also be seen that the control voltage has a very small effect on the buckling loads for hybrid laminated plates; this is because the piezoelectric layer is much thinner than the FGM substrate. Very high voltages will be able to influence

TABLE 3.3

Comparisons of Buckling Loads P_{cr} (kN) for Uniaxial Compressed, $(P/FGM)_S$ Plates with a Substrate Made of ZrO_2/Ti-6Al-4V and with Unloaded Edges Immovable under Uniform Temperature Rise and Three Sets of Electrical Loading Conditions ($b/h = 20$, $a/b = 1.0$, $T_0 = 300$ K)

ΔT (in K)	$V_U = V_L =$ (in V)	$N = 0$	$N = 0.2$	$N = 0.5$	$N = 1.0$	$N = 2.0$	$N = 5.0$
$(P/FGM)_S$, TID, $(m, n) = (1, 1)$							
0	−500	64.0000	72.9739	80.8184	87.2604	91.9643	94.6043
	0	63.4356	72.4094	80.2538	86.6957	91.3996	94.0395
	+500	62.8712	71.8449	79.6892	86.1310	90.8348	93.4747
100	−500	57.5279	64.4956	70.0364	73.9762	76.0389	75.9364
	0	56.9635	63.9311	69.4718	73.4115	75.4741	75.3716
	+500	56.3991	63.3666	68.9072	72.8468	74.9094	74.8068
200	−500	51.0557	56.0172	59.2544	60.6920	60.1135	57.2686
	0	50.4913	55.4527	58.6898	60.1273	59.5487	56.7037
	+500	49.9269	54.8882	58.1252	59.5626	58.9839	56.1389
$(P/FGM)_S$, TD-F, $(m, n) = (1, 1)$							
100	−500	54.8204	59.5851	62.9913	64.9500	65.2654	63.7199
	0	54.2561	59.0207	62.4268	64.3854	64.7007	63.1551
	+500	53.6917	58.4563	61.8623	63.8208	64.1360	62.5904
200	−500	45.9786	45.1848	42.4283	37.8905	31.5562	23.3350
	0	45.4144	44.6205	41.8639	37.3260	30.9916	22.7704
	+500	44.8502	44.0561	41.2995	36.7615	30.4270	22.2057
$(P/FGM)_S$, TD, $(m, n) = (1, 1)$							
100	−500	54.5302	59.2967	62.7049	64.6657	64.9834	63.4404
	0	53.9709	58.7372	62.1453	64.1060	64.4237	62.8805
	+500	53.4117	58.1778	61.5857	63.5464	63.8639	62.3206
200	−500	45.4104	44.6291	41.8861	37.3619	31.0410	22.8328
	0	44.8561	44.0746	41.3314	36.8071	30.4860	22.2777
	+500	44.3018	43.5201	40.7768	36.2523	29.9311	21.7227

the buckling response of the hybrid laminated plate. However, such high voltages cannot be applied, because they lead to a breakdown in the material properties.

Then Tables 3.5 and 3.6 present the buckling loads P_{cr} for the same two types of hybrid plates with unloaded edges movable subjected to uniaxial compression under three sets of temperature rise ($\Delta T = 0$, 100, 200 K). The results show that the buckling load is decreased with increase in temperature, but is increased as volume fraction index N increases at the same temperature. The numerical results also confirm that the control voltage has no effect on the buckling loads of hybrid laminated plates when the unloaded edges are movable.

TABLE 3.4

Comparisons of Buckling Loads P_{cr} (kN) for Uniaxial Compressed, $(FGM/P)_S$ Plates with a Substrate Made of ZrO_2/Ti-6Al-4V and with Unloaded Edges Immovable under Uniform Temperature Rise and Three Sets of Electrical Loading Conditions ($b/h = 20$, $a/b = 1.0$, $T_0 = 300$ K)

ΔT (in K)	V_M (in V)	$N=0$	$N=0.2$	$N=0.5$	$N=1.0$	$N=2.0$	$N=5.0$
$(FGM/P)_S$, TID, $(m, n) = (1, 1)$							
0	−500	70.7813	82.1124	92.1400	100.5459	106.9247	110.8339
	0	70.2169	81.5479	91.5754	99.9812	106.3599	110.2691
	+500	69.6525	80.9834	91.0108	99.4165	105.7951	109.7042
100	−500	64.3091	73.6340	81.3580	87.2617	90.9992	92.1660
	0	63.7448	73.0695	80.7934	86.6970	90.4345	91.6012
	+500	63.1804	72.5050	80.2288	86.1323	89.8697	91.0364
200	−500	57.8370	65.1557	70.5760	73.9775	75.0738	73.4981
	0	57.2726	64.5912	70.0113	73.4128	74.5090	72.9333
	+500	56.7082	64.0267	69.4467	72.8481	73.9443	72.3685
$(FGM/P)_S$, TD-F, $(m, n) = (1, 1)$							
100	−500	60.6917	67.4217	72.6482	76.2438	77.9534	77.4619
	0	60.1274	66.8572	72.0837	75.6792	77.3888	76.8972
	+500	59.5631	66.2928	71.5192	75.1146	76.8241	76.3325
200	−500	50.9401	51.8613	50.6936	47.5837	42.4644	35.1598
	0	50.3759	51.2970	50.1292	47.0192	41.8998	34.5951
	+500	49.8116	50.7327	49.5647	46.4547	41.3352	34.0305
$(FGM/P)_S$, TD, $(m, n) = (1, 1)$							
100	−500	60.6634	67.3937	72.6214	76.2187	77.9306	77.4421
	0	60.1041	66.8343	72.0618	75.6590	77.3708	76.8822
	+500	59.5448	66.2749	71.5022	75.0993	76.8111	76.3223
200	−500	50.8947	51.8260	50.6700	47.5728	42.4668	35.1762
	0	50.3404	51.2715	50.1153	47.0180	41.9118	34.6211
	+500	49.7861	50.7170	49.5606	46.4632	41.3569	34.0661

Figures 3.3 and 3.4 give, respectively, the postbuckling load–deflection and load–shortening curves for $(P/FGM)_S$ and $(FGM/P)_S$ hybrid laminated plates ($b/h = 40$, $N = 0.2$) with unloaded edges immovable subjected to uniaxial compression and three sets of electrical loading, $V_U = V_L$ (or V_M) $= -500$, 0, $+500$ V, and under $\Delta T = 0$ and 100 K. It is evident that the buckling loads reduce as the temperature increases, and the postbuckling path becomes lower. It can be found that the control voltage has a small effect on the postbuckling behavior of the plate. It can be seen that minus control voltages increase the buckling load and decrease the postbuckled deflection at the same temperature rise, whereas the plus control voltages decrease the buckling load and induce more large postbuckled deflections.

TABLE 3.5

Comparisons of Buckling Loads P_{cr} (kN) for Uniaxial Compressed, Si_3N_4/SUS304 Plates with Piezoelectric Actuators and with Unloaded Edges Movable Subjected to Temperature Rise ($b/h = 20$, $a/b = 1.0$, $T_0 = 300$ K)

	ΔT (in K)	$N=0.0$	$N=0.2$	$N=0.5$	$N=1.0$	$N=2.0$	$N=5.0$
$(P/FGM)_S$, $(m, n) = (1, 1)$							
TID		146.8044	167.9184	186.3678	201.5086	212.5516	218.7312
TD-FGM	100	144.8755	165.3297	183.2028	197.8707	208.5686	214.5550
	200	141.2601	161.6779	179.5191	194.1613	204.8411	210.8181
TD	100	144.5326	164.9874	182.8609	197.5289	208.2266	214.2126
	200	140.5758	160.9947	178.8367	193.4790	204.1585	210.1347
$(FGM/P)_S$, $(m, n) = (1, 1)$							
TID		176.6633	203.2696	226.8185	246.5632	261.5528	270.7491
TD-FGM	100	174.1111	199.8871	222.7014	241.8298	256.3509	265.2588
	200	169.3277	195.0592	217.8333	236.9282	251.4242	260.3173
TD	100	174.1070	199.8824	222.6962	341.8245	256.3456	265.2538
	200	169.3195	195.0497	217.8229	236.9173	251.4135	260.3075

Figures 3.5 and 3.6 show the effect of the volume fraction index N ($= 0.2$, 1.0, and 5.0) on the postbuckling behavior of $(P/FGM)_S$ and $(FGM/P)_S$ hybrid laminated plates ($b/h = 40$) with unloaded edges immovable subjected to uniaxial compression and three sets of electrical loading, and under

TABLE 3.6

Comparisons of Buckling Loads P_{cr} (kN) for Uniaxial Compressed, ZrO_2/Ti-6Al-4V Plates with Piezoelectric Actuators and with Unloaded Edges Movable Subjected to Temperature Rise Temperature Rise ($b/h = 20$, $a/b = 1.0$, $T_0 = 300$ K)

	ΔT (in K)	$N=0.0$	$N=0.2$	$N=0.5$	$N=1.0$	$N=2.0$	$N=5.0$
$(P/FGM)_S$, $(m, n) = (1, 1)$							
TID		81.2698	92.7595	102.8020	111.0477	117.0674	120.4437
TD-FGM	100	77.6612	87.2275	95.5896	102.4555	107.4673	110.2784
	200	74.0524	82.3901	89.6789	95.6635	100.0321	102.4828
TD	100	77.3189	86.8857	95.2482	102.1141	107.1257	109.9364
	200	73.3692	81.7080	88.9973	94.9821	99.3503	101.8001
$(FGM/P)_S$, $(m, n) = (1, 1)$							
TID		89.9576	104.4662	117.3045	128.0650	136.2290	141.2302
TD-FGM	100	85.1835	97.2669	107.9604	116.9225	123.7195	127.8799
	200	80.4093	90.9440	100.2675	108.0808	114.0052	117.6293
TD	100	85.1797	97.2626	107.9557	116.9175	123.7146	127.8754
	200	80.4018	90.9355	100.2582	108.0712	113.9956	117.6204

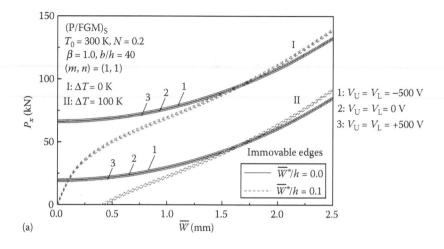

(a)

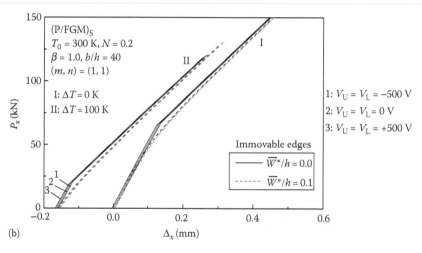

(b)

FIGURE 3.3
Thermopiezoelectric effects on the postbuckling behavior of (P/FGM)$_S$ plates with unload edges immovable: (a) load deflection; (b) load shortening.

$\Delta T = 100$ K. It can be seen that the increase of the volume fraction index N yields an increase of the buckling load and postbuckling strength.

Figures 3.7 and 3.8 show the effect of temperature rise ΔT $(= 0, 100,$ and 200 K) on the postbuckling behavior of (P/FGM)$_S$ and (FGM/P)$_S$ hybrid laminated plates ($b/h = 40$) with unloaded edges movable and with $N = 0.2$ and 2.0 subjected to uniaxial compression. It can be seen that both buckling load and postbuckling strength are decreased with increase in temperature. It can also be seen that the buckling load of hybrid laminated plates with immovable unloaded edges is lower than that of the plate with movable

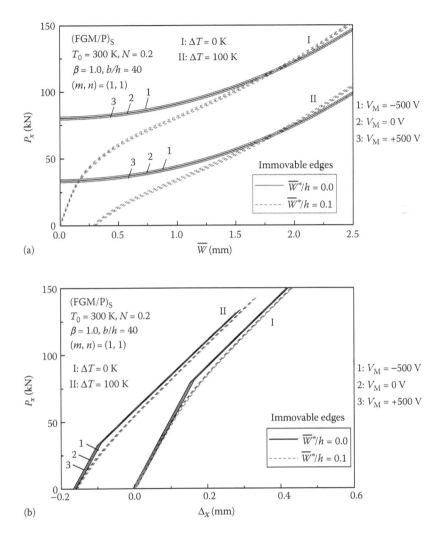

FIGURE 3.4
Thermopiezoelectric effects on the postbuckling behavior of (FGM/P)$_S$ plates with unload edges immovable: (a) load deflection; (b) load shortening.

unloaded edges under the same loading conditions, compare Figures 3.3 and 3.7, and Figures 3.4 and 3.8. In contrast, the postbuckling load-carrying capacity of the plate with immovable unloaded edges is larger than that of the plate with movable unloaded edges when the deflection $\overline{W}$ is sufficiently large.

It is appreciated that in Figures 3.3 through 3.8 $\overline{W}^*/h = 0.1$ denotes the dimensionless maximum initial geometric imperfection of the plate, and results only for $Si_3N_4/SUS304$ hybrid laminated plate under TD case.

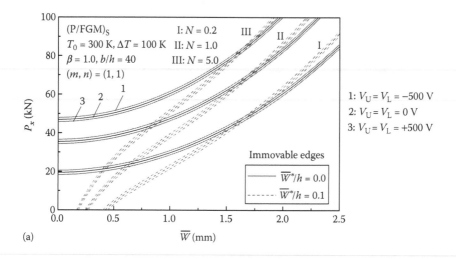

(a)

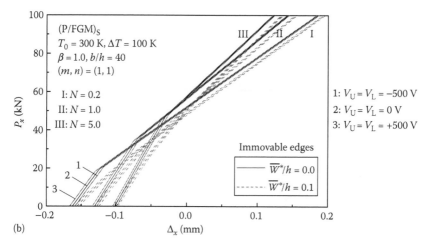

(b)

FIGURE 3.5
Effects of volume fraction index N on the postbuckling behavior of $(P/FGM)_S$ plates with unload edges immovable: (a) load deflection; (b) load shortening.

3.3   Thermal Postbuckling Behavior of FGM Plates with Piezoelectric Actuators

We now consider thermal buckling problem of these two types of hybrid laminated plate. The temperature field is assumed to be a parabolic distribution in the XY-plane of the plate, but uniform through the plate thickness, i.e.,

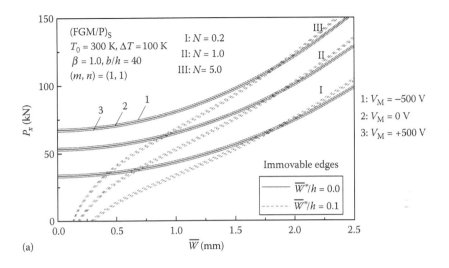

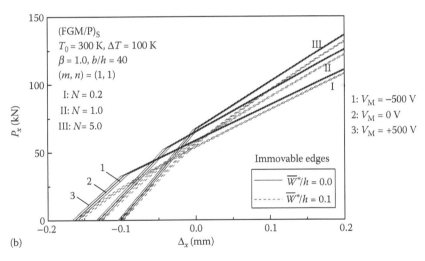

FIGURE 3.6
Effects of volume fraction index N on the postbuckling behavior of (FGM/P)$_S$ plates with unload edges immovable: (a) load deflection; (b) load shortening.

$$T(X, Y) = T_1 + T_2 \left[1 - \left(\frac{2X - a}{a} \right)^2 \right] \left[1 - \left(\frac{2Y - b}{b} \right)^2 \right] \qquad (3.36)$$

where
  T_1 is the uniform temperature rise
  T_2 is the temperature gradient

All four edges of the plate are assumed to be simply supported with no in-plane displacements. The boundary conditions become

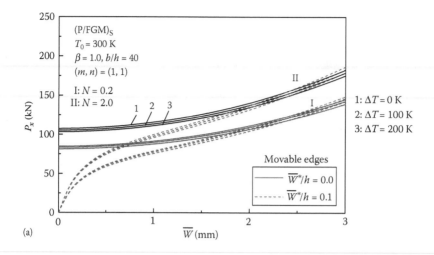

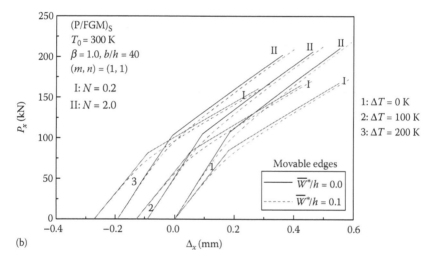

FIGURE 3.7
Effects of temperature rise on the postbuckling behavior of $(P/FGM)_S$ plates with unload edges movable: (a) load deflection; (b) load shortening.

$X = 0, a$:

$$\overline{W} = \overline{\Psi}_y = 0 \tag{3.37a}$$

$$\overline{U} = 0 \tag{3.37b}$$

$$\overline{N}_{xy} = 0, \quad \overline{M}_x = \overline{P}_x = 0 \tag{3.37c}$$

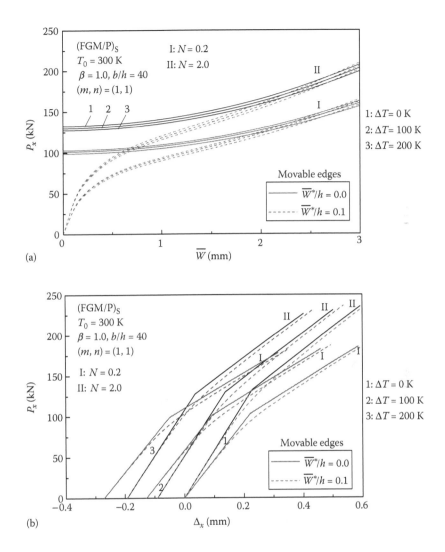

FIGURE 3.8
Effects of temperature rise on the postbuckling behavior of (FGM/P)$_S$ plates with unload edges movable: (a) load deflection; (b) load shortening.

$Y = 0, b$:

$$\overline{W} = \overline{\Psi}_x = 0 \tag{3.37d}$$

$$\overline{V} = 0 \tag{3.37e}$$

$$\overline{N}_{xy} = 0, \quad \overline{M}_y = \overline{P}_y = 0 \tag{3.37f}$$

Introducing dimensionless quantities of Equation 2.8, and

$$(\gamma_{T1}, \gamma_{T2}, \gamma_{P1}, \gamma_{P2}) = \left(A_x^T, A_y^T, B_x^P, B_y^P\right)a^2/\pi^2[D_{11}^* D_{22}^*]^{1/2}, \quad \lambda_T = \alpha_0 \Delta T \quad (3.38)$$

where α_0 is an arbitrary reference value, and

$$\alpha_{11} = a_{11}\alpha_0, \quad \alpha_{22} = a_{22}\alpha_0 \tag{3.39}$$

In Equation 3.39, $A_x^T (= A_y^T)$ and $B_x^E (= B_y^E)$ are defined by

$$\begin{bmatrix} A_x^T \\ A_y^T \end{bmatrix} \Delta T = -\sum_k \int_{t_{k-1}}^{t_k} \begin{bmatrix} A_x \\ A_y \end{bmatrix}_k \Delta T(X, Y)dZ \tag{3.40a}$$

$$\begin{bmatrix} B_x^E \\ B_y^E \end{bmatrix} \Delta V = -\sum_k \int_{t_{k-1}}^{t_k} \begin{bmatrix} B_x \\ B_y \end{bmatrix}_k \frac{V_k}{h_P} dZ \tag{3.40b}$$

where $\Delta T = T_2 - T_0$ for the in-plane parabolic temperature variation.

The nonlinear Equations 3.2 through 3.5 may then be written in dimensionless form as

$$L_{11}(W) - L_{12}(\Psi_x) - L_{13}(\Psi_y) = \gamma_{14}\beta^2 L(W + W^*, F) \tag{3.41}$$

$$L_{21}(F) - \frac{32}{\pi^2}\lambda_T C_1 = -\frac{1}{2}\gamma_{24}\beta^2 L(W + 2W^*, W) \tag{3.42}$$

$$L_{31}(W) + L_{32}(\Psi_x) - L_{33}(\Psi_y) = 0 \tag{3.43}$$

$$L_{41}(W) - L_{42}(\Psi_x) + L_{43}(\Psi_y) = 0 \tag{3.44}$$

where all nondimensional linear operators $L_{ij}()$ and nonlinear operator $L()$ are defined by Equation 2.14.

The boundary conditions expressed by Equation 3.37 become

$x = 0, \pi$:

$$W = \Psi_y = 0 \tag{3.45a}$$

$$\delta_x = 0 \tag{3.45b}$$

$$F_{,xy} = M_x = P_x = 0 \tag{3.45c}$$

$y = 0, \pi$:

$$W = \Psi_x = 0 \tag{3.45d}$$

$$\delta_y = 0 \tag{3.45e}$$

$$F_{,xy} = M_y = P_y = 0 \tag{3.45f}$$

in which

$$\delta_x = -\frac{1}{4\pi^2\beta^2\gamma_{24}} \int_0^\pi \int_0^\pi \left\{ \left[\gamma_{24}^2\beta^2\frac{\partial^2 F}{\partial y^2} - \gamma_5\frac{\partial^2 F}{\partial x^2} \right] - \frac{1}{2}\gamma_{24}\left(\frac{\partial W}{\partial x}\right)^2 - \gamma_{24}\frac{\partial W}{\partial x}\frac{\partial W^*}{\partial x} \right.$$

$$\left. + (\gamma_{24}^2\gamma_{T1} - \gamma_5\gamma_{T2})\lambda_T C_3 + (\gamma_{24}^2\gamma_{P1} - \gamma_5\gamma_{P2})\Delta V \right\} dx\, dy \tag{3.46a}$$

$$\delta_y = -\frac{1}{4\pi^2\beta^2\gamma_{24}} \int_0^\pi \int_0^\pi \left\{ \left[\frac{\partial^2 F}{\partial x^2} - \gamma_5\beta^2\frac{\partial^2 F}{\partial y^2} \right] - \frac{1}{2}\gamma_{24}\beta^2\left(\frac{\partial W}{\partial y}\right)^2 - \gamma_{24}\beta^2\frac{\partial W}{\partial y}\frac{\partial W^*}{\partial y} \right.$$

$$\left. + (\gamma_{T2} - \gamma_5\gamma_{T1})\lambda_T C_3 + (\gamma_{P2} - \gamma_5\gamma_{P1})\Delta V \right\} dy\, dx \tag{3.46b}$$

Note that in Equations 3.42 and 3.46, for the in-plane nonuniform parabolic temperature loading case,

$$C_1 = \beta^2(\gamma_{24}^2\gamma_{T1} - \gamma_5\gamma_{T2})(x/\pi - x^2/\pi^2) + (\gamma_{T2} - \gamma_5\gamma_{T1})(y/\pi - y^2/\pi^2),$$
$$C_3 = T_1/T_2 + 16(x/\pi - x^2/\pi^2)(y/\pi - y^2/\pi^2),$$

and for the uniform temperature loading case,

$$C_1 = 0, \quad C_3 = 1.0$$

By using a two-step perturbation technique, we obtain up to fourth-order asymptotic solutions

$$W = \varepsilon\left[A_{11}^{(1)}\sin mx \sin ny \right]$$
$$+ \varepsilon^3\left[A_{13}^{(3)}\sin mx \sin 3ny + A_{31}^{(3)}\sin 3mx \sin ny \right] + O(\varepsilon^5) \tag{3.47}$$

$$F = -B_{00}^{(0)}\left(\frac{y^2}{2} - C_5\frac{y^5}{120} + C_5\frac{y^6}{360\pi}\right) - b_{00}^{(0)}\left(\frac{x^2}{2} - C_6\frac{x^5}{120} + C_6\frac{x^5}{360\pi}\right)$$
$$+ \varepsilon^2\left[-B_{00}^{(2)}\left(\frac{y^2}{2} - C_5\frac{y^5}{120} + C_5\frac{y^6}{360\pi}\right) - b_{00}^{(2)}\left(\frac{x^2}{2} - C_6\frac{x^5}{120} + C_6\frac{x^5}{360\pi}\right) \right.$$
$$\left. + B_{20}^{(2)}\cos 2mx + B_{02}^{(2)}\cos 2ny \right]$$
$$+ \varepsilon^4\left[-B_{00}^{(4)}\left(\frac{y^2}{2} - C_5\frac{y^5}{120} + C_5\frac{y^6}{360\pi}\right) - b_{00}^{(4)}\left(\frac{x^2}{2} - C_6\frac{x^5}{120} + C_6\frac{x^5}{360\pi}\right) \right.$$
$$+ B_{20}^{(4)}\cos 2mx + B_{02}^{(4)}\cos 2ny + B_{22}^{(4)}\cos 2mx \cos 2ny + B_{40}^{(4)}\cos 4mx$$
$$\left. + B_{04}^{(4)}\cos 4ny + B_{24}^{(4)}\cos 2mx \cos 4ny + B_{42}^{(4)}\cos 4mx \cos 2ny \right] + O(\varepsilon^5) \tag{3.48}$$

$$\Psi_x = \varepsilon\left[C_{11}^{(1)}\cos mx \sin ny\right]$$

$$+ \varepsilon^3\left[C_{13}^{(3)}\cos mx \sin 3ny + C_{31}^{(3)}\cos 3mx \sin ny\right] + O(\varepsilon^5) \quad (3.49)$$

$$\Psi_y = \varepsilon\left[D_{11}^{(1)}\sin mx \cos ny\right]$$

$$+ \varepsilon^3\left[D_{13}^{(3)}\sin mx \cos 3ny + D_{31}^{(3)}\sin 3mx \cos ny\right] + O(\varepsilon^5) \quad (3.50)$$

Note that for the uniform temperature loading case, it is just necessary to take $C_5 = C_6 = 0$ in Equation 3.48, so that the asymptotic solutions have a similar form as expressed by Equations 3.28 through 3.31.

Next, upon substitution of Equations 3.47 through 3.50 into the boundary conditions $\delta_x = 0$ and $\delta_y = 0$, one has

$$\beta^2 B_{00}^{(0)} + \varepsilon^2\beta^2 B_{00}^{(2)} + \varepsilon^4\beta^2 B_{00}^{(4)} + \cdots = \lambda_T C_7 - \frac{1}{8}\frac{m^2 + \gamma_5 n^2\beta^2}{\gamma_{24}^2 - \gamma_5^2}(1 + 2\mu)\left(A_{11}^{(1)}\varepsilon\right)^2 \quad (3.51a)$$

$$b_{00}^{(0)} + \varepsilon^2 b_{00}^{(2)} + \varepsilon^4 b_{00}^{(4)} + \cdots = \lambda_T C_8 - \frac{1}{8}\frac{\gamma_5 m^2 + \gamma_{24}^2 n^2\beta^2}{\gamma_{24}^2 - \gamma_5^2}(1 + 2\mu)\left(A_{11}^{(1)}\varepsilon\right)^2 \quad (3.51b)$$

By adding $B_{00}^{(j)}$ and $b_{00}^{(j)}$ $(i = 0, 2, 4, \ldots)$, one has

$$\gamma_{14}\left[\left(\beta^2 B_{00}^{(0)} + \varepsilon^2\beta^2 B_{00}^{(2)} + \varepsilon^4\beta^2 B_{00}^{(4)} + \cdots\right)m^2 C_9 + \left(b_{00}^{(0)} + \varepsilon^2 b_{00}^{(2)} + \varepsilon^4 b_{00}^{(4)} + \cdots\right)n^2\beta^2 C_{10}\right]$$

$$= \frac{S_{11}}{(1+\mu)}\frac{1}{16(1+\mu)}\Theta_{22}\left(A_{11}^{(1)}\varepsilon\right)^2 - \frac{1}{256}C_{11}C_{44}\left(A_{11}^{(1)}\varepsilon\right)^4 + \cdots \quad (3.52)$$

From Equations 3.51 and 3.52, the thermal postbuckling equilibrium path can be written as

$$\lambda_T = \lambda_T^{(0)} + \lambda_T^{(2)}W_m^2 + \lambda_T^{(4)}W_m^4 + \cdots \quad (3.53)$$

in which W_m is the dimensionless form of maximum deflection, which is assumed to be at the point $(x, y) = (\pi/2m, \pi/2n)$. It is noted that $\lambda_T^{(i)}$ $(i = 0, 2, 4, \ldots)$ are all functions of temperature and position and given in detail in Appendix G.

To obtain numerical results, it is necessary to solve Equation 3.53 by an iterative numerical procedure with the following steps:

1. Begin with $\overline{W}/h = 0$.

2. Assume elastic constants and the thermal expansion coefficients are constant, i.e., at $T_0 = 300$ K. The thermal buckling load for the plate of temperature-independent material is obtained.

3. Use the temperature determined in the previous step, the temperature-dependent material properties may be decided from Equations 1.4 and 3.35, and the thermal buckling load is obtained again.

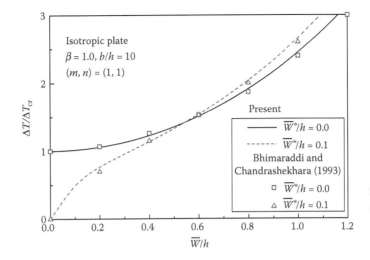

FIGURE 3.9

Comparisons of thermal postbuckling load–deflection curves for isotropic plates subjected to uniform temperature rise.

4. Repeat step (3) until the thermal buckling temperature converges.

5. Specify the new value of $\overline{W}/h$, and repeat steps (2)–(4) until the thermal postbuckling temperature converges.

The thermal postbuckling load–deflection curves for perfect and imperfect, isotropic square plates ($b/h = 10$, $\nu = 0.3$) subjected to a uniform temperature rise are compared in Figure 3.9 with the analytical solutions of Bhimaraddi and Chandrashekhara (1993). The dimensionless critical temperature $\lambda_T^* = \alpha_{22}\Delta T \times 10^4$ for these two theories are identical and $\lambda_T^* = 119.783$. In addition, the thermal postbuckling load–deflection curves for an isotropic square plate ($b/h = 40$) subjected to nonuniform parabolic temperature loading with $T_0/T_1 = 1.0$ are compared in Figure 3.10 with the finite difference method solutions of Kamiya and Fukui (1982). These two comparisons show that the results from the present method are in good agreement with the existing results, thus verifying the reliability and accuracy of the present method.

Tables 3.7 through 3.12 give thermal buckling loads ΔT_{cr} (in K) for perfect, $(P/FGM)_S$ and $(FGM/P)_S$ hybrid laminated plates with different values of the volume fraction index N ($= 0.0, 0.2, 0.5, 1.0, 2.0$, and 5.0) subjected to a uniform temperature rise ($T_2 = 0$) and nonuniform parabolic temperature variation with three thermal load ratio T_1/T_2 ($= 0.0, 0.5$, and 1.0). It can be found that the buckling temperature of $(P/FGM)_S$ plate is lower than that of $(FGM/P)_S$ plate under the same loading condition. As in the case of axial compression, for the hybrid plates with $Si_3N_4/SUS304$ substrate, a fully metallic plate ($N = 0$) has lowest buckling temperature and that the buckling temperature increases as the volume fraction index N increases. It can be seen

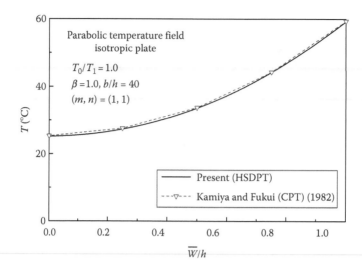

FIGURE 3.10

Comparisons of thermal postbuckling load–deflection curves for an isotropic plate subjected to nonuniform parabolic temperature loading.

TABLE 3.7

Comparisons of Buckling Temperature ΔT (K) for FGM Hybrid Plates with $Si_3N_4/SUS304$ Substrate under Uniform Temperature Rise and Three Sets of Electrical Loading Conditions ($b/h = 20$, $a/b = 1.0$, $T_0 = 300$ K)

$V_U(=V_L)$ (in V)	$N=0.0$	$N=0.2$	$N=0.5$	$N=1.0$	$N=2.0$	$N=5.0$
$(P/FGM)_S$, TID						
-500	190.8831	219.4916	250.1407	282.0667	314.3084	345.7490
0	188.3028	216.8932	247.4695	279.2786	311.3615	342.5981
$+500$	185.7225	214.2948	244.7983	276.4905	308.4147	339.4472
$(P/FGM)_S$, TD-F						
-500	172.2518	195.3290	219.4843	244.0116	267.8233	289.8148
0	170.0603	193.1553	217.2865	241.7598	265.4975	287.3994
$+500$	167.8666	190.9797	215.0868	239.5061	263.1693	284.9810
$(P/FGM)_S$, TD						
-500	171.4933	194.4750	218.5124	242.9146	266.5796	288.4171
0	169.4166	192.4294	216.4595	240.8252	264.4368	286.2063
$+500$	167.3354	190.3797	214.4023	238.7315	262.2893	283.9898
V_M	$N=0.0$	$N=0.2$	$N=0.5$	$N=1.0$	$N=2.0$	$N=5.0$
$(FGM/P)_S$, TID`						
-500	229.1825	265.1534	303.8535	344.5095	386.0890	427.2245
0	226.6022	262.5550	301.1823	341.7214	383.1422	424.0736
$+500$	224.0218	259.9565	298.5111	338.9334	380.1953	420.9227

TABLE 3.7 (continued)

Comparisons of Buckling Temperature ΔT (K) for FGM Hybrid Plates with Si_3N_4/SUS304 Substrate under Uniform Temperature Rise and Three Sets of Electrical Loading Conditions ($b/h = 20$, $a/b = 1.0$, $T_0 = 300$ K)

V_M	$N=0.0$	$N=0.2$	$N=0.5$	$N=1.0$	$N=2.0$	$N=5.0$
$(FGM/P)_S$, TD-F						
−500	202.5830	230.7437	260.2219	290.1927	319.4639	346.4602
0	200.4486	228.6269	258.0835	288.0067	317.2147	344.1400
+500	198.3126	226.5088	255.9439	285.8196	314.9637	341.8174
$(FGM/P)_S$, TD						
−500	202.4387	230.5792	260.0317	289.9734	319.2144	346.1820
0	200.4283	228.6011	258.0510	287.9671	317.1681	344.0878
+500	198.4141	226.6195	256.0670	285.9752	315.1177	341.9890

TABLE 3.8

Comparisons of Buckling Temperature ΔT (K) for FGM Hybrid Plates with ZrO_2/Ti-6Al-4V Substrate under Uniform Temperature Rise and Three Sets of Electrical Loading Conditions ($b/h = 20$, $a/b = 1.0$, $T_0 = 300$ K)

$V_U(=V_L)$ (in V)	$N=0.0$	$N=0.2$	$N=0.5$	$N=1.0$	$N=2.0$	$N=5.0$
$(P/FGM)_S$, TID						
−500	361.0100	313.6715	272.8378	238.9043	209.9199	184.1798
0	352.2897	307.0133	267.6012	234.6535	206.3736	181.1541
+500	343.5695	300.3551	262.3647	230.4026	202.8273	178.1284
$(P/FGM)_S$, TD-F						
−500	339.2903	230.5015	187.5004	161.2948	142.2591	126.8038
0	329.9423	226.2742	184.5547	158.9790	140.3220	125.1165
+500	320.6644	222.0210	181.5875	156.6469	138.3721	123.4188
$(P/FGM)_S$, TD						
−500	332.9886	228.5898	186.3837	160.5363	141.7007	126.3721
0	324.7029	224.6504	183.6053	158.3339	139.8469	124.7491
+500	316.4470	220.6800	180.8029	156.1134	137.9791	123.1148
V_M	$N=0.0$	$N=0.2$	$N=0.5$	$N=1.0$	$N=2.0$	$N=5.0$
$(FGM/P)_S$, TID						
−500	398.6698	352.4184	310.5889	274.8633	243.6992	215.4438
0	389.9496	345.7601	305.3524	270.6125	240.1529	212.4182
+500	381.2293	339.1019	300.1158	266.3616	236.6066	209.3925
$(FGM/P)_S$, TD-F						
−500	360.4985	246.6464	202.8977	175.8595	156.4976	140.4670
0	351.4150	242.6464	200.1012	173.8893	154.6699	138.8813
+500	342.4201	238.6230	197.2875	171.6830	152.8321	137.2873
$(FGM/P)_S$, TD						
−500	359.3701	246.3116	202.7124	175.7311	156.4056	140.3956
0	351.2241	242.5951	200.0737	173.8714	154.6572	138.8722
+500	343.1091	238.8494	197.4161	171.7710	152.8982	137.3400

TABLE 3.9

Comparisons of Buckling Temperature ΔT (K) for $(P/FGM)_S$ Plates with $Si_3N_4/SUS304$ Substrate under Nonuniform Parabolic Temperature Variation and Three Sets of Electrical Loading Conditions ($b/h = 20$, $a/b = 1.0$, $T_0 = 300$ K)

T_1/T_2	$V_U (= V_L)$ (in V)	$N = 0.0$	$N = 0.2$	$N = 1.0$	$N = 2.0$	$N = 5.0$
$(P/FGM)_S$, TID
0.0	−500	334.1945	384.2758	493.8177	550.2584	605.2963
	0	329.6768	379.7267	488.9366	545.0994	599.7801
	+500	325.1593	375.1775	484.0555	539.9403	594.2639
0.5	−500	178.2000	204.9059	263.3194	293.4167	322.7661
	0	175.7911	202.4802	260.7166	290.6657	319.8246
	+500	173.3822	200.0544	258.1139	287.9147	316.8832
1.0	−500	121.4908	139.6983	179.5235	200.0434	220.0532
	0	119.8485	138.0446	177.7491	198.1678	218.0478
	+500	118.2062	136.3908	175.9746	196.2923	216.0424
$(P/FGM)_S$, TD-F
0.0	−500	284.4204	321.6889	399.2022	435.4759	466.3745
	0	280.8110	318.0988	395.4951	431.6948	462.5440
	+500	277.2018	314.5095	391.7880	427.9120	458.7091
0.5	−500	161.7491	183.5069	229.4070	252.0241	272.9636
	0	159.6876	181.4618	227.2878	249.8317	270.6824
	+500	157.6241	179.4149	225.1668	247.9120	268.3984
1.0	−500	113.3601	128.9681	162.1571	178.6939	194.3458
	0	111.8991	127.5158	160.6434	177.1206	192.6956
	+500	110.4367	126.0622	159.1284	175.5458	191.0437
$(P/FGM)_S$, TD
0.0	−500	282.4131	319.4363	396.3263	432.2831	462.8696
	0	279.1118	316.1897	393.0523	428.9790	459.5585
	+500	275.8039	312.9370	389.7715	425.6670	456.2370
0.5	−500	161.0838	182.7577	228.5001	250.9276	271.7272
	0	159.1237	180.8258	226.5236	248.8976	269.6281
	+500	157.1594	178.8899	224.4809	246.8633	267.5240
1.0	−500	113.0288	128.5932	161.6644	178.1279	193.6974
	0	111.6182	127.1975	160.2241	176.6385	192.1429
	+500	110.2051	125.7995	158.7813	175.1463	190.5851

that an increase is about +69% for the $(P/FGM)_S$ plate, and about +72% for the $(FGM/P)_S$ one under uniform temperature rise, and about +65% for the $(P/FGM)_S$ plate, and about +68% for the $(FGM/P)_S$ one under nonuniform parabolic temperature variation with $T_1/T_2 = 0.0$, from $N = 0$ to $N = 5$ under TD case. It can also be seen that the buckling temperature decreases when the temperature dependency is put into consideration. The percentage decrease is about −15% for the $(P/FGM)_S$ plate and about −17% for the $(FGM/P)_S$ one from TID case to TD case under uniform temperature

TABLE 3.10

Comparisons of Buckling Temperature ΔT (K) for (FGM/P)$_S$ Plates with Si$_3$N$_4$/SUS304 Substrate under Nonuniform Parabolic Temperature Variation and Three Sets of Electrical Loading Conditions ($b/h = 20$, $a/b = 1.0$, $T_0 = 300$ K)

T_1/T_2	V_M (in V)	$N=0.0$	$N=0.2$	$N=1.0$	$N=2.0$	$N=5.0$
(FGM/P)$_S$, TID						
0.0	−500	401.2482	464.2183	603.1371	675.9244	747.9340
	0	396.7307	459.6692	598.2560	670.7654	742.4178
	+500	392.2130	455.1200	593.3749	665.6063	736.9016
0.5	−500	213.9545	247.5333	321.6120	360.4261	398.8257
	0	211.5456	245.1076	319.0093	357.6752	395.8842
	+500	209.1367	242.6819	316.4065	354.9242	392.9428
1.0	−500	145.8671	168.7604	219.2658	245.7285	271.9086
	0	144.2248	176.1066	217.4913	243.8530	269.9032
	+500	142.5825	165.4528	215.7168	241.9774	267.8978
(FGM/P)$_S$, TD-F						
0.0	−500	330.7502	376.1384	471.0853	515.2603	552.2033
	0	327.2496	372.6424	467.4599	511.5790	548.5238
	+500	323.7512	369.1491	463.8368	507.8985	544.8420
0.5	−500	190.4493	216.9529	273.1703	300.9339	326.7027
	0	188.4398	214.9614	271.1117	298.8130	324.5098
	+500	186.4289	212.9685	269.0516	296.6906	322.3148
1.0	−500	134.1980	153.3095	194.0825	214.5766	234.1622
	0	132.7667	151.8893	192.6086	213.0498	232.5667
	+500	131.3342	150.4680	191.1337	211.5216	230.9696
(FGM/P)$_S$, TD						
0.0	−500	330.4118	375.7495	470.5706	514.6824	551.5844
	0	327.2365	372.6227	467.4229	511.5351	548.4777
	+500	324.0568	369.4919	464.2708	508.3820	545.3630
0.5	−500	190.3281	216.8136	272.9858	300.7233	326.4663
	0	188.4281	214.9458	271.0864	298.7827	324.4752
	+500	186.5247	213.0748	269.1836	296.8385	322.4798
1.0	−500	134.1359	153.2376	193.9849	214.4635	234.0341
	0	132.7595	151.8797	192.5931	213.0307	232.5446
	+500	131.3808	150.5197	191.1989	211.5955	231.0524

rise and the same volume fraction distribution $N=2$. Also the thermal buckling loads under TD-F and TD cases are very close under the same volume fraction distribution and the same thermal loading condition. In contrast, it is seen that the buckling temperature of hybrid plates with ZrO$_2$/Ti-6Al-4V substrate is decreased as the volume fraction index N increases. It can also be seen that the control voltage has a small effect on the thermal buckling loads for these hybrid laminated plates.

TABLE 3.11

Comparisons of Buckling Temperature ΔT (K) for $(P/FGM)_S$ Plates with ZrO_2/Ti-6Al-4V Substrate under Nonuniform Parabolic Temperature Variation and Three Sets of Electrical Loading Conditions ($b/h = 20$, $a/b = 1.0$, $T_0 = 300$ K)

T_1/T_2	$V_U (= V_L)$ (in V)	$N = 0.0$	$N = 0.2$	$N = 1.0$	$N = 2.0$	$N = 5.0$
$(P/FGM)_S$, TID						
0.0	−500	632.1551	549.2456	418.3071	367.5503	322.4765
	0	616.8853	537.5869	410.8641	361.3410	317.1789
	+500	601.6155	525.9282	403.4211	355.1317	311.8813
0.5	−500	337.0529	292.8512	223.0412	195.9794	171.9472
	0	328.9113	286.6349	219.0726	192.6686	169.1224
	+500	320.7697	280.4187	215.1040	189.3578	166.2977
1.0	−500	229.7849	199.6515	152.0597	133.6106	117.2267
	0	224.2344	195.4136	149.3541	131.3534	115.3009
	+500	218.6839	191.1756	146.6485	129.0962	113.3752
$(P/FGM)_S$, TD-F						
0.0	−500	598.0353	343.6654	235.8257	208.3855	186.4033
	0	577.4709	337.7048	232.6350	205.7157	184.0908
	+500	557.6385	331.6966	229.4191	203.0257	181.7600
0.5	−500	317.2877	218.7738	153.5554	135.3958	120.7423
	0	308.6649	214.7361	151.3301	133.5351	119.1132
	+500	300.0934	210.6746	149.0898	131.6627	117.4751
1.0	−500	218.7285	161.5482	115.1498	101.4113	90.0399
	0	213.0890	158.4637	113.4055	99.9548	88.7771
	+500	207.4749	155.3648	111.6517	98.4907	87.5082
$(P/FGM)_S$, TD						
0.0	−500	574.8693	339.7483	234.0814	207.3249	185.5407
	0	558.7789	334.3736	231.4190	204.8110	183.3573
	+500	542.9512	328.9448	228.4113	202.2762	181.1540
0.5	−500	311.8759	217.0459	152.8709	134.8918	120.3626
	0	304.1632	213.2708	150.7486	133.1068	118.7900
	+500	296.4737	209.4670	148.6098	131.3090	117.2077
1.0	−500	216.2538	160.5520	114.7444	101.1136	89.8113
	0	211.0199	157.6210	113.0615	99.7021	88.5831
	+500	205.7910	154.6718	111.3681	98.2823	87.3481

Figure 3.11 shows the effect of material properties on the thermal post-buckling load–deflection curves for $(P/FGM)_S$ and $(FGM/P)_S$ hybrid laminated plates ($N = 2.0$) subjected to a uniform temperature rise ($T_2 = 0$) and nonuniform parabolic temperature variation ($T_1/T_2 = 0.0$), respectively, and with three sets of electrical loading, $V_U = V_L$ (or V_M) = −500, 0, +500 V, and under two cases of thermoelastic material properties, i.e., TID and TD. It can be seen that the thermal postbuckling equilibrium path becomes lower

TABLE 3.12

Comparisons of Buckling Temperature ΔT (K) for $(FGM/P)_S$ Plates with $ZrO_2/$Ti-6Al-4V Substrate under Nonuniform Parabolic Temperature Distribution and Three Sets of Electrical Loading Conditions ($b/h = 20$, $a/b = 1.0$, $T_0 = 300$ K)

T_1/T_2	V_M (in V)	$N = 0.0$	$N = 0.2$	$N = 1.0$	$N = 2.0$	$N = 5.0$
$(FGM/P)_S$, TID						
0.0	−500	698.1001	617.0921	481.2691	426.6948	377.2160
	0	682.8303	605.4334	473.8261	420.4855	371.9184
	+500	667.5605	593.7747	466.3831	414.2762	366.6208
0.5	−500	372.2135	329.0262	256.6125	227.5155	201.1348
	0	364.0719	322.8099	252.6440	224.2047	198.3100
	+500	355.9304	316.5936	248.6754	220.8939	195.4853
1.0	−500	253.7555	224.3138	174.9472	155.1106	137.1256
	0	248.2050	220.0759	172.2415	152.8534	135.1999
	+500	242.6546	215.8379	169.5359	150.5962	133.2741
$(FGM/P)_S$, TD-F						
0.0	−500	614.2082	362.4850	253.1466	225.7529	203.2276
	0	595.3040	356.8670	250.2042	223.2697	201.0907
	+500	576.9683	351.2095	247.2384	220.7703	198.9386
0.5	−500	337.9622	234.5947	167.8398	149.1455	133.6344
	0	329.5637	230.7643	165.7353	147.3920	132.1239
	+500	321.2380	226.9125	163.6189	145.6291	130.6051
1.0	−500	235.5359	175.0072	127.0871	112.8256	100.9174
	0	229.9974	172.0518	125.4258	111.4420	99.7203
	+500	224.4839	169.0833	123.7565	110.0521	98.4001
$(FGM/P)_S$, TD						
0.0	−500	610.6959	361.8958	252.9243	225.5958	203.0985
	0	595.0753	356.8408	250.1978	223.2656	201.0885
	+500	579.6954	351.7393	247.4456	220.9188	199.0620
0.5	−500	337.0280	234.3057	167.7308	149.0652	133.5706
	0	329.4402	230.7321	165.7247	147.3848	132.1188
	+500	321.8768	227.1318	163.7057	145.6942	130.6580
1.0	−500	235.0930	174.8349	127.0197	112.7761	100.8794
	0	229.9281	172.0281	125.4175	111.4363	99.7164
	+500	224.7690	169.2046	123.8067	110.0897	98.4306

when the temperature-dependent properties are taken into account. It can also be seen that the FGM hybrid laminated plate under a uniform temperature rise has a lower postbuckling temperature than does a plate under a nonuniform parabolic temperature variation. It is found that minus control voltages increase the buckling temperature and decrease the postbuckled deflection, whereas the plus control voltages decrease the buckling temperature and induce more large postbuckled deflections. Since the effect of the

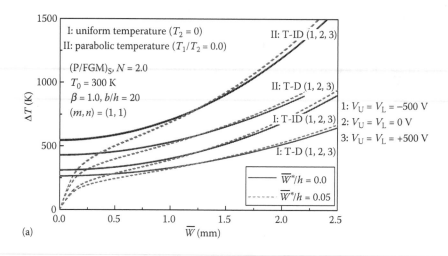

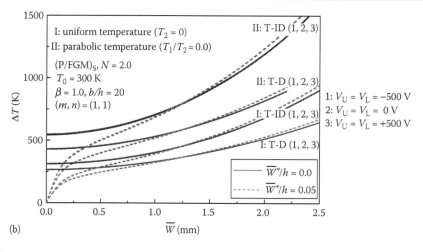

FIGURE 3.11

Effect of material properties on the thermal postbuckling behavior of FGM hybrid laminated plates under in-plane temperature variation: (a) $(P/FGM)_S$ plate; (b) $(FGM/P)_S$ plate.

control voltages is small, these three curves, referred to as 1, 2, and 3 in the figure, are very close.

Figure 3.12 shows the effect of the volume fraction index N ($= 0.2, 1.0$, and 5.0) on the thermal postbuckling behavior of $(P/FGM)_S$ and $(FGM/P)_S$ hybrid laminated plates subjected to a nonuniform parabolic temperature variation with $T_1/T_2 = 0.5$ and three sets of electrical loading. It can be seen that the increase of the volume fraction index N yields an increase of the buckling temperature and thermal postbuckling strength.

Figure 3.13 shows the effect of different values of the thermal load ratio T_1/T_2 ($= 0.0, 0.5, 1.0$) on the postbuckling behavior of $(P/FGM)_S$ and

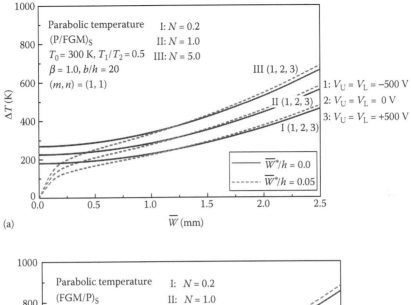

(a)

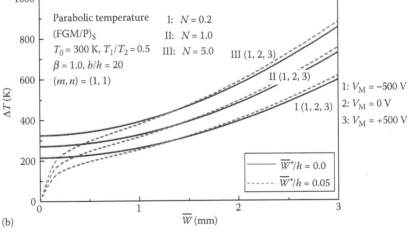

(b)

FIGURE 3.12
Effect of volume fraction index N on the thermal postbuckling behavior of FGM hybrid laminated plates under nonuniform parabolic temperature variation: (a) $(P/FGM)_S$ plate; (b) $(FGM/P)_S$ palte.

$(FGM/P)_S$ hybrid laminated plates subjected to a nonuniform parabolic temperature variation and three sets of electrical loading. It can be found that the postbuckling strength is decreased by increasing T_1/T_2, when the volume fraction index $N = 2.0$.

As mentioned in Section 3.2, in Figures 3.11 through 3.13, $\overline{W}^*/h = 0.05$ denotes the dimensionless maximum initial geometric imperfection of the plate.

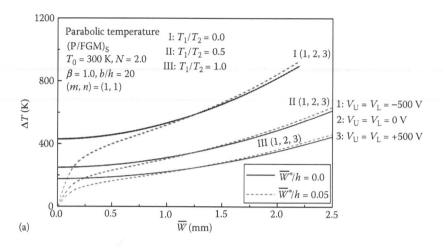

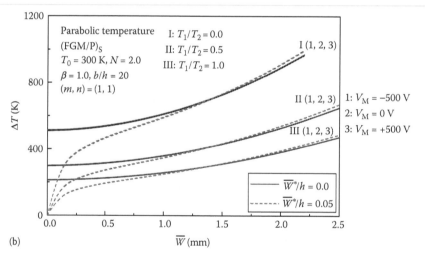

FIGURE 3.13

Effect of thermal load ratio T_1/T_2 on the postbuckling of FGM hybrid laminated plates under nonuniform parabolic temperature variation: (a) $(P/FGM)_S$ plate; (b) $(FGM/P)_S$ plate.

3.4 Postbuckling of Sandwich Plates with FGM Face Sheets in Thermal Environments

Finally, we consider a rectangular sandwich plate which consists of one homogeneous substrate and two face sheets made of functionally graded materials and is midplane symmetric, as shown in Figure 3.14. The length, width, and total thickness of the sandwich plate are a, b, and h. The thickness

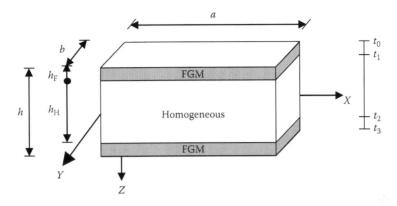

FIGURE 3.14
Configuration of a sandwich plate.

of each FGM face sheet is h_F, while the thickness of the homogeneous substrate is h_H. The FGM face sheet is made from a mixture of ceramics and metals, the mixing ratio of which is varied continuously and smoothly in the Z-direction. It is assumed that the effective Young's modulus E_f, thermal expansion coefficient α_f, and thermal conductivity κ_f of FGM face sheets are functions of temperature, so that E_f, α_f, and κ_f are both temperature- and position-dependent. The Poisson ratio ν_f depends weakly on temperature change and is assumed to be a constant. We assume the volume fraction V_m follows a simple power law. According to rule of mixture, we have

$$E_f(Z, T) = [E_m(T) - E_c(T)]\left(\frac{Z - t_0}{t_1 - t_0}\right)^N + E_c(T) \qquad (3.54a)$$

$$\alpha_f(Z, T) = [\alpha_m(T) - \alpha_c(T)]\left(\frac{Z - t_0}{t_1 - t_0}\right)^N + \alpha_c(T) \qquad (3.54b)$$

$$\kappa_f(Z, T) = [\kappa_m(T) - \kappa_c(T)]\left(\frac{Z - t_0}{t_1 - t_0}\right)^N + \kappa_c(T) \qquad (3.54c)$$

It is evident that when $Z = t_0$, $E_f = E_c$, $\alpha_f = \alpha_c$ and $\kappa_f = \kappa_c$, and when $Z = t_1$, $E_f = E_m$, $\alpha_f = \alpha_m$, and $\kappa_f = \kappa_m$.

The temperature field is assumed to be uniform over the plate surface but varying along the thickness direction due to heat conduction which can be determined by solving a steady-state heat transfer equation

$$-\frac{d}{dZ}\left[\kappa \frac{dT}{dZ}\right] = 0 \qquad (3.55)$$

where

$$
\kappa = \begin{cases} \kappa_f & (t_0 \leq Z < t_1) \\ \kappa_H & (t_1 \leq Z \leq t_2) \\ \kappa_f & (t_2 < Z \leq t_3) \end{cases} \tag{3.56a}
$$

$$
T = \begin{cases} T_F & (t_0 \leq Z \leq t_1) \\ T_H & (t_1 \leq Z \leq t_2) \\ \tilde{T}_F & (t_2 \leq Z \leq t_3) \end{cases} \tag{3.56b}
$$

where κ_H is the thermal conductivity of the homogeneous substrate. Equation 3.55 is solved by imposing the boundary conditions $T = T_U$ at $Z = t_0$ and $T = T_L$ at $Z = t_3$, and the continuity conditions

$$
T_F(t_1) = T_H(t_1) = T_{m1}, \quad T_H(t_2) = \tilde{T}_F(t_2) = T_{m2} \tag{3.57a}
$$

$$
\kappa_m \frac{dT_F}{dZ}\bigg|_{Z=t_1} = \kappa_H \frac{dT_H}{dZ}\bigg|_{Z=t_1}, \quad \kappa_H \frac{dT_H}{dZ}\bigg|_{Z=t_2} = \kappa_m \frac{d\tilde{T}_F}{dZ}\bigg|_{Z=t_2} \tag{3.57b}
$$

where
    T_U and T_L are the temperatures at top and bottom surfaces of the plate
    T_{m1} and T_{m2} are the temperatures at $Z = t_1$ and t_2 interfaces, respectively

The solutions of Equations 3.55 through 3.57, by means of polynomial series, are

$$
T_F = T_U + (T_{m1} - T_U)\eta(Z) \tag{3.58a}
$$

$$
T_H = \frac{1}{h_H}[(T_{m1}t_2 - T_{m2}t_1) + (T_{m2} - T_{m1})Z] \tag{3.58b}
$$

$$
\tilde{T}_F(Z) = T_L + (T_{m2} - T_L)\tilde{\eta}(Z) \tag{3.58c}
$$

in which

$$
\eta(Z) = \frac{1}{C}\left[\left(\frac{Z-t_0}{t_1-t_0}\right) - \frac{\kappa_{mc}}{(N+1)\kappa_c}\left(\frac{Z-t_0}{t_1-t_0}\right)^{N+1} + \frac{\kappa_{mc}^2}{(2N+1)\kappa_c^2}\left(\frac{Z-t_0}{t_1-t_0}\right)^{2N+1} \right.
$$
$$
\left. - \frac{\kappa_{mc}^3}{(3N+1)\kappa_c^3}\left(\frac{Z-t_0}{t_1-t_0}\right)^{3N+1} + \frac{\kappa_{mc}^4}{(4N+1)\kappa_c^4}\left(\frac{Z-t_0}{t_1-t_0}\right)^{4N+1} - \frac{\kappa_{mc}^5}{(5N+1)\kappa_c^5}\left(\frac{Z-t_0}{t_1-t_0}\right)^{5N+1} \right] \tag{3.59a}
$$

$$
\tilde{\eta}(Z) = \frac{1}{C}\left[\left(\frac{Z-t_3}{t_2-t_3}\right) - \frac{\kappa_{mc}}{(N+1)\kappa_c}\left(\frac{Z-t_3}{t_2-t_3}\right)^{N+1} + \frac{\kappa_{mc}^2}{(2N+1)\kappa_c^2}\left(\frac{Z-t_3}{t_2-t_3}\right)^{2N+1} \right.
$$
$$
\left. - \frac{\kappa_{mc}^3}{(3N+1)\kappa_c^3}\left(\frac{Z-t_3}{t_2-t_3}\right)^{3N+1} + \frac{\kappa_{mc}^4}{(4N+1)\kappa_c^4}\left(\frac{Z-t_3}{t_2-t_3}\right)^{4N+1} - \frac{\kappa_{mc}^5}{(5N+1)\kappa_c^5}\left(\frac{Z-t_3}{t_2-t_3}\right)^{5N+1} \right] \tag{3.59b}
$$

$$C = 1 - \frac{\kappa_{mc}}{(N+1)\kappa_c} + \frac{\kappa_{mc}^2}{(2N+1)\kappa_c^2} - \frac{\kappa_{mc}^3}{(3N+1)\kappa_c^3} + \frac{\kappa_{mc}^4}{(4N+1)\kappa_c^4}$$
$$- \frac{\kappa_{mc}^5}{(5N+1)\kappa_c^5} \tag{3.59c}$$

$$G = 1 - \frac{\kappa_{mc}}{\kappa_c} + \frac{\kappa_{mc}^2}{\kappa_c^2} - \frac{\kappa_{mc}^3}{\kappa_c^3} + \frac{\kappa_{mc}^4}{\kappa_c^4} - \frac{\kappa_{mc}^5}{\kappa_c^5} \tag{3.59d}$$

where $\kappa_{mc} = \kappa_m - \kappa_c$, and

$$T_{m1} = \frac{((\kappa_H/h_H) + (\kappa_m G/h_F C))T_U + (\kappa_H/h_H)T_L}{(\kappa_m G/h_F C) + (2\kappa_H/h_H)},$$
$$T_{m2} = \frac{((\kappa_H/h_H) + (\kappa_m G/h_F C))T_L + (\kappa_H/h_H)T_U}{(\kappa_m G/h_F C) + (2\kappa_H/h_H)} \tag{3.60}$$

The plate is assumed to be geometrically imperfect, and is subjected to a compressive edge load in the X-direction and/or heat conduction. Two cases of compressive postbuckling under thermal environments and of thermal postbuckling due to heat conduction are considered. Since the heat conduction is considered in the present case, the general von Kármán-type equations (Equations 1.33 through 1.36) can be written in the simple form as

$$\tilde{L}_{11}(\overline{W}) - \tilde{L}_{12}(\overline{\Psi}_x) - \tilde{L}_{13}(\overline{\Psi}_y) + \tilde{L}_{14}(\overline{F}) - \tilde{L}_{15}(\overline{N}^T) - \tilde{L}_{16}(\overline{M}^T) = \tilde{L}(\overline{W} + \overline{W}^*, \overline{F}) \tag{3.61}$$

$$\tilde{L}_{21}(\overline{F}) + \tilde{L}_{22}(\overline{\Psi}_x) + \tilde{L}_{23}(\overline{\Psi}_y) - \tilde{L}_{24}(\overline{W}) - \tilde{L}_{25}(\overline{N}^T) = -\frac{1}{2}\tilde{L}(\overline{W} + 2\overline{W}^*, \overline{W}) \tag{3.62}$$

$$\tilde{L}_{31}(\overline{W}) + \tilde{L}_{32}(\overline{\Psi}_x) - \tilde{L}_{33}(\overline{\Psi}_y) + \tilde{L}_{34}(\overline{F}) - \tilde{L}_{35}(\overline{N}^T) - \tilde{L}_{36}(\overline{S}^T) = 0 \tag{3.63}$$

$$\tilde{L}_{41}(\overline{W}) - \tilde{L}_{42}(\overline{\Psi}_x) + \tilde{L}_{43}(\overline{\Psi}_y) + \tilde{L}_{44}(\overline{F}) - \tilde{L}_{45}(\overline{N}^T) - \tilde{L}_{46}(\overline{S}^T) = 0 \tag{3.64}$$

where all linear operators $\tilde{L}_{ij}()$ and nonlinear operator $\tilde{L}()$ are defined by Equation 1.37. The forces and moments caused by elevated temperature are defined by Equations 3.8 and 3.9.

All four edges of the plate are assumed to be simply supported. Depending upon the in-plane behavior at the edges, two cases, case 1 (for the compressive buckling problem) and case 2 (for the thermal buckling problem), will be considered. They are

Case 1: The edges are simply supported and freely movable in the in-plane directions. In addition, the plate is subjected to uniaxial compressive edge loads.

Case 2: All four edges are simply supported with no in-plane displacements, i.e., prevented from moving in the X- and Y-directions.

For both cases the associated boundary conditions can be expressed by $X = 0$, a:

$$\overline{W} = \overline{\Psi}_y = 0 \tag{3.65a}$$

$$\overline{M}_x = \overline{P}_x = 0 \tag{3.65b}$$

$$\int_0^b \overline{N}_x dY + P = 0 \text{ (for compressive buckling problem)} \tag{3.65c}$$

$$\overline{U} = 0 \text{ (for thermal buckling problem)} \tag{3.65d}$$

$Y = 0$, b:

$$\overline{W} = \overline{\Psi}_x = 0 \tag{3.65e}$$

$$\overline{M}_y = \overline{P}_y = 0 \tag{3.65f}$$

$$\int_0^a \overline{N}_y dX = 0 \text{ (for compressive buckling problem)} \tag{3.65g}$$

$$\overline{V} = 0 \text{ (for thermal buckling problem)} \tag{3.65h}$$

where
  P is a compressive edge load in the X-direction
  $\overline{M}_x$ and $\overline{M}_y$ are the bending moments
  $\overline{P}_x$ and $\overline{P}_y$ are the higher order moments, defined by Equation 3.12

The condition expressing the immovability condition, $\overline{U} = 0$ (on $X = 0$, a) and $\overline{V} = 0$ (on $Y = 0$, b), is fulfilled on the average sense as

$$\int_0^b \int_0^a \frac{\partial \overline{U}}{\partial X} dX \, dY = 0, \quad \int_0^a \int_0^b \frac{\partial \overline{V}}{\partial Y} dY \, dX = 0 \tag{3.66}$$

This condition in conjunction with Equation 3.67 below provides thermally induced compressive stresses acting on the edges $X = 0$, a and $Y = 0$, b.
  The average end-shortening relationships are

$$\frac{\Delta_x}{a} = -\frac{1}{ab} \int_0^b \int_0^a \frac{\partial \overline{U}}{\partial X} dX \, dY$$

$$= -\frac{1}{ab} \int_0^b \int_0^a \left\{ \left[A_{11}^* \frac{\partial^2 \overline{F}}{\partial Y^2} + A_{12}^* \frac{\partial^2 \overline{F}}{\partial X^2} + \left(B_{11}^* - \frac{4}{3h^2} E_{11}^* \right) \frac{\partial \overline{\Psi}_x}{\partial X} + \left(B_{12}^* - \frac{4}{3h^2} E_{12}^* \right) \frac{\partial \overline{\Psi}_y}{\partial Y} \right. \right.$$

$$\left. - \frac{4}{3h^2} \left(E_{11}^* \frac{\partial^2 \overline{W}}{\partial X^2} + E_{12}^* \frac{\partial^2 \overline{W}}{\partial Y^2} \right) \right] - \frac{1}{2} \left(\frac{\partial \overline{W}}{\partial X} \right)^2 - \frac{\partial \overline{W}}{\partial X} \frac{\partial \overline{W}^*}{\partial X} - \left(A_{11}^* \overline{N}_x^T + A_{12}^* \overline{N}_y^T \right) \right\} dX \, dY$$

$$\tag{3.67a}$$

$$\frac{\Delta_y}{b} = -\frac{1}{ab} \int_0^a \int_0^b \frac{\partial \overline{V}}{\partial Y} \, dY \, dX$$

$$= -\frac{1}{ab} \int_0^a \int_0^b \left\{ \left[A_{22}^* \frac{\partial^2 \overline{F}}{\partial X^2} + A_{12}^* \frac{\partial^2 \overline{F}}{\partial Y^2} + \left(B_{21}^* - \frac{4}{3h^2} E_{21}^* \right) \frac{\partial \overline{\Psi}_x}{\partial X} + \left(B_{22}^* - \frac{4}{3h^2} E_{22}^* \right) \frac{\partial \overline{\Psi}_y}{\partial Y} \right. \right.$$

$$\left. \left. - \frac{4}{3h^2} \left(E_{21}^* \frac{\partial^2 \overline{W}}{\partial X^2} + E_{22}^* \frac{\partial^2 \overline{W}}{\partial Y^2} \right) \right] - \frac{1}{2} \left(\frac{\partial \overline{W}}{\partial Y} \right)^2 - \frac{\partial \overline{W}}{\partial Y} \frac{\partial \overline{W}^*}{\partial Y} - \left(A_{12}^* \overline{N}_x^{\mathrm{T}} + A_{22}^* \overline{N}_y^{\mathrm{T}} \right) \right\} dY \, dX$$

$$\tag{3.67b}$$

where Δ_x and Δ_y are plate end-shortening displacements in the X- and Y-directions.

Introducing dimensionless quantities of Equation 2.8, and

$$(\gamma_{T1}, \gamma_{T2}) = \left(A_x^{\mathrm{T}}, A_y^{\mathrm{T}} \right) a^2 / \pi^2 [D_{11}^* D_{22}^*]^{1/2}$$

$$(\gamma_{T3}, \gamma_{T4}, \gamma_{T6}, \gamma_{T7}) = \left(D_x^{\mathrm{T}}, D_y^{\mathrm{T}}, F_x^{\mathrm{T}}, F_y^{\mathrm{T}} \right) a^2 / \pi^2 h^2 D_{11}^* \tag{3.68}$$

$$\lambda_x = Pb/4\pi^2 [D_{11}^* D_{22}^*]^{1/2}, \quad \lambda_{\mathrm{T}} = \alpha_0 \Delta T$$

where α_0 is an arbitrary reference value, defined by Equation 3.39. Also we let

$$\begin{bmatrix} A_x^{\mathrm{T}} & D_x^{\mathrm{T}} & F_x^{\mathrm{T}} \\ A_y^{\mathrm{T}} & D_y^{\mathrm{T}} & F_y^{\mathrm{T}} \end{bmatrix} \Delta T = -\sum_k \int_{t_{k-1}}^{t_k} \begin{bmatrix} A_x \\ A_y \end{bmatrix}_k (1, Z, Z^3) \Delta T(Z) dZ \tag{3.69}$$

where ΔT is a constant and is defined by $\Delta T = T_{\mathrm{U}} - T_0$ for the heat conduction.

The nonlinear equations (Equations 3.61 through 3.64) may then be written in dimensionless form as

$$L_{11}(W) - L_{12}(\Psi_x) - L_{13}(\Psi_y) + \gamma_{14} L_{14}(F) - L_{16}(M^{\mathrm{T}}) = \gamma_{14} \beta^2 L(W + W^*, F) \tag{3.70}$$

$$L_{21}(F) + \gamma_{24} L_{22}(\Psi_x) + \gamma_{24} L_{23}(\Psi_y) - \gamma_{24} L_{24}(W)$$

$$= -\frac{1}{2} \gamma_{24} \beta^2 L(W + 2W^*, W) \tag{3.71}$$

$$L_{31}(W) + L_{32}(\Psi_x) - L_{33}(\Psi_y) + \gamma_{14} L_{34}(F) - L_{36}(S^{\mathrm{T}}) = 0 \tag{3.72}$$

$$L_{41}(W) - L_{42}(\Psi_x) + L_{43}(\Psi_y) + \gamma_{14} L_{44}(F) - L_{46}(S^{\mathrm{T}}) = 0 \tag{3.73}$$

where all nondimensional linear operators $L_{ij}()$ and nonlinear operator $L()$ are defined by Equation 2.14.

The boundary conditions expressed by Equation 3.65 become

$x = 0, \pi$:

$$W = \Psi_y = 0 \tag{3.74a}$$

$$M_x = P_x = 0 \tag{3.74b}$$

$$\frac{1}{\pi} \int_0^\pi \beta^2 \frac{\partial^2 F}{\partial y^2} dy + 4\lambda_x \beta^2 = 0 \text{ (for compressive buckling problem)} \tag{3.74c}$$

$$\delta_x = 0 \text{ (for thermal buckling problem)} \tag{3.74d}$$

$y = 0, \pi$:

$$W = \Psi_x = 0 \tag{3.74e}$$

$$M_y = P_y = 0 \tag{3.74f}$$

$$\int_0^\pi \frac{\partial^2 F}{\partial x^2} dx = 0 \text{ (for compressive buckling problem)} \tag{3.74g}$$

$$\delta_y = 0 \text{ (for thermal buckling problem)} \tag{3.74h}$$

and the unit end-shortening relationships become

$$
\begin{aligned}
\delta_x = -\frac{1}{4\pi^2 \beta^2 \gamma_{24}} \int_0^\pi \int_0^\pi & \left\{ \left[\gamma_{24}^2 \beta^2 \frac{\partial^2 F}{\partial y^2} - \gamma_5 \frac{\partial^2 F}{\partial x^2} + \gamma_{24} \left(\gamma_{511} \frac{\partial \Psi_x}{\partial x} + \gamma_{233}\beta \frac{\partial \Psi_y}{\partial y} \right) \right. \right. \\
& \left. - \gamma_{24} \left(\gamma_{611} \frac{\partial^2 W}{\partial x^2} + \gamma_{244}\beta^2 \frac{\partial^2 W}{\partial y^2} \right) \right] - \frac{1}{2} \gamma_{24} \left(\frac{\partial W}{\partial x} \right)^2 - \gamma_{24} \frac{\partial W}{\partial x} \frac{\partial W^*}{\partial x} \\
& \left. + (\gamma_{24}^2 \gamma_{T1} - \gamma_5 \gamma_{T2})\lambda_T \right\} dx \, dy
\end{aligned} \tag{3.75a}
$$

$$
\begin{aligned}
\delta_y = -\frac{1}{4\pi^2 \beta^2 \gamma_{24}} \int_0^\pi \int_0^\pi & \left\{ \left[\frac{\partial^2 F}{\partial x^2} - \gamma_5 \beta^2 \frac{\partial^2 F}{\partial y^2} + \gamma_{24} \left(\gamma_{220} \frac{\partial \Psi_x}{\partial x} + \gamma_{522}\beta \frac{\partial \Psi_y}{\partial y} \right) \right. \right. \\
& \left. - \gamma_{24} \left(\gamma_{240} \frac{\partial^2 W}{\partial x^2} + \gamma_{622}\beta^2 \frac{\partial^2 W}{\partial y^2} \right) \right] - \frac{1}{2} \gamma_{24}\beta^2 \left(\frac{\partial W}{\partial y} \right)^2 \\
& \left. - \gamma_{24}\beta^2 \frac{\partial W}{\partial y} \frac{\partial W^*}{\partial y} + (\gamma_{T2} - \gamma_5 \gamma_{T1})\lambda_T \right\} dy \, dx
\end{aligned} \tag{3.75b}
$$

Since the material properties are assumed to be functions of T and Z, and the temperature is also assumed to be function of Z, even for the midplane symmetric plate the coupling between transverse bending and in-plane stretching which is given in terms of B_{ij}^* and E_{ij}^* ($i, j = 1, 2, 6$) is still existed when the plate is subjected to heat conduction. In contrast, when the plate is subjected to a uniform temperature rise, all B_{ij}^* and E_{ij}^* are zero-valued. Hence, for the heat conduction case, a nonlinear static problem must be first solved to determine the prebuckling deflection W_i caused by temperature field (Shen 2007a), and then replaced W^* with $W_T^* = W^* + W_i$ in Equations 3.70 and 3.71. Note that for uniform temperature loading case, that is, $T_U = T_L$, $W_i = 0$ and $W_T^* = W^*$.

By using a two-step perturbation technique, the asymptotic solutions are obtained in the same form of Equations 3.28 through 3.31. Upon substitution of these solutions into the boundary conditions (Equations 3.74c and 3.75a), or into the boundary conditions (Equations 3.74d and 3.74h), the compressive postbuckling equilibrium path can be written as

$$\lambda_p = \lambda_p^{(0)} + \lambda_p^{(2)} W_m^2 + \lambda_p^{(4)} W_m^4 + \cdots \tag{3.76a}$$

$$\delta_x = \delta_x^{(0)} + \delta_x^{(2)} W_m^2 + \delta_x^{(4)} W_m^4 + \cdots \tag{3.76b}$$

and the thermal postbuckling equilibrium path can be written as

$$\lambda_T = \lambda_T^{(0)} + \lambda_T^{(2)} W_m^2 + \lambda_T^{(4)} W_m^4 + \cdots \tag{3.77}$$

in which W_m is the dimensionless form of maximum deflection, and $\lambda_p^{(i)}, \delta_p^{(i)}$, and $\lambda_T^{(i)}$ ($i = 0, 2, 4, \ldots$) are all functions of temperature and position and given in detail in Appendix H.

For numerical illustrations, two sets of material mixture, as shown in Section 3.2, for FGM face sheets are considered. The material properties are assumed to be nonlinear function of temperature of Equation 1.4, and typical values, in the present case, can be found in Tables 1.1 through 1.3. The same metal is selected for the homogeneous substrate, and the material properties of which are also assumed to be nonlinear function of temperature of Equation 1.4. Poisson's ratio is assumed to be a constant, that is, $\nu_f = 0.28$ for (Si$_3$N$_4$/SUS304) face sheets and $\nu_H = 0.3$ for (SUS304) substrate, also $\nu_f = \nu_H = 0.29$ for both (ZrO$_2$/Ti-6Al-4V) face sheets and (Ti-6Al-4V) substrate. For these examples, the plate geometric parameter $a/b = 1$, $b/h = 20$, and the thickness of the FGM face sheet $h_F = 1$ mm whereas the thickness of the homogeneous substrate is taken to be $h_H = 4$, 6, and 8 mm, so that the substrate-to-face sheet thickness ratio $h_H/h_F = 4$, 6, and 8, respectively. It should be appreciated that in all figures $\overline{W}^*/h = 0.1$ (for the compressive postbuckling) or 0.05 (for the thermal postbuckling) denotes the dimensionless maximum initial geometric imperfection of the plate.

Tables 3.13 and 3.14 present the compressive buckling loads P_{cr} (in kN) for perfect, sandwich plates with different values of the substrate-to-face sheet thickness ratio h_H/h_F ($= 4$, 6, and 8) and with different values of the volume fraction index N ($= 0.0$, 0.2, 0.5, 1.0, 2.0, and 5.0) of FGM face sheets subjected to uniaxial compression under three sets of uniform

TABLE 3.13

Comparisons of Buckling Loads P_{cr} (kN) for Uniaxial Compressed, (FGM/SUS304/FGM) Sandwich Plates under Uniform Temperature Rise ($b/h = 20$, $a/b = 1.0$, $T_0 = 300$ K)

h_H/h_F	ΔT (in K)	$N=0$	$N=0.2$	$N=0.5$	$N=1.0$	$N=2.0$	$N=5.0$
TID, $(m, n) = (1, 1)$							
8.0		3681.317	3859.934	4034.281	4203.058	4364.358	4515.310
6.0		2353.399	2492.362	2627.142	2756.482	2878.550	2990.619
4.0		1321.721	1421.209	1516.661	1606.887	1690.181	1764.056
TD, $(m, n) = (1, 1)$							
8.0	100	3628.076	3801.115	3970.019	4133.529	4289.799	4436.045
	200	3528.290	3701.026	3869.631	4032.846	4188.832	4334.807
	300	3381.958	3559.376	3732.538	3900.154	4060.333	4210.224
6.0	100	2319.363	2453.986	2584.559	2709.864	2828.126	2936.704
	200	2255.571	2389.958	2520.300	2645.377	2763.422	2871.797
	300	2162.024	2300.053	2433.914	2562.358	2683.572	2794.848
4.0	100	1302.606	1398.987	1491.460	1578.872	1659.571	1731.144
	200	1266.779	1362.991	1455.298	1542.551	1623.099	1694.538
	300	1214.241	1313.058	1407.855	1497.453	1580.159	1663.508

TABLE 3.14

Comparisons of Buckling Loads P_{cr} (kN) for Uniaxial Compressed, (FGM/Ti-6Al-4V/FGM) Sandwich Plates under Uniform Temperature Rise ($b/h = 20$, $a/b = 1.0$, $T_0 = 300$ K)

h_H/h_F	ΔT (in K)	$N=0$	$N=0.2$	$N=0.5$	$N=1.0$	$N=2.0$	$N=5.0$
TID, $(m, n) = (1, 1)$							
8.0		1872.292	1970.214	2065.789	2158.303	2246.714	2329.446
6.0		1198.267	1274.447	1348.327	1419.219	1486.120	1547.536
4.0		674.0251	728.5607	780.8779	830.3261	875.9709	916.4501
TD, $(m, n) = (1, 1)$							
8.0	100	1772.731	1854.270	1933.864	2010.920	2084.566	2153.490
	200	1673.170	1744.246	1813.632	1880.810	1945.019	2005.116
	300	1573.609	1639.296	1703.425	1765.512	1824.857	1880.401
6.0	100	1134.548	1197.983	1259.513	1318.563	1374.296	1425.467
	200	1070.829	1126.125	1179.764	1231.246	1279.841	1324.460
	300	1007.110	1058.214	1107.789	1155.370	1200.284	1241.523
4.0	100	638.1832	683.5961	727.1699	768.3617	806.3907	840.1191
	200	602.3412	641.9277	679.9152	715.8294	748.9890	778.4004
	300	566.4993	603.0850	638.1940	617.3875	702.0355	729.2194

temperature rise ($\Delta T = 100$, 200, 300 K). Two kinds of sandwich plates, (FGM/SUS304/FGM) and (FGM/Ti-6Al-4V/FGM), are considered. It can be found that the buckling load of (FGM/Ti-6Al-4V/FGM) sandwich plate is lower than that of (FGM/SUS304/FGM) one. It can be seen that a fully metallic plate ($N=0$) has lowest buckling load and that the buckling load increases as the volume fraction index N increases. It is found that the increase is about $+36\%$ for the (FGM/SUS304/FGM) sandwich plate, and about $+29\%$ for the (FGM/Ti-6Al-4V/FGM) one with $h_H/h_F = 4$, from $N=0$ to $N=5$, under temperature change $\Delta T = 300$ K. It can also be seen that the temperature reduces the buckling load when the temperature dependency is put into consideration. The percentage decrease is about -6.5% for the (FGM/SUS304/FGM) sandwich plate and about -20% for the (FGM/Ti-6Al-4V/FGM) one with $h_H/h_F = 4$ from temperature changes from $\Delta T = 100$ K to $\Delta T = 300$ K under the same volume fraction distribution $N=2$.

Then Tables 3.15 and 3.16 present thermal buckling temperature ΔT_{cr} (in K) for the same two kinds of sandwich plates with different values of the volume fraction index N ($= 0.0, 0.2, 0.5, 1.0, 2.0,$ and 5.0) subjected to a uniform temperature rise. Note that, for the thermal buckling problem, it is necessary to solve Equation 3.77 by an iterative numerical procedure, as previously shown in Section 3.3. The results confirm that the buckling temperature is reduced when the TD properties are taken into account. Now the buckling

TABLE 3.15

Comparisons of Buckling Temperature ΔT (K) for (FGM/SUS304/FGM) Sandwich Plates under Uniform Temperature Rise ($b/h = 20$, $a/b = 1.0$, $T_0 = 300$ K)

h_H/h_F		$N=0$	$N=0.2$	$N=0.5$	$N=1.0$	$N=2.0$	$N=5.0$
8.0	TID	503.4949	513.6594	524.5022	535.7767	547.2765	558.7750
	TD	481.9259	490.0950	498.7635	507.7332	516.8405	525.9066
6.0	TID	503.5508	515.9382	529.1391	542.8510	556.8042	570.6840
	TD	481.9711	491.9193	502.4538	513.3313	524.3387	535.2285
4.0	TID	503.7081	519.5416	536.3883	553.8568	571.5649	589.0330
	TD	482.0982	494.7982	508.2006	521.9910	535.8672	549.4501

TABLE 3.16

Comparisons of Buckling Temperature ΔT (K) for (FGM/Ti-6Al-4V/FGM) Sandwich Plates under Uniform Temperature Rise ($b/h = 20$, $a/b = 1.0$, $T_0 = 300$ K)

h_H/h_F		$N=0$	$N=0.2$	$N=0.5$	$N=1.0$	$N=2.0$	$N=5.0$
8.0	TID	654.3356	650.9921	644.7942	637.1305	628.6901	619.7468
	TD	621.3503	591.1686	567.3193	548.8891	533.7027	521.1356
6.0	TID	654.3357	649.6246	641.5507	631.8915	621.4725	610.5759
	TD	621.3504	585.1382	558.8517	538.8232	523.0273	510.2279
4.0	TID	654.3357	646.9642	635.6199	622.7022	609.1954	595.3411
	TD	621.3504	576.1403	546.6052	525.0375	508.7258	495.3070

temperature of (FGM/Ti-6Al-4V/FGM) sandwich plate is larger than that of (FGM/SUS304/FGM) one. It is found that the increase is about +14% for the (FGM/SUS304/FGM) sandwich plate, and about −20% for the (FGM/Ti- 6Al-4V/FGM) one with $h_H/h_F = 4$, from $N = 0$ to $N = 5$, under TD case.

Figure 3.15 shows the effect of material properties on the postbuckling behavior of the (FGM/SUS304/FGM) sandwich plate with $h_H/h_F = 4$

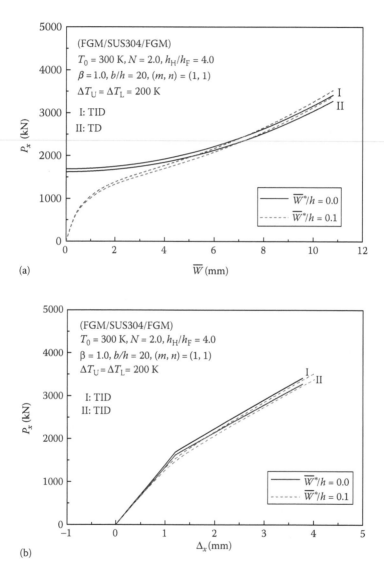

(a)

(b)

FIGURE 3.15
Effects of material properties on the postbuckling behavior of (FGM/SUS304/FGM) sandwich plates subjected to uniaxial compression under uniform temperature field: (a) load deflection; (b) load shortening.

and $N = 2.0$ subjected to uniaxial compression under uniform temperature field $\Delta T_U = \Delta T_L = 200$ K, and under two cases of thermoelastic material properties, i.e., TID and TD. It can be seen that the postbuckling equilibrium path becomes lower when the TD properties are taken into account.

Figure 3.16 shows temperature changes on the postbuckling behavior of the same plate subjected to uniaxial compression under heat conduction

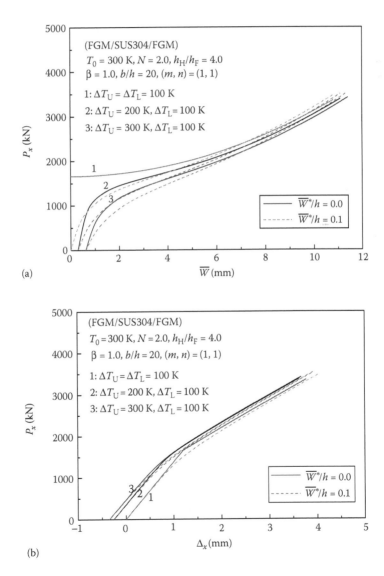

(a)

(b)

FIGURE 3.16
Thermal effects on the postbuckling behavior of (FGM/SUS304/FGM) sandwich plates subjected to uniaxial compression: (a) load deflection; (b) load shortening.

and TD case. It can be seen that the postbuckling strength is decreased with increase in temperature. It can be seen that the initial deflection is not zero-valued and an initial extension occurs (curves 2 and 3 in the figure) when the heat conduction is put into consideration. The results confirm that no buckling loads exist and the plate will buckle at the onset of edge compression when under heat conduction, and the shape of the load–deflection curves for perfect plates appears similar to that for plates with an initial geometric imperfection. Note that $\Delta T_U = \Delta T_L = 100$ K means uniform temperature field, so that the buckling load still exists (curve 1 in the figure).

Figure 3.17 shows the effect of the volume fraction index N ($= 0.2, 2.0,$ and 5.0) on the postbuckling behavior of the (FGM/SUS304/FGM) sandwich plate with $h_H/h_F = 4$ subjected to uniaxial compression under heat conduction $\Delta T_U = 200$ K, $\Delta T_L = 100$ K and TD case. It can be seen that the initial deflection is decreased, but the initial extension is almost the same when the volume fraction index N is increased. It is found that the increase of the volume fraction index N yields an increase of postbuckling strength.

Figures 3.18 and 3.19 are thermal postbuckling results for the (FGM/SUS304/FGM) sandwich plate analogous to the compressive postbuckling results of Figures 3.15 and 3.17, which are for the thermal loading case of heat conduction with $T_L = 300$ K. The results confirm that for the case of heat conduction, the thermal postbuckling path is no longer of the bifurcation type.

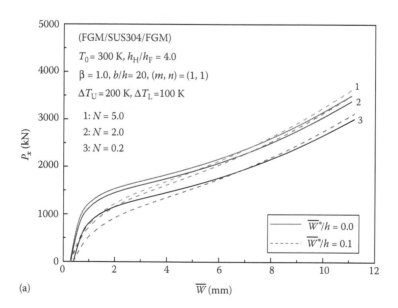

(a)

FIGURE 3.17
Effects of volume fraction index N on the postbuckling behavior of (FGM/SUS304/FGM) sandwich plates subjected to uniaxial compression under heat conduction: (a) load deflection;

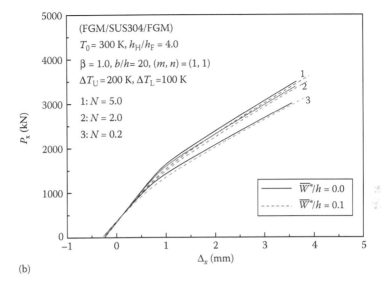

(b)

FIGURE 3.17 (continued)
(b) load shortening.

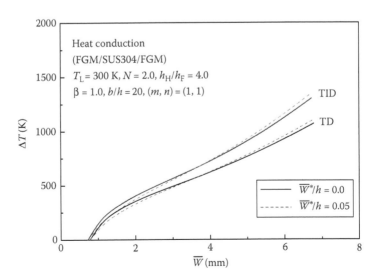

FIGURE 3.18
Effects of material properties on the thermal postbuckling behavior of (FGM/SUS304/FGM) sandwich plates under heat conduction.

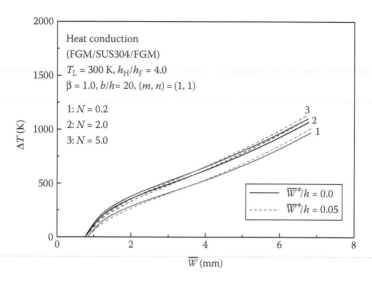

FIGURE 3.19
Effects of volume fraction index N on the thermal postbuckling behavior of (FGM/SUS304/FGM) sandwich plates under heat conduction.

References

Bhimaraddi A. and Chandrashekhara K. (1993), Nonlinear vibrations of heated anti-symmetric angle-ply laminated plates, *International Journal of Solids and Structures*, **30**, 1255–1268.

Birman V. (1995), Stability of functionally graded hybrid composite plates, *Composites Engineering*, **5**, 913–921.

Chen X.L. and Liew K.M. (2004), Buckling of rectangular functionally graded material plates subjected to nonlinearly distributed in-plane edge loads, *Smart Materials and Structures*, **13**, 1430–1437.

Chen X.L., Zhao Z.Y., and Liew K.M. (2008), Stability of piezoelectric FGM rectangular plates subjected to non-uniformly distributed load, heat and voltage, *Advances in Engineering Software*, **39**, 121–131.

Dym C.L. (1974), *Stability Theory and its Applications to Structural Mechanics*, Noordhoff, Leyden.

Feldman E. and Aboudi J. (1997), Buckling analysis of functionally graded plates subjected to uniaxial loading, *Composite Structures*, **38**, 29–36.

Ganapathi M. and Prakash T. (2006), Thermal buckling of simply supported functionally graded skew plates, *Composite Structures*, **74**, 247–250.

Ganapathi M., Prakash T., and Sundararajan N. (2006), Influence of functionally graded material on buckling of skew plates under mechanical loads, *Journal of Engineering Mechanics ASCE*, **132**, 902–905.

Javaheri R. and Eslami M.R. (2002a), Buckling of functionally graded plates under in-plane compressive loading, *ZAMM*, **82**, 277–283.

Javaheri R. and Eslami M.R. (2002b), Thermal buckling of functionally graded plates, *AIAA Journal*, **40**, 162–169.

Javaheri R. and Eslami M.R. (2002c), Thermal buckling of functionally graded plates based on higher order theory, *Journal of Thermal Stresses*, **25**, 603–625.

Kamiya N. and Fukui A. (1982), Finite deflection and postbuckling behavior of heated rectangular plates with temperature-dependent properties, *Nuclear Engineering and Design*, **72**, 415–420.

Leissa A.W. (1986), Conditions for laminated plates to remain flat under inplane loading, *Composite Structures*, **6**, 261–270.

Li S.-R., Zhang J.-H., and Zhao Y.-G. (2007), Nonlinear thermomechanical post-buckling of circular FGM plate with geometric imperfection, *Thin-Walled Structures*, **45**, 528–536.

Liew K.M., Yang J., and Kitipornchai S. (2003), Postbuckling of piezoelectric FGM plates subject to thermo-electro-mechanical loading, *International Journal of Solids and Structures*, **40**, 3869–3892.

Liew K.M., Yang J., and Kitipornchai S. (2004), Thermal post-buckling of laminated plates comprising functionally graded materials with temperature-dependent properties, *Journal of Applied Mechanics ASME*, **71**, 839–850.

Ma L.S. and Wang T.J. (2003a), Axisymmetric post-buckling of a functionally graded circular plate subjected to uniformly distributed radial compression, *Materials Science Forum*, **423–425**, 719–724.

Ma L.S. and Wang T.J. (2003b), Nonlinear bending and post-buckling of a functionally graded circular plate under mechanical and thermal loadings, *International Journal of Solids and Structures*, **40**, 3311–3330.

Ma L.S. and Wang T.J. (2004), Relationships between axisymmetric bending and buckling solutions of FGM circular plates based on third-order plate theory and classical plate theory, *International Journal of Solids and Structures*, **41**, 85–101.

Na K.-S. and Kim J.-H. (2004), Three-dimensional thermal buckling analysis of functionally graded materials, *Composites Part B*, **35**, 429–437.

Na K.-S. and Kim J.-H. (2006a), Thermal postbuckling investigations of functionally graded plates using 3-D finite element method, *Finite Elements in Analysis and Design*, **42**, 749–756.

Na K.-S. and Kim J.-H. (2006b), Three-dimensional thermomechanical buckling analysis of functionally graded composite plates, *Composite Structures*, **73**, 413–422.

Naei M.H., Masoumi A., and Shamekhi A. (2007), Buckling analysis of circular functionally graded material plate having variable thickness under uniform compression by finite-element method, *Proceedings of the Institution of Mechanical Engineers Part C—Journal of Mechanical Engineering Science*, **221**, 1241–1247.

Najafizadeh M.M. and Eslami M.R. (2002a), Buckling analysis of circular plates of functionally graded materials under uniform radial compression, *International Journal of Mechanical Sciences*, **44**, 2479–2493.

Najafizadeh M.M. and Eslami M.R. (2002b), First-order-theory-based thermoelastic stability of functionally graded material circular plates, *AIAA Journal*, **40**, 1444–1450.

Najafizadeh M.M. and Heydari H.R. (2004a), Refined theory for thermoelastic stability of functionally graded circular plates, *Journal of Thermal Stresses*, **27**, 857–880.

Najafizadeh M.M. and Heydari H.R. (2004b), Thermal buckling of functionally graded circular plates based on higher order shear deformation plate theory, *European Journal of Mechanics A/Solids*, **23**, 1085–1100.

Najafizadeh M.M. and Heydari H.R. (2008), An exact solution for buckling of functionally graded circular plates based on higher order shear deformation plate theory under uniform radial compression, *International Journal of Mechanical Sciences*, **50**, 603–612.

Oh I.K., Han J.H., and Lee I. (2000), Postbuckling and vibration characteristics of piezolaminated composite plate subjected to thermo-piezoelectric loads, *Journal of Sound and Vibration*, **233**, 19–40.

Park J.S. and Kim J.-H. (2006), Thermal postbuckling and vibration analyses of functionally graded plates, *Journal of Sound and Vibration*, **289**, 77–93.

Prakash T., Singha M.K., and Ganapathi M. (2008), Thermal postbuckling analysis of FGM skew plates, *Engineering Structures*, **30**, 22–32.

Qatu M.S. and Leissa A.W. (1993), Buckling or transverse deflections of unsymmetrically laminated plates subjected to in-plane loads, *AIAA Journal*, **31**, 189–194.

Shariat B.A.S. and Eslami M.R. (2005), Effect of initial imperfections on thermal buckling of functionally graded plates, *Journal of Thermal Stresses*, **28**, 1183–1198.

Shariat B.A.S. and Eslami M.R. (2006), Thermal buckling of imperfect functionally graded plates, *International Journal of Solids and Structures*, **43**, 4082–4096.

Shen H.-S. (2002), Nonlinear bending response of functionally graded plates subjected to transverse loads and in thermal environments, *International Journal of Mechanical Sciences*, **44**, 561–584.

Shen H.-S. (2005), Postbuckling of FGM plates with piezoelectric actuators under thermo-electro-mechanical loadings, *International Journal of Solids and Structures*, **42**, 6101–6121.

Shen H.-S. (2007a), Nonlinear thermal bending response of FGM plates due to heat conduction, *Composites Part B*, **38**, 201–215.

Shen H.-S. (2007b), Thermal postbuckling behavior of shear deformable FGM plates with temperature-dependent properties, *International Journal of Mechanical Sciences*, **49**, 466–478.

Shen H.-S. and Li S.-R. (2008), Postbuckling of sandwich plates with FGM face sheets and temperature-dependent properties, *Composites Part B*, **39**, 332–344.

Shukla K.K., Kumar K.V.R., Pandey R., and Nath Y. (2007), Postbuckling response of functionally graded rectangular plates subjected to thermo-mechanical loading, *International Journal of Structural Stability and Dynamics*, **7**, 519–541.

Woo J., Meguid S.A., Stranart J.C., and Liew K.M. (2005), Thermomechanical postbuckling analysis of moderately thick functionally graded plates and shallow shells, *International Journal of Mechanical Sciences*, **47**, 1147–1171.

Wu L. (2004), Thermal buckling of simply supported moderately thick rectangular FGM plate, *Composite Structures*, **64**, 211–218.

Wu T.-L., Shukla K.K., and Huang J.H. (2007), Post-buckling analysis of functionally graded rectangular plates, *Composite Structures*, **81**, 1–10.

Yamaki N. (1961), Experiments on the postbuckling behavior of square plates loaded in edge compression, *Journal of Applied Mechanics ASME*, **28**, 238–244.

Yang J., Liew K.M., and Kitipornchai S. (2006), Imperfection sensitivity of the postbuckling behavior of higher-order shear deformable functionally graded plates, *International Journal of Solids and Structures*, **43**, 5247–5266.

Yang J. and Shen H.-S. (2003), Nonlinear analysis of functionally graded plates under transverse and in-plane loads, *International Journal of Non-Linear Mechanics*, **38**, 467–482.

4

Nonlinear Vibration of Shear Deformable FGM Plates

4.1 Introduction

The external loads applied to a plate invariably change with time. If the variations in time are small and occur over an extended interval, the inertial effects may be neglected and the behavior of the plate can be approximately determined from considerations of equilibrium and material properties. However, in some modern engineering constructions, like aircraft, missiles, and launch vehicles, rapid time variations of loadings occur that must be considered in formulating structural design. In such cases, inertial effects must be taken into account and the dynamic behavior of the plate must be treated as a function of time.

Many studies have been reported on the free and forced vibration of FGM plates, for example, Yang and Shen (2001, 2002), Cheng and Reddy (2003), Vel and Batra (2004), Kim (2005), Elishakoff et al. (2005), Prakash and Ganapathi (2006), Efraim and Eisenberger (2007), and Li et al. (2008). According to mixture rules, Abrate (2006) found that the natural frequencies of FGM plates are proportional to those of the corresponding homogeneous plate and concluded that only one case is needed to determine the proportionality constant, and direct analysis is unnecessary.

In most conditions of severe environments, when the plate deflection-to-thickness ratio is greater than 0.4, the nonlinearity is very important and the nonlinear dynamic equations of plates are required to perform the analysis. Praveen and Reddy (1998) analyzed the nonlinear static and dynamic response of functionally graded ceramic–metal plates in a steady temperature field and subjected to dynamic transverse loads by finite element method (FEM). Reddy (2000) developed both theoretical and finite element formulations for thick FGM plates according to the HSDPT, and studied the nonlinear dynamic response of FGM plates subjected to suddenly applied uniform pressure. Yang et al. (2003) presented a large amplitude vibration analysis of an initially stressed FGM plate with surface-bonded piezoelectric layers by using a semianalytical method based on 1D differential quadrature and Galerkin technique. Chen (2005), Chen et al. (2006), Fung and Chen (2006), and Chen and Tan (2007) studied the large amplitude vibration of an initially stressed FGM plate with or without initial geometric imperfections. In their studies, the initial stress was taken to be a combination of pure bending stress and an extensional stress in the plane of the plate, and the formulations were based on the FSDPT and classical plate theory (CPT), respectively. Also, Allahverdizadeh et al. (2008a–c) studied the nonlinear free and forced vibration of FGM circular thin plates based on the CPT. However, in the references cited above (Praveen and Reddy 1998, Reddy 2000, Yang et al. 2003, Chen 2005, Chen et al. 2006, Chen and Tan 2007, Allahverdizadeh et al. 2008a–c), the material properties were either assumed to be independent of temperature or considered in a constant temperature environment ($T = 300$ K). Since FGMs always serve in the high-temperature environments, the materials properties of FGM plates must be temperature- and position-dependent. Kitipornchai et al. (2004) and Yang and Huang (2007) studied nonlinear free and forced vibration of imperfect FGM laminated plates with various boundary conditions and with temperature-dependent material properties, respectively. On the other hand, ceramics and the metals used in FGM do store different amounts of heat, and therefore the heat conduction usually occurs. This leads to a nonuniform distribution of temperature through the plate thickness, but it is not accounted for in the above study. Also recently, Huang and Shen (2004, 2006) provided nonlinear free and forced vibration analysis of shear deformable FGM plates without or with surface-bonded piezoelectric layers in thermal environments. Xia and Shen (2008) provided small- and large-amplitude vibration analysis of compressively and thermally postbuckled sandwich plates with FGM face sheets in thermal environments. In these studies, heat conduction and temperature-dependent material properties were both taken into account. Sundararajan et al. (2005) calculated frequencies for nonlinear free flexural vibration of functionally graded rectangular and skew plates under thermal environments. In their studies, the material properties were based on the Mori–Tanaka scheme, and a remarkable synergism between the Mori–Tanaka scheme and the rule of mixture was found.

―――――――――

4.2   Nonlinear Vibration of FGM Plates in Thermal Environments

Here, we consider an FGM plate of length a, width b, and thickness h, which is made from a mixture of ceramics and metals. We assume that the composition is varied from the top to the bottom surface, i.e., the top surface ($Z = h/2$) of the plate is ceramic-rich, whereas the bottom surface ($Z = -h/2$) is metal-rich. Note that in Sections 4.2 and 4.3 the Z is in the direction of the upward normal to the middle surface. A simple power law exponent of the volume fraction distribution is used to provide a measure of the amount of ceramic and metal in the FGM. In the present case, the volume fraction of ceramic, instead of V_m, is defined by

$$V_c(Z) = \left(\frac{2Z + h}{2h}\right)^N \tag{4.1}$$

It is assumed that the effective Young's modulus E_f and thermal expansion coefficient α_f of the FGM plate are temperature dependent, whereas the mass density ρ_f and thermal conductivity κ_f are independent of the temperature. Poisson's ratio ν_f depends weakly on temperature change and is assumed to be a constant. As long as the volume fraction is known, the effective material properties can be obtained easily using the rule of mixture

$$E_f(Z, T) = [E_c(T) - E_m(T)]\left(\frac{2Z + h}{2h}\right)^N + E_m(T) \tag{4.2a}$$

$$\alpha_f(Z, T) = [\alpha_c(T) - \alpha_m(T)]\left(\frac{2Z + h}{2h}\right)^N + \alpha_m(T) \tag{4.2b}$$

$$\rho_f(Z) = (\rho_c - \rho_m)\left(\frac{2Z + h}{2h}\right)^N + \rho_c \tag{4.2c}$$

$$\kappa(Z) = (\kappa_c - \kappa_m)\left(\frac{2Z + h}{2h}\right)^N + \kappa_m \tag{4.2d}$$

We assume that the temperature variation occurs in the thickness direction only and 1D temperature field is assumed to be constant in the XY plane of the plate. In such a case, the temperature distribution along the thickness can be obtained by solving a steady-state heat transfer equation:

$$-\frac{d}{dZ}\left[\kappa(Z)\frac{dT}{dZ}\right] = 0 \tag{4.3}$$

This equation is solved by imposing boundary condition of $T = T_U$ at $Z = h/2$ and $T = T_L$ at $Z = -h/2$. The solution of this equation, by means of polynomial series, is

$$T(Z) = T_L + (T_U - T_L)\eta(Z) \tag{4.4}$$

where T_U and T_L are the temperatures at top and bottom surfaces of the plate, and

$$\eta(Z) = \frac{1}{C}\left[\left(\frac{2Z+h}{2h}\right) - \frac{\kappa_{cm}}{(N+1)\kappa_m}\left(\frac{2Z+h}{2h}\right)^{N+1} + \frac{\kappa_{cm}^2}{(2N+1)\kappa_m^2}\left(\frac{2Z+h}{2h}\right)^{2N+1}\right.$$

$$- \frac{\kappa_{cm}^3}{(3N+1)\kappa_m^3}\left(\frac{2Z+h}{2h}\right)^{3N+1} + \frac{\kappa_{cm}^4}{(4N+1)\kappa_m^4}\left(\frac{2Z+h}{2h}\right)^{4N+1}$$

$$\left. - \frac{\kappa_{cm}^5}{(5N+1)\kappa_m^5}\left(\frac{2Z+h}{2h}\right)^{5N+1}\right] \tag{4.5a}$$

$$C = 1 - \frac{\kappa_{cm}}{(N+1)\kappa_m} + \frac{\kappa_{cm}^2}{(2N+1)\kappa_m^2} - \frac{\kappa_{cm}^3}{(3N+1)\kappa_m^3}$$

$$+ \frac{\kappa_{cm}^4}{(4N+1)\kappa_m^4} - \frac{\kappa_{cm}^5}{(5N+1)\kappa_m^5} \tag{4.5b}$$

where $\kappa_{cm} = \kappa_c - \kappa_m$.

The plate is assumed to be geometrically perfect, and is subjected to a transverse dynamic load $q(X, Y, t)$ in thermal environments. Hence, the general von Kármán-type equations can be expressed by Equations 1.33 through 1.36, and the forces, moments, and higher order moments caused by elevated temperature are defined by Equation 1.28.

It is assumed that all four edges are simply supported with no in-plane displacements. In such a case, the boundary conditions can be expressed by Equation 2.6 for immovable edges.

Introducing dimensionless quantities Equation 2.8, and

$$\tau = \frac{\pi t}{a}\sqrt{\frac{E_0}{\rho_0}}, \quad \gamma_{170} = -\frac{I_1 E_0 a^2}{\pi^2 \rho_0 D_{11}^*}, \quad \gamma_{171} = \frac{4E_0(I_5 I_1 - I_4 I_2)}{3\rho_0 h^2 I_1 D_{11}^*},$$

$$(\gamma_{80}, \gamma_{90}, \gamma_{10}) = (-\bar{I}_8, \bar{I}_5, -\bar{I}_3)\frac{E_0}{\rho_0 D_{11}^*} \tag{4.6}$$

where E_0 and ρ_0 are the reference values of E_m and ρ_m, respectively, at room temperature ($T_0 = 300$ K) and $A_x^T(=A_y^T)$, $D_x^T(=D_y^T)$, and $F_x^T(=F_y^T)$ are redefined by

$$\begin{bmatrix} A_x^T & D_x^T & F_x^T \\ A_y^T & D_y^T & F_y^T \end{bmatrix} T_1 = -\int_{-h/2}^{h/2}\begin{bmatrix} A_x \\ A_y \end{bmatrix}(1, Z, Z^3)\Delta T(Z)dZ \tag{4.7}$$

where $T_1 = (T_U + T_L - 2T_0)/2$.

The nonlinear motion equations can then be written in dimensionless form as

$$L_{11}(W) - L_{12}(\Psi_x) - L_{13}(\Psi_y) + \gamma_{14}L_{14}(F) - L_{16}(M^T)$$

$$= \gamma_{14}\beta^2 L(W, F) + L_{17}(\ddot{W}) + \gamma_{80}\left(\frac{\partial\ddot{\Psi}_x}{\partial x} + \beta\frac{\partial\ddot{\Psi}_y}{\partial y}\right) + \lambda_q \tag{4.8}$$

$$L_{21}(F) + \gamma_{24}L_{22}(\Psi_x) + \gamma_{24}L_{23}(\Psi_y) - \gamma_{24}L_{24}(W) = -\frac{1}{2}\gamma_{24}\beta^2 L(W,W) \quad (4.9)$$

$$L_{31}(W) + L_{32}(\Psi_x) - L_{33}(\Psi_y) + \gamma_{14}L_{34}(F) - L_{36}(S^T) = \gamma_{90}\frac{\partial \ddot{W}}{\partial x} + \gamma_{10}\ddot{\Psi}_x \quad (4.10)$$

$$L_{41}(W) - L_{42}(\Psi_x) + L_{43}(\Psi_y) + \gamma_{14}L_{44}(F) - L_{46}(S^T) = \gamma_{90}\beta\frac{\partial \ddot{W}}{\partial y} + \gamma_{10}\ddot{\Psi}_y \quad (4.11)$$

where

$$L_{17}() = \gamma_{170} + \gamma_{171}\frac{\partial^2}{\partial x^2} + \gamma_{171}\beta^2\frac{\partial^2}{\partial y^2} \quad (4.12)$$

and all other nondimensional linear operators $L_{ij}()$ and nonlinear operator $L()$ are defined by Equation 2.14.

The boundary conditions can be written in dimensionless form as

$x = 0, \pi$:

$$W = \Psi_y = 0 \quad (4.13a)$$

$$\int_0^\pi \int_0^\pi \left[\gamma_{24}^2\beta^2\frac{\partial^2 F}{\partial y^2} - \gamma_5\frac{\partial^2 F}{\partial x^2} + \gamma_{24}\left(\gamma_{511}\frac{\partial \Psi_x}{\partial x} + \gamma_{233}\beta\frac{\partial \Psi_y}{\partial y}\right) - \gamma_{24}\left(\gamma_{611}\frac{\partial^2 W}{\partial x^2} + \gamma_{244}\beta^2\frac{\partial^2 W}{\partial y^2}\right) \right.$$
$$\left. -\frac{1}{2}\gamma_{24}\left(\frac{\partial W}{\partial x}\right)^2 + (\gamma_{24}^2\gamma_{T1} - \gamma_5\gamma_{T2})T_1 \right] dx\, dy = 0 \quad (4.13b)$$

$y = 0, \pi$:

$$W = \Psi_x = 0 \quad (4.13c)$$

$$\int_0^\pi \int_0^\pi \left[\frac{\partial^2 F}{\partial x^2} - \gamma_5\beta^2\frac{\partial^2 F}{\partial y^2} + \gamma_{24}\left(\gamma_{220}\frac{\partial \Psi_x}{\partial x} + \gamma_{522}\beta\frac{\partial \Psi_y}{\partial y}\right) - \gamma_{24}\left(\gamma_{240}\frac{\partial^2 W}{\partial x^2} + \gamma_{622}\beta^2\frac{\partial^2 W}{\partial y^2}\right) \right.$$
$$\left. -\frac{1}{2}\gamma_{24}\beta^2\left(\frac{\partial W}{\partial y}\right)^2 + (\gamma_{T2} - \gamma_5\gamma_{T1})T_1 \right] dy\, dx = 0 \quad (4.13d)$$

We assume that the solutions of Equations 4.8 through 4.11 can be expressed as

$$W(x,y,\tau) = W^*(x,y) + \tilde{W}_x(x,y,\tau)$$
$$\Psi_x(x,y,\tau) = \Psi_x^*(x,y) + \tilde{\Psi}_x(x,y,\tau)$$
$$\Psi_y(x,y,\tau) = \Psi_y^*(x,y) + \tilde{\Psi}_y(x,y,\tau)$$
$$F(x,y,\tau) = F^*(x,y) + \tilde{F}(x,y,\tau)$$

$$(4.14)$$

where
$W^*(x, y)$ is an initial deflection due to initial thermal bending moment
$\tilde{W}(x, y, \tau)$ is an additional deflection

$\Psi_x^*(x, y)$, $\Psi_y^*(x, y)$, and $F^*(x, y)$ are the midplane rotations and stress function corresponding to $W^*(x, y)$. $\tilde{\Psi}_x(x, y, \tau)$, $\tilde{\Psi}_y(x, y, \tau)$, and $\tilde{F}(x, y, \tau)$ are defined analogously to $\Psi_x^*(x, y)$, $\Psi_y^*(x, y)$, and $F^*(x, y)$, but is for $\tilde{W}(x, y, \tau)$.

Due to the bending–stretching coupling effect in the FGM plate, the thermal preload will bring about deflections and bending curvatures which have significant influences on the plate vibration characteristics. To account for this effect, the previbration solutions $W^*(x, y)$, $\Psi_x^*(x, y)$, $\Psi_y^*(x, y)$, and $F^*(x, y)$ are sought at the first step from the following nonlinear equations:

$$L_{11}(W^*) - L_{12}(\Psi_x^*) - L_{13}(\Psi_y^*) + \gamma_{14}L_{14}(F^*) - L_{16}(M^T)$$
$$+ \gamma_{14}\beta^2 \left(p_x \frac{\partial^2 W^*}{\partial x^2} + p_y \frac{\partial^2 W^*}{\partial y^2} \right) = \gamma_{14}\beta^2 L(W^*, F^*) \tag{4.15}$$

$$L_{21}(F^*) + \gamma_{24}L_{22}(\Psi_x^*) + \gamma_{24}L_{23}(\Psi_y^*) - \gamma_{24}L_{24}(W^*) = -\frac{1}{2}\gamma_{24}\beta^2 L(W^*, W^*) \tag{4.16}$$

$$L_{31}(W^*) + L_{32}(\Psi_x^*) - L_{33}(\Psi_y^*) + \gamma_{14}L_{34}(F^*) - L_{36}(S^T) = 0 \tag{4.17}$$

$$L_{41}(W^*) - L_{42}(\Psi_x^*) + L_{43}(\Psi_y^*) + \gamma_{14}L_{44}(F^*) - L_{46}(S^T) = 0 \tag{4.18}$$

In Equation 4.15, p_x and p_y are edge compressive stresses induced by temperature rise with edge restraints. The solutions of Equations 4.15 through 4.18 can be assumed to be as

$$W^*(x, y) = \sum_{k=1,3,\dots} \sum_{l=1,3,\dots} w_{kl} \sin kx \sin ly$$
$$\Psi_x^*(x, y) = \sum_{k=1,3,\dots} \sum_{l=1,3,\dots} (\psi_x)_{kl} \cos kx \sin ly$$
$$\Psi_y^*(x, y) = \sum_{k=1,3,\dots} \sum_{l=1,3,\dots} (\psi_y)_{kl} \sin kx \cos ly \tag{4.19}$$
$$F^*(x, y) = -\frac{1}{2} \left(B_{00}^{(0)} y^2 + b_{00}^{(0)} x^2 \right) + \sum_{k=1,3,\dots} \sum_{l=1,3,\dots} f_{kl} \sin kx \sin ly$$

We then expand the constant thermal bending moments in the double Fourier sine series as

$$\begin{bmatrix} M_x^T & S_x^T \\ M_y^T & S_y^T \end{bmatrix} = - \begin{bmatrix} M_x^{(0)} & S_x^{(0)} \\ M_y^{(0)} & S_y^{(0)} \end{bmatrix} \sum_{k=1,3,\dots} \sum_{l=1,3,\dots} \frac{1}{kl} \sin kx \sin ly \tag{4.20}$$

Substituting Equations 4.19 and 4.20 into Equations 4.15 through 4.18 and applying Galerkin procedure to Equations 4.15 and 4.16, W_{kl}, $(\psi_x)_{kl}$, $(\psi_y)_{kl}$, and f_{kl} can easily be determined. The detailed expressions are given in Appendix I.

Then an initially stressed FGM plate is under consideration and $\tilde{W}(x, y, \tau)$, $\tilde{\Psi}_x(x, y, \tau)$, $\tilde{\Psi}_y(x, y, \tau)$, and $\tilde{F}(x, y, \tau)$ satisfy the nonlinear equations:

$$L_{11}(\tilde{W}) - L_{12}(\tilde{\Psi}_x) - L_{13}(\tilde{\Psi}_y) + \gamma_{14}L_{14}(\tilde{F}) = \gamma_{14}\beta^2[L(\tilde{W} + W^*, \tilde{F}) + L(\tilde{W}, F^*)]$$

$$+ L_{17}(\ddot{\tilde{W}}) + \gamma_{80}\left(\frac{\partial \ddot{\tilde{\Psi}}_x}{\partial x} + \beta\frac{\partial \ddot{\tilde{\Psi}}_y}{\partial y}\right) + \lambda_q \tag{4.21}$$

$$L_{21}(\tilde{F}) + \gamma_{24}L_{22}(\tilde{\Psi}_x) + \gamma_{24}L_{23}(\tilde{\Psi}_y) - \gamma_{24}L_{24}(\tilde{W}) = -\frac{1}{2}\gamma_{24}\beta^2 L(\tilde{W} + 2W^*, \tilde{W}) \tag{4.22}$$

$$L_{31}(\tilde{W}) + L_{32}(\tilde{\Psi}_x) - L_{33}(\tilde{\Psi}_y) + \gamma_{14}L_{34}(\tilde{F}) = \gamma_{90}\frac{\partial \ddot{\tilde{W}}}{\partial x} + \gamma_{10}\ddot{\tilde{\Psi}}_x \tag{4.23}$$

$$L_{41}(\tilde{W}) - L_{42}(\tilde{\Psi}_x) + L_{43}(\tilde{\Psi}_y) + \gamma_{14}L_{44}(\tilde{F}) = \gamma_{90}\beta\frac{\partial \ddot{\tilde{W}}}{\partial y} + \gamma_{10}\ddot{\tilde{\Psi}}_y \tag{4.24}$$

The initial conditions are assumed to be

$$\tilde{W}\Big|_{\tau=0} = \frac{\partial \tilde{W}}{\partial \tau}\Big|_{\tau=0} = 0 \tag{4.25a}$$

$$\tilde{\Psi}_x\Big|_{\tau=0} = \frac{\partial \tilde{\Psi}_x}{\partial \tau}\Big|_{\tau=0} = 0 \tag{4.25b}$$

$$\tilde{\Psi}_y\Big|_{\tau=0} = \frac{\partial \tilde{\Psi}_y}{\partial \tau}\Big|_{\tau=0} = 0 \tag{4.25c}$$

A perturbation technique is now used to solve Equations 4.21 through 4.24. The essence of this procedure, in the present case, is to assume that

$$\tilde{W}(x, y, \tilde{\tau}, \varepsilon) = \sum_{j=1} \varepsilon^j W_j(x, y, \tilde{\tau})$$

$$\tilde{F}(x, y, \tilde{\tau}, \varepsilon) = \sum_{j=1} \varepsilon^j F_j(x, y, \tilde{\tau})$$

$$\tilde{\Psi}_x(x, y, \tilde{\tau}, \varepsilon) = \sum_{j=1} \varepsilon^j \Psi_{xj}(x, y, \tilde{\tau}) \tag{4.26}$$

$$\tilde{\Psi}_y(x, y, \tilde{\tau}, \varepsilon) = \sum_{j=1} \varepsilon^j \Psi_{yj}(x, y, \tilde{\tau})$$

$$\lambda_q(x, y, \tilde{\tau}, \varepsilon) = \sum_{j=1} \varepsilon^j \lambda_j(x, y, \tilde{\tau})$$

where ε is a small perturbation parameter. Here, we introduce an important parameter $\tilde{\tau} = \varepsilon\tau$ to improve perturbation procedure for solving nonlinear dynamic problem.

Substituting Equation 4.26 into Equations 4.21 through 4.24 and collecting the terms of the same order of ε, we obtain a set of perturbation equations which can be written, for example, as

$O(\varepsilon^0)$:

$$L_{14}(f_0) = 0 \tag{4.27a}$$

$$L_{21}(f_0) = 0 \tag{4.27b}$$

$$L_{34}(f_0) = 0 \tag{4.27c}$$

$$L_{44}(f_0) = 0 \tag{4.27d}$$

$O(\varepsilon^1)$:

$$L_{11}(w_1) - L_{12}(\psi_{x1}) - L_{13}(\psi_{y1}) + \gamma_{14}L_{14}(f_1)$$
$$= \gamma_{14}\beta^2[L(w_1 + W^*, f_0) + L(w_1, F^*)] + \lambda_1 \tag{4.28a}$$

$$L_{21}(f_1) + \gamma_{24}L_{22}(\psi_{x1}) + \gamma_{24}L_{23}(\psi_{y1}) - \gamma_{24}L_{24}(w_1) = 0 \tag{4.28b}$$

$$L_{31}(w_1) + L_{32}(\psi_{x1}) - L_{33}(\psi_{y1}) + \gamma_{14}L_{34}(f_1) = 0 \tag{4.28c}$$

$$L_{41}(w_1) - L_{42}(\psi_{x1}) + L_{43}(\psi_{y1}) + \gamma_{14}L_{44}(f_1) = 0 \tag{4.28d}$$

$O(\varepsilon^2)$:

$$L_{11}(w_2) - L_{12}(\psi_{x2}) - L_{13}(\psi_{y2}) + \gamma_{14}L_{14}(f_2)$$
$$= \gamma_{14}\beta^2[L(w_2, f_0 + F^*) + L(w_1 + W^*, f_1)] + \lambda_2 \tag{4.29a}$$

$$L_{21}(f_2) + \gamma_{24}L_{22}(\psi_{x2}) + \gamma_{24}L_{23}(\psi_{y2}) - \gamma_{24}L_{24}(w_2)$$
$$= -\frac{1}{2}\gamma_{24}\beta^2 L(w_1 + 2W^*, w_1) \tag{4.29b}$$

$$L_{31}(w_2) + L_{32}(\psi_{x2}) - L_{33}(\psi_{y2}) + \gamma_{14}L_{34}(f_2) = 0 \tag{4.29c}$$

$$L_{41}(w_2) - L_{42}(\psi_{x2}) + L_{43}(\psi_{y2}) + \gamma_{14}L_{44}(f_2) = 0 \tag{4.29d}$$

$O(\varepsilon^3)$:

$$L_{11}(w_3) - L_{12}(\psi_{x3}) - L_{13}(\psi_{y3}) + \gamma_{14}L_{14}(f_3)$$
$$= \gamma_{14}\beta^2[L(w_3, f_0 + F^*) + L(w_2, f_1) + L(w_1 + W^*, f_2)] + \lambda_3 \tag{4.30a}$$

$$L_{21}(f_3) + \gamma_{24}L_{22}(\psi_{x3}) + \gamma_{24}L_{23}(\psi_{y3}) - \gamma_{24}L_{24}(w_3)$$
$$= -\frac{1}{2}\gamma_{24}\beta^2 L(w_1 + 2W^*, w_2) \tag{4.30b}$$

$$L_{31}(w_3) + L_{32}(\psi_{x3}) - L_{33}(\psi_{y3}) + \gamma_{14}L_{34}(f_3) = \gamma_{90}\frac{\partial \ddot{w}_1}{\partial x} + \gamma_{10}\ddot{\psi}_{x1} \tag{4.30c}$$

$$L_{41}(w_3) - L_{42}(\psi_{x3}) + L_{43}(\psi_{y3}) + \gamma_{14}L_{44}(f_3) = \gamma_{90}\beta\frac{\partial \ddot{w}_1}{\partial y} + \gamma_{10}\ddot{\psi}_{y1} \tag{4.30d}$$

To solve these perturbation equations of each order, the amplitudes of the terms $w_j(x, y)$, $f_j(x, y)$, $\psi_{xj}(x, y)$, and $\psi_{yj}(x, y)$ can be determined step by step. As a result, we obtain asymptotic solutions, up to third order, as

$$\tilde{W}(x, y, \tau) = \varepsilon[w_1(\tau) + g_1\ddot{w}_1(\tau)]\sin mx \sin ny + (\varepsilon w_1(\tau))^3[\alpha g_{311}\sin mx \sin ny$$
$$+ g_{331}\sin 3mx \sin ny + g_{313}\sin mx \sin 3ny] + O(\varepsilon^4) \tag{4.31}$$

$$\tilde{\Psi}_x(x, y, \tau) = \varepsilon\left[g_{11}^{(1,1)}w_1(\tau) + g_2\ddot{w}_1(\tau)\right]\cos mx \sin ny + g_{12}(\varepsilon w_1(\tau))^2\sin 2mx$$
$$+ (\varepsilon w_1(\tau))^3\left[\alpha g_{11}^{(1,1)}g_{311}\cos mx \sin ny + g_{11}^{(3,1)}g_{331}\cos 3mx \sin ny\right.$$
$$\left.+ g_{11}^{(1,3)}g_{313}\cos mx \sin 3ny\right] + O(\varepsilon^4) \tag{4.32}$$

$$\tilde{\Psi}_y(x, y, \tau) = \varepsilon\left[g_{21}^{(1,1)}w_1(\tau) + g_3\ddot{w}_1(\tau)\right]\sin mx \cos ny + g_{22}(\varepsilon w_1(\tau))^2\sin 2ny$$
$$+ (\varepsilon w_1(\tau))^3\left[\alpha g_{21}^{(1,1)}g_{311}\sin mx \cos ny + g_{21}^{(3,1)}g_{331}\sin 3mx \cos ny\right.$$
$$\left.+ g_{21}^{(1,3)}g_{313}\sin mx \cos 3ny\right] + O(\varepsilon^4) \tag{4.33}$$

$$\tilde{F}(x, y, \tau) = \varepsilon\left[g_{31}^{(1,1)}w_1(\tau) + g_4\ddot{w}_1(\tau)\right]\sin mx \sin ny - (\varepsilon w_1(\tau))^2$$
$$\times\left(B_{00}^{(2)}y^2/2 + b_{00}^{(2)}x^2/2 - g_{402}\cos 2ny - g_{420}\cos 2mx\right)$$
$$+ (\varepsilon w_1(\tau))^3\left[\alpha g_{31}^{(1,1)}g_{311}\sin mx \sin ny + g_{31}^{(3,1)}g_{331}\sin 3mx \sin ny\right.$$
$$\left.+ g_{31}^{(1,3)}g_{313}\sin mx \sin 3ny\right] + O(\varepsilon^4) \tag{4.34}$$

$$\lambda_q(x, y, \tau) = \varepsilon[g_{41}w_1(\tau) + g_{43}\ddot{w}_1(\tau)]\sin mx \sin ny$$
$$+ (\varepsilon w_1(\tau))^2(g_{441}\cos 2mx + g_{442}\cos 2ny) - \gamma_{14}\beta^2(\varepsilon w_1(\tau))^2\sum_k\sum_l w_{kl}$$
$$\times\left(B_{00}^{(2)}k^2 + b_{00}^{(2)}l^2 - 4k^2n^2g_{402}\cos 2ny - 4l^2m^2g_{420}\cos 2mx\right)\sin kx \sin ly$$
$$+ \bar{\alpha}g_{42}(\varepsilon w_1(\tau))^3\sin mx \sin ny + O(\varepsilon^4) \tag{4.35}$$

Note that in Equations 4.31 through 4.35 $\tilde{\tau}$ is replaced by τ, and for the case of free vibration $\alpha = 0$, $\bar{\alpha} = 1$, otherwise $\alpha = 1$, $\bar{\alpha} = 0$. Coefficients $g_{11}^{(i,j)}$, $g_{21}^{(i,j)}$, $g_{31}^{(i,j)}$ ($i, j = 1, 3$), etc. are given in detail in Appendix J.

Multiplying Equation 4.34 by ($\sin mx \sin ny$) and integrating over the entire plate area, one has

$$g_{43}\frac{d^2(\varepsilon w_1)}{d\tau^2} + g_{41}(\varepsilon w_1) + g_{44}(\varepsilon w_1)^2 + \overline{\alpha}g_{42}(\varepsilon w_1)^3 = \overline{\lambda}_q(\tau) \tag{4.36}$$

where

$$\overline{\lambda}_q(\tau) = \frac{4}{\pi^2}\int_0^\pi\int_0^\pi \lambda_q(x, y, \tau)\sin mx \sin ny \, dx \, dy \tag{4.37}$$

4.2.1   Free Vibration

When $\overline{\alpha} = 1, \lambda_q(\tau) = 0$, Equation 4.36 becomes the free vibration equation of the plate. The nonlinear frequency of the plates can be expressed as

$$\omega_{NL} = \omega_L\left[1 + \frac{9g_{42}g_{41} - 10g_{44}^2}{12g_{41}^2}A^2\right]^{1/2} \tag{4.38}$$

where $A = \overline{W}_{max}/h$ is the amplitude to thickness ratio, and $\omega_L = [g_{41}/g_{43}]^{1/2}$ is the dimensionless linear frequency, from which the corresponding linear frequency can be expressed as $\overline{\omega}_L = \omega_L(\pi/a)(E_0/\rho_0)^{1/2}$, where E_0 and ρ_0 are defined as in Equation 4.6.

4.2.2   Forced Vibration

When the forced vibration is under consideration, we take $\overline{\alpha} = 0$. In such a case, Equation 4.36 can be rewritten as

$$\varepsilon\ddot{w}_1(\tau) + \varepsilon w_1(\tau)\omega_L^2 + \frac{g_{44}}{g_{43}}(\varepsilon w_1(\tau))^2 + O(\varepsilon^4) = \frac{\overline{\lambda}_q(\tau)}{g_{43}} \tag{4.39}$$

If zero-valued initial conditions prevail, i.e., $w_1(0) = \dot{w}_1(0) = 0$, Equation 4.39 may then be solved by using the Runge–Kutta iteration scheme (Pearson 1986):

$$(\varepsilon w_1)_{i+1} = (\varepsilon w_1)_i + \Delta\tau(\varepsilon\dot{w}_1)_i + \frac{(\Delta\tau)^2}{6}(L_1 + L_2 + L_3)$$

$$(\varepsilon\dot{w}_1)_{i+1} = (\varepsilon\dot{w}_1)_i + \frac{\Delta\tau}{6}(L_1 + 2L_2 + 2L_3 + L_4) \tag{4.40}$$

where $\Delta\tau$ is the time step, and

$$L_1 = f(\tau_i, (\varepsilon w_1))$$

$$L_2 = f\left(\tau_i + \frac{\Delta\tau}{2}, (\varepsilon w_1)_i + \frac{\Delta\tau(\varepsilon \dot{w}_1)_i}{2}\right)$$

$$L_3 = f\left(\tau_i + \frac{\Delta\tau}{2}, (\varepsilon w_1)_i + \frac{\Delta\tau(\varepsilon \dot{w}_1)_i}{2} + \frac{(\Delta\tau)^2}{4}L_1\right) \tag{4.41}$$

$$L_4 = f\left(\tau_i + \Delta\tau, (\varepsilon w_1)_i + \Delta\tau(\varepsilon \dot{w}_1)_i + \frac{(\Delta\tau)^2}{2}L_2\right)$$

where

$$f(\tau, x) = -\omega_L^2 x - \frac{g_{44}}{g_{43}}x^2 + \frac{\overline{\lambda}_q(\tau)}{g_{43}} \tag{4.42}$$

As a result, the solution of Equation 4.39 is obtained numerically. Resubstituting it into Equations 4.31 through 4.35, both displacement and stress function are determined. Next, substituting Equation 4.14 into boundary conditions (Equation 4.13), the coefficients $B_{00}^{(0)}, b_{00}^{(0)}, B_{00}^{(2)}$, and $b_{00}^{(2)}$ are then determined as given in Appendix J.

For numerical illustrations, two sets of material mixture are considered. One is zirconium oxide and titanium alloy, referred to as ZrO$_2$/Ti-6Al-4V, and the other is silicon nitride and stainless steel, referred to as Si$_3$N$_4$/SUS304. The upper surface of these two FGM plates is ceramic-rich and the lower surface is metal-rich. The thickness and side of the square plate are $h = 0.025$ m and $a = 0.2$ m, respectively. The mass density and thermal conductivity are $\rho = 3000$ kg m^{-3}, $\kappa = 1.80$ W mK^{-1} for ZrO$_2$; $\rho = 4429$ kg m^{-3}, $\kappa = 7.82$ W mK^{-1} for Ti-6Al-4V; $\rho = 2370$ kg m^{-3}, $\kappa = 9.19$ W mK^{-1} for Si$_3$N$_4$; and $\rho = 8166$ kg m^{-3}, $\kappa = 12.04$ W mK^{-1} for SUS304. Young's modulus and thermal expansion coefficient of these materials are assumed to be nonlinear function of temperature of Equation 1.4, and typical values are listed in Tables 1.1 and 1.2. Poisson's ratio ν_f is assumed to be a constant, for ZrO$_2$/Ti-6Al-4V plate $\nu = 0.3$, and for Si$_3$N$_4$/SUS304 one $\nu = 0.28$.

For the first example, we consider the nonlinear free vibration of an isotropic square plate ($a/b = 1.0$, $b/h = 10$, and $\nu = 0.3$) under different thermal loading conditions $\Delta T/T_{cr} = 0$, 0.25, 0.5, and 0.75, where $T_{cr} = 119.783/(\alpha \times 10^4)$ is the critical temperature of the plate (Bhimaraddi and Chandrashekhara 1993). The frequency parameter $\Omega = \overline{\omega}_L(a^2/h)[\rho(1 - \nu^2)/E]^{1/2}$ and nonlinear to linear frequency ratio ω_{NL}/ω_L are calculated and compared in Table 4.1 with the results of Bhimaraddi and Chandrashekhara (1993) based on the CPT, FSDPT, and HSDPT.

For the second example, we consider the free vibration of an FGM square plate made of aluminum oxide and Ti-6Al-4V. The top surface is ceramic-rich, whereas the bottom surface is metal-rich. The material properties, as given in He et al. (2001), are $E_m = 105.7$ GPa, $\nu_m = 0.2981$, $\rho_m = 4429$ kg m^{-3}

TABLE 4.1

Comparison of Natural Frequency Ω and Nonlinear to Linear Frequency Ratios for an Isotropic Square Plate under Different Thermal Loading Conditions ($a/b = 1.0$, $b/h = 10$, and $\nu = 0.3$)

$\Delta T/T_{cr}$	Sources	Ω	$\overline{W}_{max}/h$					
			0.0	0.2	0.4	0.6	0.8	1.0
0.25	HSDPT[a]	4.7624	1.000	1.027	1.105	1.222	1.368	1.535
	FSDPT[a]	4.7232	0.922	1.019	1.097	1.215	1.362	1.529
	CPT[a]	4.9380	1.037	1.063	1.138	1.252	1.395	1.559
	Present	4.7636	1.000	1.027	1.105	1.225	1.374	1.546
0.5	HSDPT	3.8884	1.000	1.041	1.153	1.318	1.517	1.739
	FSDPT	3.8405	0.988	1.029	1.143	1.309	1.509	1.732
	CPT	4.1017	1.055	1.094	1.201	1.360	1.554	1.772
	Present	3.8891	1.000	1.040	1.155	1.323	1.528	1.757
0.75	HSDPT	2.7495	1.000	1.080	1.287	1.569	1.893	2.242
	FSDPT	2.6813	0.975	1.057	1.267	1.553	1.880	2.230
	CPT	3.0437	1.107	1.180	1.372	1.640	1.953	2.293
	Present	2.7492	1.000	1.080	1.291	1.582	1.916	2.275

[a] HSDPT, FSDPT, and CPT results all from Bhimaraddi and Chandrashekhara (1993).

for Ti-6Al-4V; and $E_c = 320.24$ GPa, $\nu_c = 0.26$, $\rho_c = 3750$ kg m^{-3} for aluminum oxide. The FGM plate has $a = b = 0.4$ m and $h = 5$ mm. Table 4.2 gives the comparison of natural frequency $\bar{\omega}_L$ (in Hz) for the two special cases of isotropy, i.e., volume fraction index $N = 0$ and 2000. The FEM results of He et al. (2001) based on the CPT and seminumerical results of Yang and Shen (2002) based on HSDPT are given for direct comparison.

TABLE 4.2

Comparison of Natural Frequency $\bar{\omega}_L$ (Hz) for Simply Supported FGM Plates for the Two Special Cases of Isotropy

Mode Sequence	$N = 0$			$N = 2000$		
	He et al. (2001)	Yang and Shen (2002)	Huang and Shen (2004)	He et al. (2001)	Yang and Shen (2002)	Huang and Shen (2004)
1	144.66	143.96	144.94	268.92	261.46	271.03
2	360.53	360.07	362.04	669.40	653.14	677.04
3	360.53	360.07	362.04	669.40	653.14	677.04
4	569.89	568.88	578.78	1052.49	1044.31	1082.38
5	720.57	718.22	723.06	1338.52	1304.79	1352.24
6	720.57	718.22	723.06	1338.52	1304.79	1352.24
7	919.74	916.40	939.19	1695.23	1694.98	1756.49
8	919.74	916.40	939.19	1695.23	1694.98	1756.49
9	1225.72	1207.09	1226.19	2280.95	2214.34	2294.47
10	1225.72	1207.09	1226.19	2280.95	2214.34	2294.47

We then examine the dynamic response of an FGM square plate subjected to a uniform sudden load with $q_0 = 1.0$ MPa in thermal environments. The FGM plate is made of aluminum and alumina. The side and thickness of the square plate are 200 and 10 mm, respectively. The top surface is ceramic-rich, whereas the bottom surface is metal-rich. The temperature is varied only in the thickness direction and determined by the steady-state heat conduction equation with the boundary conditions. A stress-free temperature $T_0 = 0°C$ was taken. The material properties adopted are $E_b = 70$ GPa, $\nu_b = 0.3$, $\rho_b = 2707$ kg m^{-3}, $\alpha_b = 23.0 \times 10^{-6}$ °C^{-1}, $\kappa_b = 204$ W mK^{-1}, for aluminum; and $E_t = 380$ GPa, $\nu_t = 0.3$, $\rho_t = 3800$ kg m^{-3}, $\alpha_t = 7.4 \times 10^{-6}$ °C^{-1}, $\kappa_t = 10.4$ W mK^{-1}, for alumina. The curves of central deflection as functions of time are plotted and compared in Figure 4.1 with the FEM results of Praveen and Reddy (1998) based on FSDPT. In Figure 4.1, dimensionless central deflection and time are defined by $W = (\bar{E}_m h / q_0 a^2)$ and $\bar{t} = t[E_m / a^2 \rho_m]^{1/2}$, respectively.

Note that in these three examples the material properties are assumed to be independent of temperature. Table 4.3 gives comparisons of frequency parameter for Si$_3$N$_4$/SUS304 square plates with temperature-dependent material properties under heat conduction. The results of Huang and Shen (2004) based on HSDPT, the FEM results of Sundararajan et al. (2005) based on the Mori–Tanaka scheme, and the differential quadrature method (DQM) results of Wu et al. (2007) are also given for direct comparison. These four comparisons show that the present results agree well with existing results, and can be used as benchmark for other numerical studies.

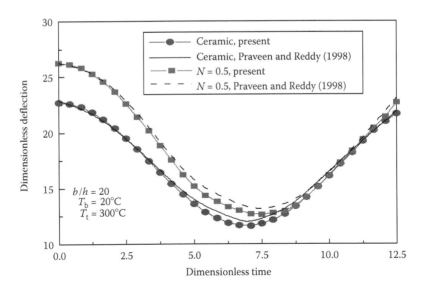

FIGURE 4.1
Comparison of central deflection versus time curves for an FGM square plate subjected to a suddenly applied uniform load and in thermal environments.

TABLE 4.3

Comparisons of Frequency Parameter for Si_3N_4/SUS304 Square Plates under Heat Conduction

	Mode	$N=0.0$	$N=0.5$	$N=1.0$	$N=2.0$
$T_U=400$ K, $T_L=300$ K					
Huang and Shen (2004)	(1,1)	12.397	8.615	7.474	6.639
	(1,2)	29.083	20.215	17.607	15.762
	(2,2)	43.835	30.530	26.590	23.786
Sundararajan et al. (2005)	(1,1)	12.311	8.276	7.302	6.572
	(1,2)	29.016	19.772	17.369	15.599
	(2,2)	44.094	30.184	26.506	23.787
Wu et al. (2007)	(1,1)	12.353	8.513	7.439	6.678
	(1,2)	29.033	20.115	17.578	15.717
	(2,2)	43.775	30.226	26.510	23.699
$T_U=600$ K, $T_L=300$ K					
Huang and Shen (2004)	(1,1)	11.984	8.269	7.171	6.398
	(1,2)	28.504	19.783	17.213	15.384
	(2,2)	43.107	29.998	26.109	23.327
Sundararajan et al. (2005)	(1,1)	11.888	7.943	6.989	6.269
	(1,2)	28.421	19.327	16.959	15.207
	(2,2)	43.343	29.629	25.997	23.303
Wu et al. (2007)	(1,1)	11.958	8.253	7.144	6.378
	(1,2)	28.433	19.468	17.116	15.375
	(2,2)	43.003	29.886	25.994	23.315

Tables 4.4 and 4.5 show the effect of volume fraction index N on the natural frequency parameter of ZrO_2/Ti-6Al-4V and Si_3N_4/SUS304 plates under three thermal loading conditions: case 1, $T_L=300$ K, $T_U=300$ K; case 2, $T_L=300$ K, $T_U=400$ K; and case 3, $T_L=300$ K, $T_U=600$ K. Temperature-dependent

TABLE 4.4

Natural Frequency Parameter $\Omega = \bar{\omega}_L (a^2/h) \left[\rho_0 (1-\nu^2)/E_0\right]^{1/2}$ for ZrO_2/Ti-6Al-4V Square Plates in Thermal Environments

	Mode				
	(1,1)	(1,2)	(2,2)	(1,3)	(2,3)
$T_L=300$ K, $T_U=300$ K					
ZrO_2	8.273	19.261	28.962	34.873	43.070
0.5	7.139	16.643	25.048	30.174	37.288
1.0	6.657	15.514	23.345	28.120	34.747
2.0	6.286	14.625	21.978	26.454	32.659
Ti-6Al-4V	5.400	12.571	18.903	22.762	28.111

TABLE 4.4 (continued)

Natural Frequency Parameter $\Omega = \bar{\omega}_L (a^2/h)\left[\rho_0(1 - \nu^2)/E_0\right]^{1/2}$ for $ZrO_2/Ti\text{-}6Al\text{-}4V$ Square Plates in Thermal Environments

| | \multicolumn{5}{c}{Mode} | | | | |
	(1, 1)	(1, 2)	(2, 2)	(1, 3)	(2, 3)
$T_L = 300\ K,\ T_U = 400\ K,\ temperature\ dependent$					
ZrO_2	7.868	18.659	28.203	34.015	42.045
0.5	6.876	16.264	24.578	29.651	36.664
1.0	6.437	15.202	22.956	27.696	34.236
2.0	6.101	14.372	21.653	26.113	32.239
Ti-6Al-4V	5.322	12.455	18.766	22.603	27.921
$T_L = 300\ K,\ T_U = 400\ K,\ temperature\ independent$					
ZrO_2	8.122	19.193	28.986	34.958	43.190
0.5	7.154	16.644	25.136	30.136	37.476
1.0	6.592	15.531	23.442	28.273	34.936
2.0	6.238	14.655	22.078	26.605	32.840
Ti-6Al-4V	5.389	12.620	19.104	22.905	28.261
$T_L = 300\ K,\ T_U = 600\ K,\ temperature\ dependent$					
ZrO_2	6.685	16.986	26.073	31.567	39.212
0.5	6.123	15.169	23.166	28.041	34.789
1.0	5.819	14.287	21.768	26.342	32.660
2.0	5.612	13.611	20.652	24.961	30.904
Ti-6Al-4V	5.118	12.059	18.175	21.898	27.045
$T_L = 300\ K,\ T_U = 600\ K,\ temperature\ independent$					
ZrO_2	7.686	18.749	28.527	34.472	42.713
0.5	6.776	16.367	24.859	30.044	37.201
1.0	6.362	15.308	23.216	28.036	34.714
2.0	6.056	14.474	21.896	26.435	32.664
Ti-6Al-4V	5.284	12.511	18.902	22.784	28.168

TABLE 4.5

Natural Frequency Parameter $\Omega = \bar{\omega}_L (a^2/h)\left[\rho_0(1 - \nu^2)/E_0\right]^{1/2}$ for $Si_3N_4/SUS304$ Square Plates in Thermal Environments

| | \multicolumn{5}{c}{Mode} | | | | |
	(1, 1)	(1, 2)	(2, 2)	(1, 3)	(2, 3)
$T_L = 300\ K,\ T_U = 300\ K$					
Si_3N_4	12.495	29.131	43.845	52.822	65.281
0.5	8.675	20.262	30.359	36.819	45.546
1.0	7.555	17.649	26.606	32.081	39.692
2.0	6.777	15.809	23.806	28.687	35.466
SUS304	5.405	12.602	18.967	22.850	28.239

(continued)

TABLE 4.5 (continued)

Natural Frequency Parameter $\Omega = \bar{\omega}_L(a^2/h)\left[\rho_0(1-\nu^2)/E_0\right]^{1/2}$ for Si$_3$N$_4$/SUS304 Square Plates in Thermal Environments

	Mode				
	(1, 1)	(1, 2)	(2, 2)	(1, 3)	(2, 3)
$T_L = 300$ K, $T_U = 400$ K, *temperature dependent*					
Si$_3$N$_4$	12.397	29.083	43.835	52.822	65.310
0.5	8.615	20.215	30.530	36.824	45.575
1.0	7.474	17.607	26.590	32.088	39.721
2.0	6.693	15.762	23.786	28.686	35.491
SUS304	5.311	12.539	18.959	22.828	28.246
$T_L = 300$ K, $T_U = 400$ K, *temperature independent*					
Si$_3$N$_4$	12.382	29.243	44.072	53.105	65.559
0.5	8.641	20.316	30.682	37.007	45.802
1.0	7.514	17.694	26.717	32.242	39.908
2.0	6.728	15.836	23.893	28.816	35.648
SUS304	5.335	12.587	19.008	22.908	28.344
$T_L = 300$ K, $T_U = 600$ K, *temperature dependent*					
Si$_3$N$_4$	11.984	28.504	43.107	51.998	64.358
0.5	8.269	19.783	29.998	36.239	44.901
1.0	7.171	17.213	26.109	31.557	39.114
2.0	6.398	15.384	23.327	28.185	34.918
SUS304	4.971	12.089	18.392	22.221	27.557
$T_L = 300$ K, $T_U = 600$ K, *temperature independent*					
Si$_3$N$_4$	12.213	28.976	43.797	52.821	65.365
0.5	8.425	20.099	30.458	36.781	45.572
1.0	7.305	17.486	26.506	31.970	39.692
2.0	6.523	15.632	23.685	28.609	35.436
SUS304	5.104	12.342	18.763	22.658	28.084

and temperature-independent material properties (values at fixed temperature 300 K) are both taken into account. In these two tables $\Omega = \bar{\omega}_L(a^2/h)[\rho_0(1-\nu^2)/E_0]^{1/2}$, where E_0 and ρ_0 are the reference values of E_m and ρ_m at $T_0 = 300$ K. Then Tables 4.6 and 4.7 show, respectively, the effects of volume fraction index N and temperature field on the nonlinear to linear frequency ratios ω_{NL}/ω_L of the same two FGM plates. It can be seen that the natural frequency of the FGM plate decreases with the increase of volume fraction index N, but it has a small effect on the nonlinear to linear frequency ratios. On the other hand, the temperature rise decreases the natural frequencies but increases the nonlinear to linear frequency ratios. The results show that the FGM plate will have lower natural frequency and slightly higher nonlinear to linear frequency ratios when the temperature-dependent material properties are taken into account.

TABLE 4.6

Effect of Volume Fraction Index N on the Nonlinear to Linear Frequency Ratio ω_{NL}/ω_L of FGM Square Plates in Thermal Environments ($T_L = 300$ K, $T_U = 400$ K)

	$\overline{W}_{max}/h$					
	0.0	0.2	0.4	0.6	0.8	1.0
ZrO_2/Ti-$6Al$-$4V$						
ZrO_2	1.000	1.023	1.087	1.186	1.312	1.461
0.5	1.000	1.023	1.087	1.186	1.312	1.460
1.0	1.000	1.022	1.086	1.183	1.310	1.455
2.0	1.000	1.022	1.084	1.179	1.302	1.444
Ti-6Al-4V	1.000	1.022	1.083	1.177	1.300	1.440
$Si_3N_4/SUS304$						
Si_3N_4	1.000	1.022	1.084	1.181	1.303	1.446
0.5	1.000	1.022	1.084	1.181	1.302	1.444
1.0	1.000	1.022	1.084	1.180	1.301	1.442
2.0	1.000	1.022	1.082	1.176	1.299	1.440
SUS304	1.000	1.022	1.082	1.172	1.296	1.438

TABLE 4.7

Effect of Temperature Field on the Nonlinear to Linear Frequency Ratio ω_{NL}/ω_L of FGM Square Plates ($N = 2.0$)

	$\overline{W}_{max}/h$					
	0.0	0.2	0.4	0.6	0.8	1.0
ZrO_2/Ti-$6Al$-$4V$						
$T_L = 300$ K, $T_U = 300$ K	1.000	1.021	1.082	1.176	1.296	1.436
$T_L = 300$ K, $T_U = 400$ K						
Temperature dependent	1.000	1.022	1.084	1.179	1.302	1.444
Temperature independent	1.000	1.022	1.083	1.178	1.300	1.441
$T_L = 300$ K, $T_U = 600$ K						
Temperature dependent	1.000	1.024	1.091	1.194	1.325	1.477
Temperature independent	1.000	1.023	1.087	1.183	1.314	1.462
$Si_3N_4/SUS304$						
$T_L = 300$ K, $T_U = 300$ K	1.000	1.021	1.081	1.174	1.293	1.432
$T_L = 300$ K, $T_U = 400$ K						
Temperature dependent	1.000	1.022	1.082	1.176	1.299	1.440
Temperature independent	1.000	1.021	1.082	1.175	1.255	1.437
$T_L = 300$ K, $T_U = 600$ K						
Temperature dependent	1.000	1.023	1.088	1.188	1.315	1.463
Temperature independent	1.000	1.023	1.087	1.187	1.313	1.460

Figures 4.2 and 4.3 show, respectively, the effect of volume fraction index N on the dynamic response of ZrO_2/Ti-6Al-4V and Si_3N_4/SUS304 plates under thermal environmental condition $T_L = 300$ K and $T_U = 400$ K. It can

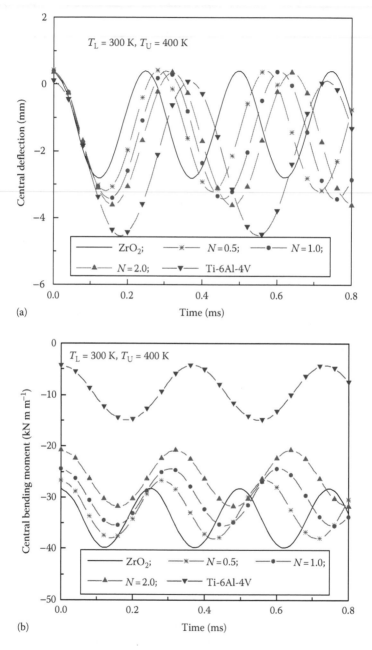

(a)

(b)

FIGURE 4.2

Effect of volume fraction index N on the dynamic response of ZrO_2/Ti-6Al-4V square plate subjected to a suddenly applied uniform load and in thermal environments: (a) central deflection versus time; (b) bending moment versus time.

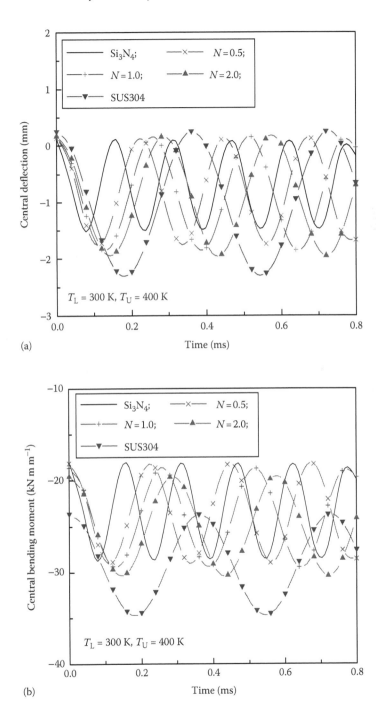

FIGURE 4.3
Effect of volume fraction index N on the dynamic response of Si_3N_4/SUS304 square plate subjected to a suddenly applied uniform load and in thermal environments: (a) central deflection versus time; (b) bending moment versus time.

be seen that the plate deflections are increased by increasing the volume fraction index N. The bending moment is decreased for the ZrO_2/Ti-6Al-4V plate, but it is increased for the $Si_3N_4/SUS304$ plate when the volume fraction index N is increased.

Figures 4.4 and 4.5 show, respectively, the effect of thermal environmental conditions on the dynamic response of ZrO_2/Ti-6Al-4V and $Si_3N_4/SUS304$ plates with $N=2.0$. The results show that both central deflections and bending moments are increased with the increase in temperature. It is also seen that the greater the temperature rise is, the greater will be the thermally induced initial bending moments.

It is appreciated that in Figures 4.2 through 4.5 the deflection mode $(m,n)=(1,1)$ was used and in Equation 4.19 k and l are taken as 1, 3, and 5, and in Equations 4.41 and 4.42 $\Delta\tau=2$ μs is taken as the time step for Runge–Kutta iteration method. The dynamic load is assumed to be a suddenly applied uniform load with $q_0=-50$ MPa.

4.3   Nonlinear Vibration of FGM Plates with Piezoelectric Actuators in Thermal Environments

We now consider the nonlinear free and forced vibration of FGM hybrid laminated plates. The plate is assumed to be made of a substrate FGM layer with surface-bonded piezoelectric layers. The substrate FGM layer is made of the combined ceramic and metallic materials with continuously varying mix-ratios comprising ceramic and metal. The length, width, and total thickness of the hybrid laminated plate are a, b, and h. The thickness of the FGM layer is h_f, while the thickness of the piezoelectric layer is h_p. We assume that the material composition varies smoothly from the upper to the lower surface of the FGM layer, such that the upper surface $(Z=h_2)$ of the FGM layer is ceramic-rich, and the lower surface $(Z=h_1)$ is metal-rich.

As in the case of Section 4.2, we assume the effective Young's modulus E_f and thermal expansion coefficient α_f of the FGM layer are temperature dependent, whereas the thermal conductivity κ_f and mass density ρ_f are independent to the temperature. Poisson's ratio ν_f depends weakly on temperature change and is assumed to be a constant. We assume that the volume fraction V_c follows a simple power law. Now Equation 4.2 may be rewritten as

$$E_f(Z,T) = [E_c(T) - E_m(T)]\left(\frac{Z-h_1}{h_2-h_1}\right)^N + E_m(T) \tag{4.43a}$$

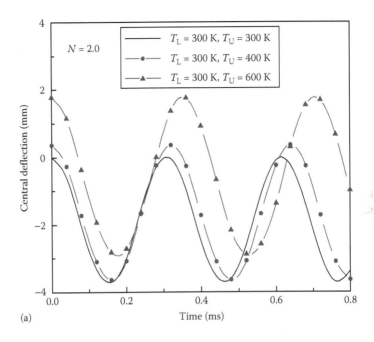

(a)

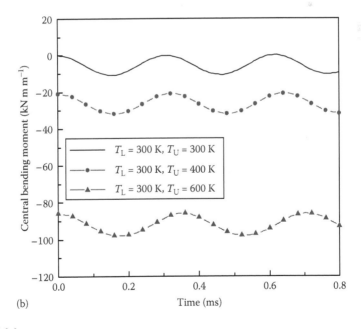

(b)

FIGURE 4.4
Effect of temperature field on the dynamic response of ZrO$_2$/Ti-6Al-4V square plate subjected to a suddenly applied uniform load: (a) central deflection versus time; (b) bending moment versus time.

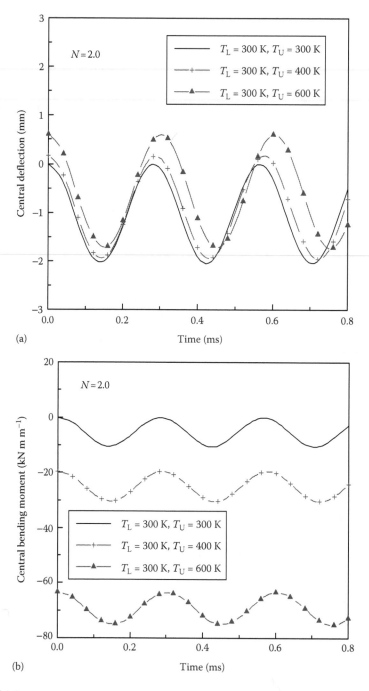

FIGURE 4.5
Effect of temperature field on the dynamic response of $Si_3N_4/SUS304$ square plate subjected to a suddenly applied uniform load: (a) central deflection versus time; (b) bending moment versus time.

$$\alpha_f(Z, T) = [\alpha_c(T) - \alpha_m(T)]\left(\frac{Z - h_1}{h_2 - h_1}\right)^N + \alpha_m(T) \qquad (4.43b)$$

$$\rho_f(Z) = (\rho_c - \rho_m)\left(\frac{Z - h_1}{h_2 - h_1}\right)^N + \rho_m \qquad (4.43c)$$

$$\kappa_f(Z) = (\kappa_c - \kappa_m)\left(\frac{Z - h_1}{h_2 - h_1}\right)^N + \kappa_m \qquad (4.43d)$$

It is evident that when $Z = h_1$, $E_f = E_m(T_m)$ and $\alpha_f = \alpha_m(T_m)$, and when $Z = h_2$, $E_f = E_c(T_c)$ and $\alpha_f = \alpha_c(T_c)$. Furthermore, E_f and α_f are both temperature- and position-dependent.

The temperature field is assumed to be the same as in Section 4.2. Then the temperature distribution along the thickness can be obtained by solving a steady-state heat transfer equation:

$$-\frac{d}{dZ}\left[\kappa(Z)\frac{dT}{dZ}\right] = 0 \qquad (4.44)$$

where

$$\kappa(Z) = \begin{cases} \kappa_p & (h_0 < Z < h_1) \\ \kappa_f(Z) & (h_1 < Z < h_2) \\ \kappa_p & (h_2 < Z < h_3) \end{cases} \qquad (4.45a)$$

$$T(Z) = \begin{cases} \tilde{T}_p(Z) & (h_0 \leq Z \leq h_1) \\ T_f(Z) & (h_1 \leq Z \leq h_2) \\ T_p(Z) & (h_2 \leq Z \leq h_3) \end{cases} \qquad (4.45b)$$

where κ_p is the thermal conductivity of the piezoelectric layer. Equation 4.44 is solved by imposing the boundary conditions $T = T_U$ at $Z = h_3$ and $T = T_L$ at $Z = h_0$, and the continuity conditions

$$\tilde{T}_p(h_1) = T_f(h_1) = T_m, \quad T_p(h_2) = T_f(h_2) = T_c \qquad (4.46a)$$

$$\kappa_p \frac{d\tilde{T}_p(Z)}{dZ}\bigg|_{Z=h_1} = \kappa_m \frac{dT_f(Z)}{dZ}\bigg|_{Z=h_1}, \quad \kappa_p \frac{dT_p(Z)}{dZ}\bigg|_{Z=h_2} = \kappa_c \frac{dT_f(Z)}{dZ}\bigg|_{Z=h_2} \qquad (4.46b)$$

The solution of Equations 4.44 through 4.46, by means of polynomial series, is

$$\tilde{T}_p(Z) = \frac{1}{h_p}[(T_L h_1 - T_m h_0) + (T_m - T_L)Z] \qquad (4.47a)$$

$$T_f(Z) = T_m + (T_c - T_m)\eta(Z) \tag{4.47b}$$

$$T_p(Z) = \frac{1}{h_p}[(T_c h_3 - T_U h_2) + (T_U - T_c)Z] \tag{4.47c}$$

where

$$\eta(Z) = \frac{1}{C}\left[\left(\frac{Z - h_1}{h_2 - h_1}\right) - \frac{\kappa_{mc}}{(N+1)\kappa_c}\left(\frac{Z - h_1}{h_2 - h_1}\right)^{N+1} + \frac{\kappa_{mc}^2}{(2N+1)\kappa_c^2}\left(\frac{Z - h_1}{h_2 - h_1}\right)^{2N+1}\right.$$

$$- \frac{\kappa_{mc}^3}{(3N+1)\kappa_c^3}\left(\frac{Z - h_1}{h_2 - h_1}\right)^{3N+1} + \frac{\kappa_{mc}^4}{(4N+1)\kappa_c^4}\left(\frac{Z - h_1}{h_2 - h_1}\right)^{4N+1}$$

$$\left. - \frac{\kappa_{mc}^5}{(5N+1)\kappa_c^5}\left(\frac{Z - h_1}{h_2 - h_1}\right)^{5N+1}\right] \tag{4.48a}$$

$$C = 1 - \frac{\kappa_{mc}}{(N+1)\kappa_c} + \frac{\kappa_{mc}^2}{(2N+1)\kappa_c^2} - \frac{\kappa_{mc}^3}{(3N+1)\kappa_c^3} + \frac{\kappa_{mc}^4}{(4N+1)\kappa_c^4} - \frac{\kappa_{mc}^5}{(5N+1)\kappa_c^5} \tag{4.48b}$$

$$G = 1 - \frac{\kappa_{mc}}{\kappa_c} + \frac{\kappa_{mc}^2}{\kappa_c^2} - \frac{\kappa_{mc}^3}{\kappa_c^3} + \frac{\kappa_{mc}^4}{\kappa_c^4} - \frac{\kappa_{mc}^5}{\kappa_c^5} \tag{4.48c}$$

where $\kappa_{mc} = \kappa_m - \kappa_c$, and

$$T_c = \frac{(1/h_f C)(\kappa_c G T_L + \kappa_m T_U) + (\kappa_p/h_p)T_U}{(1/h_f C)(\kappa_c G + \kappa_m) + (1/h_p)\kappa_p} \tag{4.49a}$$

$$T_m = \frac{(1/h_f C)(\kappa_c G T_L + \kappa_m T_U) + (\kappa_p/h_p)T_L}{(1/h_f C)(\kappa_c G + \kappa_m) + (1/h_p)\kappa_p} \tag{4.49b}$$

The plate is assumed to be geometrically perfect, and is subjected to a transverse dynamic load $q(X, Y, t)$ in thermal environments. Hence, the general von Kármán-type equations can be written in a similar form as expressed by Equations 1.33 through 1.36, just necessary to replace $\bar{N}^T$, $\bar{M}^T$, $\bar{S}^T$, $\bar{P}^T$ with $\bar{N}^P$, $\bar{M}^P$, $\bar{S}^P$, $\bar{P}^P$, and these equivalent thermopiezoelectric loads are defined by Equations 3.6 through 3.9.

Introducing dimensionless quantities (Equations 2.8 and 4.6), and

$$(\gamma_{T1}, \gamma_{T2}) = (A_x^T, A_y^T)a^2/\pi^2[D_{11}^* D_{22}^*]^{1/2}, \quad (\gamma_{P1}, \gamma_{P2}) = (B_x^E, B_y^E)a^2/\pi^2[D_{11}^* D_{22}^*]^{1/2}$$

$$(\gamma_{T3}, \gamma_{T4}, \gamma_{T6}, \gamma_{T7}) = (D_x^T, D_y^T, F_x^T, F_y^T)a^2/\pi^2 h^2 D_{11}^* \tag{4.50}$$

$$(\gamma_{P3}, \gamma_{P4}, \gamma_{P6}, \gamma_{P7}) = (D_x^E, D_y^E, F_x^E, F_y^E)a^2/\pi^2 h^2 D_{11}^*$$

where $B_x^E (=B_y^E)$, $D_x^E (=D_y^E)$, and $F_x^E (=F_y^E)$ are defined by

$$\begin{bmatrix} B_x^E & D_x^E & F_x^E \\ B_y^E & D_y^E & F_y^E \end{bmatrix} \Delta V = -\sum_k \int_{h_{k-1}}^{h_k} \begin{bmatrix} B_x \\ B_y \end{bmatrix}_k (1, Z, Z^3) \frac{V_k}{h_k} dZ \qquad (4.51)$$

The nonlinear motion equations can then be written in similar form as

$$L_{11}(W) - L_{12}(\Psi_x) - L_{13}(\Psi_y) + \gamma_{14}L_{14}(F) - L_{16}(M^P)$$

$$= \gamma_{14}\beta^2 L(W,F) + L_{17}(\ddot{W}) + \gamma_{80}\left(\frac{\partial \ddot{\Psi}_x}{\partial x} + \beta \frac{\partial \ddot{\Psi}_y}{\partial y}\right) + \lambda_q \qquad (4.52)$$

$$L_{21}(F) + \gamma_{24}L_{22}(\Psi_x) + \gamma_{24}L_{23}(\Psi_y) - \gamma_{24}L_{24}(W) = -\frac{1}{2}\gamma_{24}\beta^2 L(W,W) \qquad (4.53)$$

$$L_{31}(W) + L_{32}(\Psi_x) - L_{33}(\Psi_y) + \gamma_{14}L_{34}(F) - L_{36}(S^P) = \gamma_{90}\frac{\partial \ddot{W}}{\partial x} + \gamma_{10}\ddot{\Psi}_x \qquad (4.54)$$

$$L_{41}(W) - L_{42}(\Psi_x) + L_{43}(\Psi_y) + \gamma_{14}L_{44}(F) - L_{46}(S^P) = \gamma_{90}\beta\frac{\partial \ddot{W}}{\partial y} + \gamma_{10}\ddot{\Psi}_y \qquad (4.55)$$

and the boundary conditions can be written in dimensionless form as

$x = 0, \pi$:

$$W = \Psi_y = 0 \qquad (4.56a)$$

$$\int_0^\pi \int_0^\pi \left[\gamma_{24}^2\beta^2 \frac{\partial^2 F}{\partial y^2} - \gamma_5 \frac{\partial^2 F}{\partial x^2} + \gamma_{24}\left(\gamma_{511}\frac{\partial \Psi_x}{\partial x} + \gamma_{233}\beta\frac{\partial \Psi_y}{\partial y}\right) \right.$$

$$- \gamma_{24}\left(\gamma_{611}\frac{\partial^2 W}{\partial x^2} + \gamma_{244}\beta^2\frac{\partial^2 W}{\partial y^2}\right) - \frac{1}{2}\gamma_{24}\left(\frac{\partial W}{\partial x}\right)^2$$

$$\left. + (\gamma_{24}^2\gamma_{T1} - \gamma_5\gamma_{T2})T_1 + (\gamma_{24}^2\gamma_{P1} - \gamma_5\gamma_{P2})\Delta V \right] dx \, dy = 0 \qquad (4.56b)$$

$y = 0, \pi$:

$$W = \Psi_x = 0 \qquad (4.56c)$$

$$\int_0^\pi \int_0^\pi \left[\frac{\partial^2 F}{\partial x^2} - \gamma_5\beta^2 \frac{\partial^2 F}{\partial y^2} + \gamma_{24}\left(\gamma_{220}\frac{\partial \Psi_x}{\partial x} + \gamma_{522}\beta\frac{\partial \Psi_y}{\partial y}\right) \right.$$

$$- \gamma_{24}\left(\gamma_{240}\frac{\partial^2 W}{\partial x^2} + \gamma_{622}\beta^2\frac{\partial^2 W}{\partial y^2}\right) - \frac{1}{2}\gamma_{24}\beta^2\left(\frac{\partial W}{\partial y}\right)^2$$

$$\left. + (\gamma_{T2} - \gamma_5\gamma_{T1})T_1 + (\gamma_{P2} - \gamma_5\gamma_{P1})\Delta V \right] dx \, dy = 0 \qquad (4.56d)$$

All the necessary steps of the solution methodology are described in Section 4.2, and the solutions are not repeated herein for convenience.

We first examine the free vibration of an FGM square plate with symmetrically fully covered G-1195N piezoelectric layers. The substrate FGM plate is made of aluminum oxide and Ti-6Al-4V. The material properties adopted are $E_c = 320.24$ GPa, $\nu_c = 0.26$, $\rho_c = 3750$ kg m^{-3}, for aluminum oxide; $E_m = 105.70$ GPa, $\nu_m = 0.2981$, $\rho_m = 4429$ kg m^{-3}, for Ti-6Al-4V; $E_p = 63.0$ GPa, $\nu_p = 0.3$, $\rho_p = 7600$ kg m^{-3}, $d_{31} = d_{32} = 254 \times 10^{-12}$ mV^{-1}. The side and thickness of the substrate FGM square plate are 400 and 5 mm, and the thickness of each piezoelectric layer is 0.1 mm. The initial 10 frequencies of the plate as a function of the volume fraction index N are listed in Table 4.8 and compared with the FEM results of He et al. (2001) based on classical laminated plate theory (CLPT). Again, good agreement can be seen. Note that in this example the material properties are assumed to be independent of temperature.

In this section, two types of the hybrid FGM plate are considered. The first hybrid FGM plate has fully covered piezoelectric actuators on the top surface (referred to as P/FGM), and the second has two piezoelectric layers symmetrically bonded to the top and bottom surfaces (referred to as P/FGM/P). Silicon nitride (Si$_3$N$_4$) and stainless steel (SUS304) are chosen to be the

TABLE 4.8

Comparison of Natural Frequency $\bar{\omega}_L$ (Hz) for FGM Plates with Piezoelectric Actuator Bonded on the Top and Bottom Surfaces

Mode	Method	$N=0$	$N=0.5$	$N=1$	$N=5$	$N=15$	$N=100$	$N=1000$
1	He et al. (2001)	144.25	185.45	198.92	230.46	247.30	259.35	261.73
	Present	143.25	184.73	198.78	229.47	246.86	258.78	260.84
2	He et al. (2001)	359.00	462.65	495.62	573.82	615.58	645.55	651.49
	Present	358.87	461.02	494.65	571.87	613.95	643.92	649.83
3	He et al. (2001)	359.00	462.47	495.62	573.82	615.58	645.55	651.49
	Present	358.87	461.02	494.65	571.87	613.95	643.92	649.83
4	He et al. (2001)	564.10	731.12	778.94	902.04	967.78	1014.94	1024.28
	Present	563.42	727.98	778.61	899.91	964.31	1012.54	1023.72
5	He et al. (2001)	717.80	925.45	993.11	1148.12	1231.00	1290.78	1302.64
	Present	717.65	922.83	992.87	1146.87	1229.44	1288.73	1301.34
6	He et al. (2001)	717.80	925.45	993.11	1148.12	1231.00	1290.78	1302.64
	Present	717.65	922.83	992.87	1146.87	1229.44	1288.73	1301.34
7	He et al. (2001)	908.25	1180.93	1255.98	1453.32	1558.77	1634.65	1649.70
	Present	907.87	1177.34	1223.36	1451.66	1557.12	1632.18	1648.56
8	He et al. (2001)	908.25	1180.93	1255.98	1453.32	1558.77	1634.65	1649.70
	Present	907.87	1177.34	1223.36	1451.66	1557.12	1632.18	1648.56
9	He et al. (2001)	1223.14	1576.91	1697.15	1958.17	2097.91	2199.46	2219.67
	Present	1219.32	1571.65	1695.17	1956.79	2095.67	2197.47	2217.94
10	He et al. (2001)	1223.14	1576.91	1697.15	1958.17	2097.91	2199.46	2219.67
	Present	1219.32	1571.65	1695.17	1956.79	2095.67	2197.47	2217.94

constituent materials of the substrate FGM layer. The mass density ρ_f, Poisson's ratio ν_f, and thermal conductivity κ_f are 2370 kg m^{-3}, 0.24, 9.19 W mK^{-1} for Si$_3$N$_4$, and 8166 kg m^{-3}, 0.33, 12.04 W mK^{-1} for SUS304. Young's modulus E_f and thermal expansion coefficient α_f for these two constituent materials are listed in Tables 1.1 and 1.2. PZT-5A is selected for the piezoelectric layers. The material properties of which, as linear functions of temperature of Equation 3.35, are $E_{110} = E_{220} = 63$ GPa, $G_{120} = G_{130} = G_{230} = 24.2$ GPa, $\alpha_{110} = \alpha_{220} = 0.9 \times 10^{-6}$ K^{-1}, $\rho_p = 7600$ kg m^{-3}, $\nu_p = 0.3$, $\kappa_p = 2.1$ W mK^{-1}, and $d_{31} = d_{32} = 2.54 \times 10^{-10}$ mV^{-1}, and $E_{111} = -0.0005$, $E_{221} = G_{121} = G_{131} = G_{231} = -0.0002$, $\alpha_{111} = \alpha_{221} = 0.0005$. The side of the hybrid FGM plate is $a = b = 24$ mm. The thickness of the substrate FGM 1layer $h_f = 1.0$ mm, whereas the thickness of each piezoelectric layer $h_p = 0.1$ mm.

Tables 4.9 and 4.10 present the natural frequency parameter $\Omega = \bar{\omega}_L(a^2/h_f)[\rho_0/E_0]^{1/2}$ of these two types of the FGM hybrid plate with different

TABLE 4.9

Natural Frequency Parameter $\Omega = \bar{\omega}_L(a^2/h_f)[\rho_0/E_0]^{1/2}$ for the Hybrid (P/FGM) Plates under Different Sets of Thermal and Electric Loading Conditions

	$V_U = V_L =$	Si$_3$N$_4$ ($N=0$)	$N=0.5$	$N=2.0$	$N=4.0$	SUS304 ($N=\infty$)
(P/FGM), TID						
$T_U = 300$ K,	−200 V	10.726	7.885	6.334	5.920	5.194
$T_L = 300$ K	0 V	10.704	7.868	6.320	5.906	5.179
	+200 V	10.682	7.852	6.306	5.892	5.164
$T_U = 400$ K,	−200 V	10.149	7.380	5.852	5.427	4.668
$T_L = 300$ K	0 V	10.134	7.371	5.846	5.422	4.665
	+200 V	10.119	7.363	5.841	5.418	4.663
$T_U = 600$ K,	−200 V	9.237	6.667	5.277	4.903	4.324
$T_L = 300$ K	0 V	9.248	6.685	5.299	4.926	4.352
	+200 V	9.260	6.704	5.322	4.950	4.378
(P/FGM), TD-F						
$T_U = 400$ K,	−200 V	10.099	7.346	5.824	5.402	4.649
$T_L = 300$ K	0 V	10.084	7.336	5.819	5.397	4.646
	+200 V	10.070	7.328	5.814	5.393	4.645
$T_U = 600$ K,	−200 V	9.093	6.592	5.242	4.882	3.432
$T_L = 300$ K	0 V	9.109	6.614	5.266	4.907	3.277
	−200 V	9.125	6.636	5.291	4.932	3.159
(P/FGM), TD						
$T_U = 400$ K,	−200 V	10.080	7.331	5.812	5.390	4.635
$T_L = 300$ K	0 V	10.066	7.322	5.807	5.385	4.633
	−200 V	10.052	7.315	5.803	5.382	4.632
$T_U = 600$ K,	−200 V	9.042	6.557	5.216	4.858	3.183
$T_L = 300$ K	0 V	9.055	6.576	5.238	4.881	3.089
	−200 V	9.070	6.595	5.259	4.902	3.010

TABLE 4.10

Natural Frequency Parameter $\Omega = \bar{\omega}_L(a^2/h_f)[\rho_0/E_0]^{1/2}$ for the Hybrid (P/FGM/P) Plates under Different Sets of Thermal and Electric Loading Conditions

	$V_U=V_L=$	Si_3N_4 $(N=0)$	$N=0.5$	$N=2.0$	$N=4.0$	SUS304 $(N=\infty)$
(P/FGM/P), TID						
$T_U=300$ K,	-200 V	9.121	7.032	5.794	5.446	4.810
$T_L=300$ K	0 V	9.085	7.000	5.766	5.418	4.782
	$+200$ V	9.050	6.969	5.738	5.391	4.755
$T_U=400$ K,	-200 V	8.435	6.344	5.102	4.740	4.058
$T_L=300$ K	0 V	8.397	6.310	5.070	4.709	4.026
	$+200$ V	8.358	6.276	5.038	4.678	3.995
$T_U=600$ K,	-200 V	7.124	5.004	3.886	3.622	1.521
$T_L=300$ K	0 V	7.085	4.971	3.860	3.601	1.424
	$+200$ V	7.046	4.938	3.837	3.581	1.323
(P/FGM/P), TD-F						
$T_U=400$ K,	-200 V	8.372	6.295	5.059	4.698	4.022
$T_L=300$ K	0 V	8.333	6.260	5.027	4.668	3.990
	$+200$ V	8.294	6.226	4.996	4.636	3.959
$T_U=600$ K,	-200 V	6.900	4.854	3.816	2.511	1.287
$T_L=300$ K	0 V	6.863	4.825	3.797	2.370	1.172
	-200 V	6.826	4.795	3.778	2.246	1.045
(P/FGM/P), TD						
$T_U=400$ K,	-200 V	8.340	6.266	5.033	4.673	3.996
$T_L=300$ K	0 V	8.303	6.233	5.002	4.643	3.966
	-200 V	8.266	6.200	4.972	4.613	3.936
$T_U=600$ K,	-200 V	6.806	4.777	3.763	2.190	0.976
$T_L=300$ K	0 V	6.775	4.752	3.747	2.094	0.847
	-200 V	6.744	4.729	3.731	2.003	0.695

values of the volume fraction index N ($=0.0$, 0.5, 2.0, 4.0, and ∞) under different sets of thermal and electric loading conditions. Here, E_0 and ρ_0 are the reference values of SUS304 at the room temperature ($T_0=300$ K). TD represents material properties for both substrate FGM layer and piezoelectric layers are temperature dependent. TD-F represents material properties of substrate FGM layer are temperature dependent but material properties of piezoelectric layers are temperature independent, i.e., $E_{111}=E_{221}=G_{121}=G_{131}=G_{231}=\alpha_{111}=\alpha_{221}=0$ in Equation 3.35. TID represents material properties for both piezoelectric layers and substrate FGM layer are temperature independent, i.e., in a fixed temperature $T_0=300$ K for FGM layer, as previously used in Yang and Shen (2001). Six different applied voltages: $V_U=-200$ V, $V_U=0$ V, $V_U=200$ V, and $V_L=V_U=-200$ V, $V_L=V_U=0$ V, $V_L=V_U=200$ V are used, where subscripts "L" and "U" imply the low and upper piezoelectric layer. It can be seen that the natural frequency of these two plates is decreased by increasing temperature and volume fraction index N. The plus voltage decreases, but the minus voltage increases the plate natural frequency.

Tables 4.11 through 4.13 show, respectively, the effect of volume fraction index N, control voltage, and temperature field on the nonlinear to linear

TABLE 4.11

Effect of Volume Fraction Index N on Nonlinear to Linear Frequency Ratio ω_{NL}/ω_L for the Hybrid FGM Plates in Thermal Environments ($T_L = 300$ K, $T_U = 400$ K)

	$\overline{W}_{max}/h$					
	0.0	0.2	0.4	0.6	0.8	1.0
(P/FGM) ($V_U = +200$ V), TID						
Si_3N_4	1.000	1.021	1.081	1.175	1.294	1.434
0.5	1.000	1.022	1.084	1.182	1.305	1.449
2.0	1.000	1.022	1.084	1.180	1.303	1.446
4.0	1.000	1.021	1.082	1.178	1.299	1.440
SUS304	1.000	1.022	1.087	1.188	1.315	1.463
(P/FGM) ($V_U = +200$ V), TD-F						
Si_3N_4	1.000	1.021	1.081	1.175	1.296	1.435
0.5	1.000	1.022	1.085	1.182	1.307	1.451
2.0	1.000	1.022	1.084	1.181	1.305	1.448
4.0	1.000	1.021	1.083	1.179	1.301	1.442
SUS304	1.000	1.023	1.088	1.189	1.317	1.466
(P/FGM) ($V_U = +200$ V), TD						
Si_3N_4	1.000	1.021	1.082	1.176	1.296	1.436
0.5	1.000	1.022	1.085	1.183	1.307	1.452
2.0	1.000	1.022	1.084	1.182	1.306	1.449
4.0	1.000	1.021	1.083	1.179	1.301	1.444
SUS304	1.000	1.023	1.088	1.190	1.318	1.467
(P/FGM/P) ($V_L = V_U = +200$ V), TID						
Si_3N_4	1.000	1.021	1.082	1.177	1.298	1.439
0.5	1.000	1.022	1.087	1.186	1.313	1.460
2.0	1.000	1.022	1.088	1.188	1.315	1.463
4.0	1.000	1.022	1.087	1.187	1.314	1.460
SUS304	1.000	1.025	1.096	1.205	1.343	1.502
(P/FGM/P) ($V_L = V_U = +200$ V), TD-F						
Si_3N_4	1.000	1.021	1.082	1.177	1.299	1.439
0.5	1.000	1.023	1.087	1.187	1.315	1.462
2.0	1.000	1.023	1.088	1.189	1.318	1.467
4.0	1.000	1.023	1.088	1.188	1.316	1.465
SUS304	1.000	1.025	1.096	1.205	1.343	1.502

(continued)

TABLE 4.11 (continued)

Effect of Volume Fraction Index N on Nonlinear to Linear Frequency Ratio ω_{NL}/ω_L for the Hybrid FGM Plates in Thermal Environments ($T_L = 300$ K, $T_U = 400$ K)

	$\overline{W}_{max}/h$					
	0.0	0.2	0.4	0.6	0.8	1.0
$(P/FGM/P)$ $(V_L = V_U = +200$ V$)$, TD						
Si_3N_4	1.000	1.021	1.083	1.178	1.300	1.441
0.5	1.000	1.023	1.088	1.188	1.316	1.464
2.0	1.000	1.023	1.089	1.190	1.320	1.470
4.0	1.000	1.023	1.088	1.190	1.318	1.468
SUS304	1.000	1.025	1.097	1.207	1.346	1.506

TABLE 4.12

Effect of Temperature Field on Nonlinear to Linear Frequency Ratio ω_{NL}/ω_L for the Hybrid FGM Plates ($N = 0.5$)

	$\overline{W}_{max}/h$					
	0.0	0.2	0.4	0.6	0.8	1.0
(P/FGM) $(V_U = +200$ V$)$, TID						
$T_L = 300$ K, $T_U = 300$ K	1.000	1.019	1.075	1.161	1.272	1.402
$T_L = 300$ K, $T_U = 400$ K	1.000	1.022	1.084	1.182	1.306	1.450
$T_L = 300$ K, $T_U = 600$ K	1.000	1.026	1.099	1.211	1.352	1.515
(P/FGM) $(V_U = +200$ V$)$, TD-F						
$T_L = 300$ K, $T_U = 400$ K	1.000	1.022	1.085	1.182	1.307	1.451
$T_L = 300$ K, $T_U = 600$ K	1.000	1.025	1.096	1.205	1.343	1.502
(P/FGM) $(V_U = +200$ V$)$, TD						
$T_L = 300$ K, $T_U = 400$ K	1.000	1.022	1.085	1.183	1.307	1.452
$T_L = 300$ K, $T_U = 600$ K	1.000	1.025	1.096	1.206	1.345	1.504
$(P/FGM/\!/P)$ $(V_U = V_L = +200$ V$)$, TID						
$T_L = 300$ K, $T_U = 300$ K	1.000	1.018	1.071	1.153	1.259	1.384
$T_L = 300$ K, $T_U = 400$ K	1.000	1.022	1.087	1.186	1.313	1.460
$T_L = 300$ K, $T_U = 600$ K	1.000	1.035	1.134	1.282	1.464	1.670
$(P/FGM/\!/P)$ $(V_U = V_L = +200$ V$)$, TD-F						
$T_L = 300$ K, $T_U = 400$ K	1.000	1.023	1.087	1.187	1.315	1.462
$T_L = 300$ K, $T_U = 600$ K	1.000	1.034	1.128	1.271	1.447	1.645
$(P/FGM/\!/P)$ $(V_U = V_L = +200$ V$)$, TD						
$T_L = 300$ K, $T_U = 400$ K	1.000	1.023	1.088	1.188	1.316	1.464
$T_L = 300$ K, $T_U = 600$ K	1.000	1.033	1.128	1.270	1.445	1.643

TABLE 4.13

Effect of Applied Voltage on Nonlinear to Linear Frequency Ratio ω_{NL}/ω_L for the Hybrid FGM Plates in Thermal Environments ($T_L = 300$ K, $T_U = 400$ K, $N = 2.0$)

	$\overline{W}_{max}/h$					
	0.0	0.2	0.4	0.6	0.8	1.0
(P/FGM), TID						
$V_U = -200$ V	1.000	1.022	1.084	1.180	1.302	1.445
$V_U = 0$ V	1.000	1.022	1.084	1.180	1.303	1.446
$V_U = +200$ V	1.000	1.022	1.084	1.180	1.304	1.447
(P/FGM), TD-F						
$V_U = -200$ V	1.000	1.022	1.084	1.180	1.304	1.446
$V_U = 0$ V	1.000	1.022	1.084	1.181	1.304	1.447
$V_U = +200$ V	1.000	1.022	1.084	1.181	1.305	1.448
(P/FGM), TD						
$V_U = -200$ V	1.000	1.022	1.084	1.181	1.304	1.448
$V_U = 0$ V	1.000	1.022	1.084	1.181	1.305	1.448
$V_U = +200$ V	1.000	1.022	1.084	1.182	1.306	1.450
(P/FGM/P), TID						
$V_U = V_L = -200$ V	1.000	1.022	1.085	1.183	1.308	1.454
$V_U = V_L = 0$ V	1.000	1.022	1.086	1.186	1.312	1.458
$V_U = V_L = +200$ V	1.000	1.022	1.087	1.188	1.315	1.463
(P/FGM/P), TD-F						
$V_U = V_L = -200$ V	1.000	1.022	1.086	1.185	1.311	1.457
$V_U = V_L = 0$ V	1.000	1.022	1.087	1.187	1.314	1.462
$V_U = V_L = +200$ V	1.000	1.023	1.088	1.189	1.318	1.467
(P/FGM/P), TD						
$V_U = V_L = -200$ V	1.000	1.022	1.087	1.186	1.313	1.460
$V_U = V_L = 0$ V	1.000	1.023	1.088	1.188	1.316	1.465
$V_U = V_L = +200$ V	1.000	1.023	1.089	1.190	1.320	1.470

frequency ratios ω_{NL}/ω_L of these two types of FGM hybrid plates. It can be seen that the ratios ω_{NL}/ω_L increase as the volume fraction index N or temperature increases. It is noted that the control voltage only has a small effect on the frequency ratios. It is found that the decrease of natural frequency is about -46% for the P/FGM plate, and about -43% for the P/FGM/P one, from $N = 0$ to $N = 4$, in thermal environmental condition $T_L = 300$ K, $T_U = 400$ K under TD-F and TD cases. It can also be seen that the natural frequency and the nonlinear to linear frequency ratio under TD-F and TD cases are very close.

Figures 4.6 and 4.7 show the effect of temperature dependency and volume fraction index N ($= 0$, 0.5, 2.0, and ∞) on the dynamic response of P/FGM

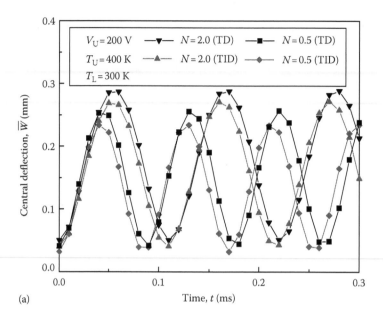

(a)

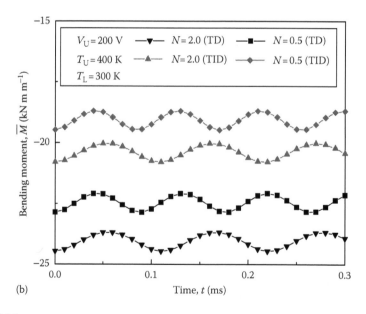

(b)

FIGURE 4.6
Effect of temperature dependency on the dynamic response of P/FGM plate subjected to a sudden load, control voltage and in thermal environments: (a) central deflection versus time; (b) central bending moment versus time.

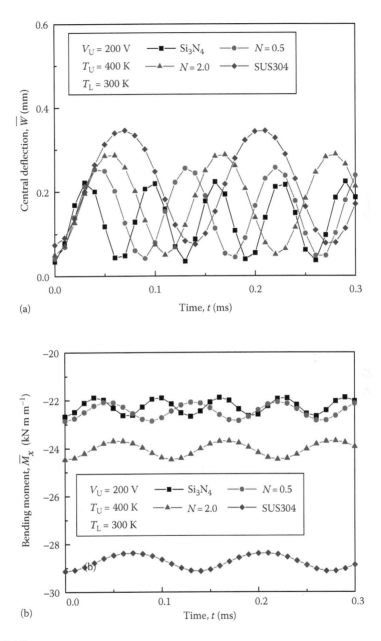

FIGURE 4.7
Effect of volume fraction index N on the dynamic response of P/FGM plate subjected to a suddenly applied uniform load, control voltage and in thermal environments: (a) central deflection versus time; (b) central bending moment versus time.

plates subjected to a sudden applied load with $q_0 = 2.0$ MPa, under electric loading condition $V_U = 200$ V and in thermal environmental condition $T_L = 300$ K, $T_U = 400$ K. The results show that the dynamic deflections of the P/FGM plate are increased by increasing volume fraction index N; this is because the stiffness of the plate becomes weaker when the volume fraction index N is increased. It can also be seen that the dynamic response becomes greater when the temperature-dependent properties are taken into account.

4.4   Vibration of Postbuckled Sandwich Plates with FGM Face Sheets in Thermal Environments

Finally, we consider the small- and large-amplitude vibrations of compressively and thermally postbuckled sandwich plates with FGM face sheets in thermal environments. The length, width, and total thickness of the sandwich plate are a, b, and h. The thickness of each FGM face sheet is h_F, while the thickness of the homogeneous substrate is h_H (see Figure 3.14). Note that in this section the Z is in the direction of the downward normal to the middle surface. The FGM face sheet is made from a mixture of ceramics and metals, the mixing ratio of which is varied continuously and smoothly in the Z-direction.

As in the case of Section 3.4, we assume the effective Young's modulus E_f, thermal expansion coefficient α_f, and thermal conductivity κ_f of FGM face sheets are functions of temperature, so that E_f, α_f, and κ_f are both temperature- and position-dependent. The Poisson ratio ν_f depends weakly on temperature change and is assumed to be a constant. Note that in this section, we assume the volume fraction V_m follows a simple power law. According to rule of mixture, we have

$$E_f(Z, T) = [E_m(T) - E_c(T)]\left(\frac{Z - t_0}{t_1 - t_0}\right)^N + E_c(T) \tag{4.57a}$$

$$\alpha_f(Z, T) = [\alpha_m(T) - \alpha_c(T)]\left(\frac{Z - t_0}{t_1 - t_0}\right)^N + \alpha_c(T) \tag{4.57b}$$

$$\kappa_f(Z, T) = [\kappa_m(T) - \kappa_c(T)]\left(\frac{Z - t_0}{t_1 - t_0}\right)^N + \kappa_c(T) \tag{4.57c}$$

$$\rho_f(Z) = (\rho_m - \rho_c)\left(\frac{Z - t_0}{t_1 - t_0}\right)^N + \rho_c \tag{4.57d}$$

The temperature field is assumed to be the same as in Section 4.2. The 1D steady-state heat transfer equation (Equation 3.55) will be solved and solutions are the same as Equations 3.56 through 3.60.

The plate is assumed to be geometrically perfect, and is subjected to a compressive edge load in the X-direction and/or thermal loading. Two cases of compressively postbuckled plates and of thermally postbuckled plates are considered. All four edges of the plate are assumed to be simply supported. Depending upon the in-plane behavior at the edges, two cases, case 1 (for the compressively buckled plate) and case 2 (for the thermally buckled plate), will be considered. The boundary conditions can be expressed by Equation 3.65 for these two cases.

Introducing dimensionless quantities of Equations 2.8 and 4.6, and

$$(\gamma_{T3}, \gamma_{T4}, \gamma_{T6}, \gamma_{T7}) = (D_x^T, D_y^T, F_x^T, F_y^T)a^2/\pi^2 h^2 D_{11}^*$$
$$\lambda_x = Pb/4\pi^2[D_{11}^* D_{22}^*]^{1/2}, \quad \lambda_T = \alpha_0 \Delta T_1 \qquad (4.58)$$

where α_0 is an arbitrary reference value, defined by Equation 3.39.

As mentioned in Section 3.4 the stretching–bending coupling, which is given in terms of B_{ij}^* and E_{ij}^* ($i,j = 1,2,6$), is still existed even for the mid-plane symmetric plate, when the plate is subjected to heat conduction. Hence the general von Kármán-type equations can be written in a similar form as expressed by Equations 1.33 through 1.36, just necessary to delete transverse applied load q, and the nonlinear motion equations can then be written in dimensionless form as

$$L_{11}(W) - L_{12}(\Psi_x) - L_{13}(\Psi_y) + \gamma_{14}L_{14}(F) - L_{16}(M^T)$$
$$= \gamma_{14}\beta^2 L(W,F) + L_{17}(\ddot{W}) + \gamma_{80}\left(\frac{\partial \ddot{\Psi}_x}{\partial x} + \beta \frac{\partial \ddot{\Psi}_y}{\partial y}\right) \qquad (4.59)$$

$$L_{21}(F) + \gamma_{24}L_{22}(\Psi_x) + \gamma_{24}L_{23}(\Psi_y) - \gamma_{24}L_{24}(W) = -\frac{1}{2}\gamma_{24}\beta^2 L(W,W) \qquad (4.60)$$

$$L_{31}(W) + L_{32}(\Psi_x) - L_{33}(\Psi_y) + \gamma_{14}L_{34}(F) - L_{36}(S^T) = \gamma_{90}\frac{\partial \ddot{W}}{\partial x} + \gamma_{10}\ddot{\Psi}_x \qquad (4.61)$$

$$L_{41}(W) - L_{42}(\Psi_x) + L_{43}(\Psi_y) + \gamma_{14}L_{44}(F) - L_{46}(S^T) = \gamma_{90}\beta\frac{\partial \ddot{W}}{\partial y} + \gamma_{10}\ddot{\Psi}_y \qquad (4.62)$$

where nondimensional linear operators $L_{ij}()$ and nonlinear operator $L()$ are defined by Equations 2.14 and 4.12, and the boundary conditions can be written in dimensionless form as

$x = 0, \pi$:

$$W = \Psi_y = 0 \qquad (4.63a)$$

$$M_x = P_x = 0 \qquad (4.63b)$$

$$\frac{1}{\pi}\int_0^\pi \beta^2 \frac{\partial^2 F}{\partial y^2}\,dy + 4\lambda_x \beta^2 = 0 \quad \text{(for compressively buckled plate)} \tag{4.63c}$$

$$\int_0^\pi \int_0^\pi \left[\gamma_{24}^2\beta^2 \frac{\partial^2 F}{\partial y^2} - \gamma_5 \frac{\partial^2 F}{\partial x^2} + \gamma_{24}\left(\gamma_{511}\frac{\partial \Psi_x}{\partial x} + \gamma_{233}\beta\frac{\partial \Psi_y}{\partial y} \right) \right.$$

$$- \gamma_{24}\left(\gamma_{611}\frac{\partial^2 W}{\partial x^2} + \gamma_{244}\beta^2\frac{\partial^2 W}{\partial y^2} \right) - \frac{1}{2}\gamma_{24}\left(\frac{\partial W}{\partial x} \right)^2$$

$$\left. + \gamma_{24}^2\gamma_{T1} - \gamma_5\gamma_{T2})\lambda_T \right]\,dx\,dy = 0 \quad \text{(for thermally buckled plate)} \tag{4.63d}$$

$y = 0, \pi$:

$$W = \Psi_x = 0 \tag{4.63e}$$

$$M_y = P_y = 0 \tag{4.63f}$$

$$\int_0^\pi \frac{\partial^2 F}{\partial x^2}\,dx = 0 \quad \text{(for compressively buckled plate)} \tag{4.63g}$$

$$\int_0^\pi \int_0^\pi \left[\frac{\partial^2 F}{\partial x^2} - \gamma_5\beta^2\frac{\partial^2 F}{\partial y^2} + \gamma_{24}\left(\gamma_{220}\frac{\partial \Psi_x}{\partial x} + \gamma_{522}\beta\frac{\partial \Psi_y}{\partial y} \right) \right.$$

$$- \gamma_{24}\left(\gamma_{240}\frac{\partial^2 W}{\partial x^2} + \gamma_{622}\beta^2\frac{\partial^2 W}{\partial y^2} \right) - \frac{1}{2}\gamma_{24}\beta^2\left(\frac{\partial W}{\partial y} \right)^2$$

$$\left. + (\gamma_{T2} - \gamma_5\gamma_{T1})\lambda_T \right]\,dy\,dx = 0 \quad \text{(for thermally buckled plate)} \tag{4.63h}$$

We assume that the solution of Equations 4.59 through 4.62 can be expressed as

$$W(x,y,\tau) = W^*(x,y) + \tilde{W}(x,y,\tau)$$
$$\Psi_x(x,y,\tau) = \Psi_x^*(x,y) + \tilde{\Psi}_x(x,y,\tau)$$
$$\Psi_y(x,y,\tau) = \Psi_y^*(x,y) + \tilde{\Psi}_y(x,y,\tau) \tag{4.64}$$
$$F(x,y,\tau) = F^*(x,y) + \tilde{F}(x,y,\tau)$$

where $W^*(x,y)$ is an initial time-independent deflection due to pre- and postbuckling equilibrium states of sandwich plates subjected to uniaxial compression and/or thermal loading. $\Psi_x^*(x,y)$, $\Psi_y^*(x,y)$, and $F^*(x,y)$ are the midplane rotations and stress function corresponding to $W^*(x,y)$. $\tilde{W}(x,y,\tau)$ is an additional time-dependent displacement which is considered to originate from the linear or nonlinear vibration of sandwich plates. $\tilde{\Psi}_x(x, y, \tau)$, $\tilde{\Psi}_y(x, y, \tau)$, and $\tilde{F}(x, y, \tau)$ are defined analogously to $\Psi_x^*(x, y)$, $\Psi_y^*(x, y)$, and $F^*(x, y)$, but is for $W(x, y, \tau)$.

Substituting Equation 4.64 into Equations 4.59 through 4.62, we obtain two sets of equations and can be solved in sequence. The first set of equations yields the particular solution of static postbuckling or thermal postbuckling deflection, and the second set of equations gives the homogeneous solution of vibration characteristics on the buckled plate.

As has been shown in Section 3.4, the prebuckling deflection caused by temperature field (Shen 2007) should be included, when the plate is subjected to heat conduction. Solutions $W^*(x, y)$, $\Psi_x^*(x, y)$, $\Psi_y^*(x, y)$, and $F^*(x, y)$ may be expressed as

$$W^* = \varepsilon \left[A_{11}^{(1)} \sin kx \sin ly \right] + \varepsilon^3 \left[A_{13}^{(3)} \sin kx \sin 3ly + A_{31}^{(3)} \sin 3k \sin ly \right] + O(\varepsilon^5) \quad (4.65)$$

$$F^* = -B_{00}^{(0)} \frac{y^2}{2} - b_{00}^{(0)} \frac{x^2}{2} + \varepsilon^2 \left[-B_{00}^{(2)} \frac{y^2}{2} - b_{00}^{(2)} \frac{x^2}{2} + B_{20}^{(2)} \cos 2kx + B_{02}^{(2)} \cos 2ly \right]$$

$$+ \varepsilon^4 \left[-B_{00}^{(4)} \frac{y^2}{2} - b_{00}^{(4)} \frac{x^2}{2} + B_{20}^{(4)} \cos 2kx + B_{02}^{(4)} \cos 2ly + B_{22}^{(4)} \cos 2kx \cos 2ly + B_{40}^{(4)} \cos 4kx \right.$$

$$\left. + B_{04}^{(4)} \cos 4ly + B_{24}^{(4)} \cos 2kx \cos 4ly + B_{42}^{(4)} \cos 4kx \cos 2ly \right] + O(\varepsilon^5) \quad (4.66)$$

$$\Psi_x^* = \varepsilon \left[C_{11}^{(1)} \cos kx \sin ly \right] + \varepsilon^3 \left[C_{13}^{(3)} \cos kx \sin 3ly + C_{31}^{(3)} \cos 3kx \sin ly \right] + O(\varepsilon^5) \quad (4.67)$$

$$\Psi_y^* = \varepsilon \left[D_{11}^{(1)} \sin kx \cos ly \right] + \varepsilon^3 \left[D_{13}^{(3)} \sin kx \cos 3ly + D_{31}^{(3)} \sin 3kx \cos ly \right] + O(\varepsilon^5) \quad (4.68)$$

As is mentioned before, all coefficients in Equations 4.65 through 4.68 are related and can be expressed in terms of $A_{11}^{(1)}$.

Then, $\tilde{W}(x, y, \tau)$, $\tilde{\Psi}_x(x, y, \tau)$, $\tilde{\Psi}_y(x, y, \tau)$, and $\tilde{F}(x, y, \tau)$ satisfy the nonlinear motion equations:

$$L_{11}(\tilde{W}) - L_{12}(\tilde{\Psi}_x) - L_{13}(\tilde{\Psi}_y) + \gamma_{14} L_{14}(\tilde{F}) = \gamma_{14} \beta^2 \left[L(\tilde{W} + W^*, \tilde{F}) + L(\tilde{W}, F^*) \right]$$

$$+ L_{17}(\ddot{\tilde{W}}) + \gamma_{80} \left(\frac{\partial \ddot{\tilde{\Psi}}_x}{\partial x} + \beta \frac{\partial \ddot{\tilde{\Psi}}_y}{\partial y} \right) \quad (4.69)$$

$$L_{21}(\tilde{F}) + \gamma_{24} L_{22}(\tilde{\Psi}_x) + \gamma_{24} L_{23}(\tilde{\Psi}_y) - \gamma_{24} L_{24}(\tilde{W}) = -\frac{1}{2} \gamma_{24} \beta^2 L(\tilde{W} + 2W^*, \tilde{W}) \quad (4.70)$$

$$L_{31}(\tilde{W}) + L_{32}(\tilde{\Psi}_x) - L_{33}(\tilde{\Psi}_y) + \gamma_{14} L_{34}(\tilde{F}) = \gamma_{90} \frac{\partial \ddot{\tilde{W}}}{\partial x} + \gamma_{10} \ddot{\tilde{\Psi}}_x \quad (4.71)$$

$$L_{41}(\tilde{W}) - L_{42}(\tilde{\Psi}_x) + L_{43}(\tilde{\Psi}_y) + \gamma_{14} L_{44}(\tilde{F}) = \gamma_{90} \beta \frac{\partial \ddot{\tilde{W}}}{\partial y} + \gamma_{10} \ddot{\tilde{\Psi}}_y \quad (4.72)$$

Using the perturbation procedure as described in Section 4.2, we obtain asymptotic solutions, up to third order, as

$$\tilde{W}(x,y,\tau) = \varepsilon[w_1(\tau) + g_1\ddot{w}_1(\tau)]\sin mx\sin ny$$
$$+ (\varepsilon w_1(\tau))^3[g_{331}\sin 3mx\sin ny + g_{313}\sin mx\sin 3ny] + O(\varepsilon^4) \quad (4.73)$$

$$\tilde{F}(x,y,\tau) = \varepsilon\left[g_{31}^{(1,1)}w_1(\tau) + g_4\ddot{w}_1(\tau)\right]\sin mx\sin ny$$
$$+ (\varepsilon w_1(\tau))^2(g_{402}\cos 2ny + g_{420}\cos 2mx) + (\varepsilon w_1(\tau))^3$$
$$\times \left[g_{31}^{(3,1)}g_{331}\sin 3mx\sin ny + g_{31}^{(1,3)}g_{313}\sin mx\sin 3ny\right] + O(\varepsilon^4) \quad (4.74)$$

$$\tilde{\Psi}_x(x,y,\tau) = \varepsilon\left[g_{11}^{(1,1)}w_1(\tau) + g_2\ddot{w}_1(\tau)\right]\cos mx\sin ny$$
$$+ (\varepsilon w_1(\tau))^2 g_{12}\sin 2mx + (\varepsilon w_1(\tau))^3$$
$$\times \left[g_{11}^{(3,1)}g_{331}\cos 3mx\sin ny + g_{11}^{(1,3)}g_{313}\cos mx\sin 3ny\right] + O(\varepsilon^4) \quad (4.75)$$

$$\tilde{\Psi}_y(x,y,\tau) = \varepsilon\left[g_{21}^{(1,1)}w_1(\tau) + g_3\ddot{w}_1(\tau)\right]\sin mx\cos ny$$
$$+ (\varepsilon w_1(\tau))^2 g_{22}\sin 2ny + (\varepsilon w_1(\tau))^3$$
$$\times \left[g_{21}^{(3,1)}g_{331}\sin 3mx\cos ny + g_{21}^{(1,3)}g_{313}\sin mx\cos 3ny\right] + O(\varepsilon^4) \quad (4.76)$$

Note that in Equations 4.73 through 4.76 coefficients $g_{11}^{(i,j)}$, $g_{21}^{(i,j)}$, $g_{31}^{(i,j)}(i,j=1,3)$, etc. are given in detail in Appendix K. Also we have

$$\varepsilon[g_{41}w_1(\tau) + g_{43}\ddot{w}_1(\tau)]\sin mx\sin ny + (\varepsilon w_1(\tau))^2(g_{441}\cos 2mx + g_{442}\cos 2ny)$$
$$+ \gamma_{14}\beta^2(\varepsilon w_1(\tau))^2\left(\varepsilon A_{11}^{(1)}\right)(4k^2n^2 g_{402}\cos 2ny + 4l^2m^2 g_{420}\cos 2mx)\sin kx\sin ly$$
$$+ \gamma_{14}\beta^2(\varepsilon w_1(\tau))^2\left(\varepsilon^3 A_{13}^{(3)}\right)(4k^2n^2 g_{402}\cos 2ny + 36l^2m^2 g_{420}\cos 2mx)\sin kx\sin 3ly$$
$$+ \gamma_{14}\beta^2(\varepsilon w_1(\tau))^2\left(\varepsilon^3 A_{31}^{(3)}\right)(36k^2n^2 g_{402}\cos 2ny + 4l^2m^2 g_{420}\cos 2mx)\sin 3kx\sin ly$$
$$+ g_{42}(\varepsilon w_1(\tau))^3\sin mx\sin ny = 0 \quad (4.77)$$

Multiplying Equation 4.77 by $(\sin mx \sin ny)$ and integrating over the plate area, we obtain

$$g_{43}\frac{d^2(\varepsilon w_1)}{d\tau^2} + g_{41}(\varepsilon w_1) + g_{44}(\varepsilon w_1)^2 + g_{42}(\varepsilon w_1)^3 = 0 \quad (4.78)$$

From Equation 4.78, the nonlinear frequency of the plates can be expressed as

$$\omega_{NL} = \omega_L\left[1 + \frac{9g_{42}g_{41} - 10g_{44}^2}{12g_{41}^2}A^2\right]^{1/2} \quad (P_x \neq P_{cr}\ \text{or}\ \Delta T \neq \Delta T_{cr}) \quad (4.79a)$$

$$\omega_{NL} = \left[\frac{3}{4} \frac{g_{42}}{g_{43}} \right]^{1/2} A \quad (P_x = P_{cr} \text{ or } \Delta T = \Delta T_{cr}) \tag{4.79b}$$

From Equation 4.79b, it can be seen that the nonlinear frequency varies linearly with the amplitude $\overline{W}_{max}/h$, when P_x/P_{cr} (or $\Delta T/\Delta T_{cr}$) = 1.

We first examine the relationship between natural frequency ratio (ω/ω_0) and in-plane compressive load ratio P/P_{cr} for an isotropic rectangular plate with aspect ratio of 3, where ω_0 is the lowest linear natural frequency of the plate and P_{cr} is the buckling load of the same plate under the uniaxial compression. The curves are plotted in Figure 4.8 and compared with the FEM result of Yang and Han (1983) based on high-order triangular membrane finite element combined with a fully conforming triangular plate bending element.

We then examine the relationship between fundamental frequencies $\overline{\omega}_L$ (in Hz) and temperature rise T (in K) for a simply supported Si_3N_4/SUS304 plate under the two special cases of isotropy, i.e., volume fraction index $N = 0$ and 2000. The top surface is ceramic-rich, whereas the bottom surface is metal-rich. The material properties are the same as listed in Tables 1.1 through 1.3. The FGM plate has $a = b = 0.3$ m and $a/h = 100$. The curves are plotted in Figure 4.9 and compared with the FEM result of Park and Kim (2006) based on the FSDPT.

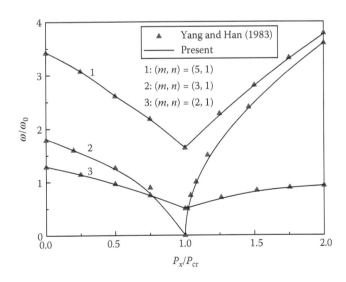

FIGURE 4.8
Comparisons of natural frequencies for the compressively postbuckled isotropic rectangular plate with aspect ratio of 3 ($\nu = 0.3$).

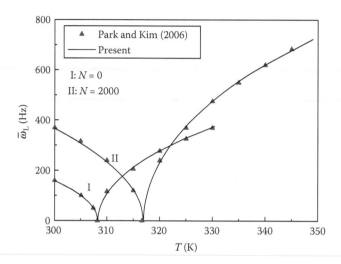

FIGURE 4.9
Comparisons of natural frequencies for the thermally buckled Si_3N_4/SUS304 square plate.

These two comparisons show that the present results agree well with existing results for both compressively and thermally buckled plates. In this section, the material mixture for FGM face sheets is considered to be silicon nitride and stainless steel, referred to as Si_3N_4/SUS304. The material properties are the same as those adopted in Section 3.4. The plate geometric parameter $a/b = 1$, $b/h = 20$, and the thickness of the FGM face sheets $h_F = 1$ mm, whereas the thickness of the homogeneous substrate is taken to be $h_H = 4$, 6, and 8 mm, so that the substrate-to-face sheet thickness ratio $h_H/h_F = 4$, 6, 8, respectively.

Figure 4.10 shows the effects of volume fraction index N on the linear fundamental frequencies $\bar{\omega}_L$ of the pre- and postbuckled sandwich plate with $h_H/h_F = 4$ under uniform or nonuniform temperature field. It can be seen that as the volume fraction index increases, the fundamental frequency increases in the prebuckling region, but decreases in the initial postbuckling region ($P_x < 3500$ kN), and in the deep postbuckling region the fundamental frequency becomes greater, when increasing in N. This is due to the fact that the plate stiffness is increased in the prebuckling region, when increasing in N, but in the initial postbuckling region the initial deflection is an important issue and the plate will have a small deflection when it has a great stiffness, further in the deep postbuckling region the effect of plate stiffness becomes more pronounced again.

Figure 4.11 shows temperature changes on the fundamental frequencies of the pre- and postbuckled sandwich plate with $h_H/h_F = 4$ and $N = 2$ under uniform or nonuniform temperature field. It can be seen that, for uniform temperature field, as the temperature is increased, fundamental frequencies

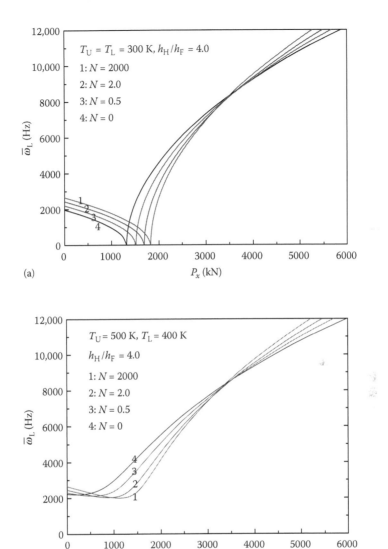

FIGURE 4.10
Effects of volume fraction index N on the fundamental frequencies of the pre- and postbuckled sandwich plate in thermal environments: (a) uniform temperature field; (b) heat conduction.

have decreased in the prebuckling region, but increased in the postbuckling region. It can also be seen that the effect of nonuniform temperature field is larger than that of uniform temperature field. When heat conduction is put into consideration, the larger top-bottom temperature difference leads to larger fundamental frequency rise.

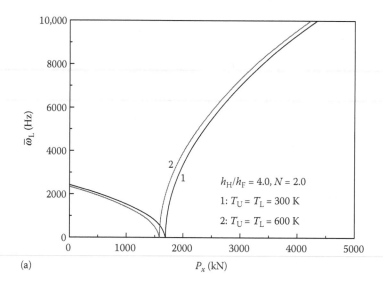

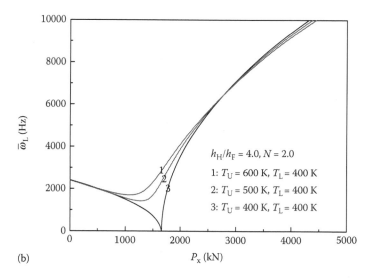

FIGURE 4.11
Effects of temperature changes on the fundamental frequencies of the pre- and postbuckled sandwich plate: (a) uniform temperature field; (b) heat conduction.

From Figures 4.10 and 4.11, it can be seen that, as the compressive load reaches the buckling load, the fundamental frequencies will drop to zero under uniform temperature field. In contrast, the fundamental frequencies do not go to zero, because no bifurcation-type buckling could occur when heat conduction is taken into account.

Figures 4.12 and 4.13 show, respectively, the effects of volume fraction index N, and temperature changes on the nonlinear frequency ratio of

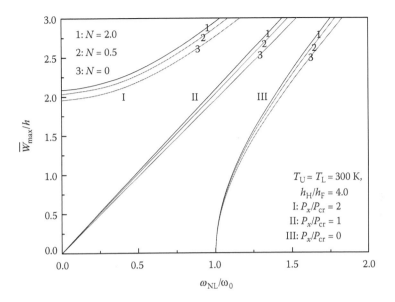

FIGURE 4.12
Effects of volume fraction index N on the nonlinear frequency ratio of the pre- and postbuckled sandwich plate.

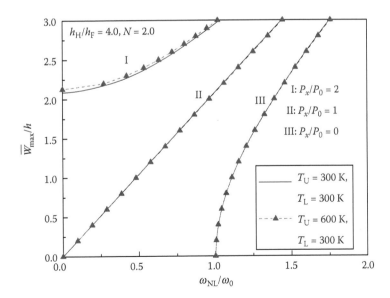

FIGURE 4.13
Effects of temperature changes on the nonlinear frequency ratio of the pre- and postbuckled sandwich plate.

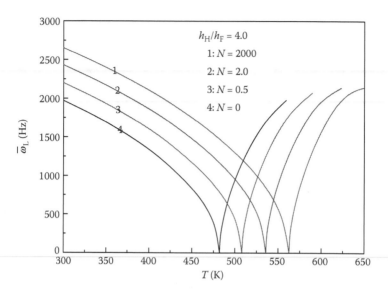

FIGURE 4.14
Effects of volume fraction index N on the fundamental frequencies of the thermally pre- and postbuckled sandwich plate.

the pre- and postbuckled sandwich plate in thermal environments. Three cases, i.e., $P_x/P_{cr} = 0$, 1, and 2, are considered. $P_x/P_{cr} = 0$ denotes no in-plane loads, $P_x/P_{cr} = 1$ denotes bifurcation buckling case, and $P_x/P_{cr} = 2$ represents a large-amplitude free vibration about a postbuckled equilibrium state. Note that in Figure 4.13 P_x/P_{cr} is replaced by P_x/P_0, where P_0 is a reference value of buckling load of the plate at $T_U = T_L = 300$ K. It can be seen that the nonlinear frequency ratio is decreased, when the volume fraction index N is increased. The temperature changes only have small effects on the nonlinear frequency ratio of the plate. Note that, in the present study, the solution is based on the assumption that the vibration of the plate is symmetric about the flat position. Another type of motion is possible in which the plate vibrates about a static buckled position on one side of the flat position. Such motion is however not considered in the present study.

Figures 4.14 and 4.15 are thermally postbuckled vibration results for the same plate analogous to the compressively postbuckled vibration results of Figures 4.10 and 4.12, which are for the thermal loading case of uniform temperature field. They lead to broadly the same conclusions as do Figures 4.10 and 4.12.

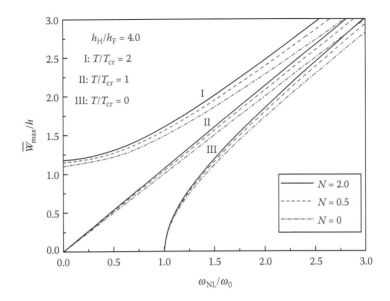

FIGURE 4.15
Effects of volume fraction index N on the nonlinear frequency ratio of the thermally pre- and postbuckled sandwich plate.

References

Abrate S. (2006), Free vibration, buckling, and static deflections of functionally graded plates, *Composites Science and Technology*, **66**, 2383–2394.

Allahverdizadeh A., Naei M.H., and Bahrami M.N. (2008a), Nonlinear free and forced vibration analysis of thin circular functionally graded plates, *Journal of Sound and Vibration*, **310**, 966–984.

Allahverdizadeh A., Naei M.H., and Bahrami M.N. (2008b), Vibration amplitude and thermal effects on the nonlinear behavior of thin circular functionally graded plates, *International Journal of Mechanical Sciences*, **50**, 445–454.

Allahverdizadeh A., Rastgo A., and Naei M.H. (2008c), Nonlinear analysis of a thin circular functionally graded plate and large deflection effects on the forces and moments, *Journal of Engineering Materials and Technology-ASME*, **130**, Article Number: 011009.

Bhimaraddi A. and Chandrashekhara K. (1993), Nonlinear vibrations of heated antisymmetric angle-ply laminated plates, *International Journal of Solids and Structures*, **30**, 1255–1268.

Chen C.-S. (2005), Nonlinear vibration of a shear deformable functionally graded plate, *Composite Structures*, **68**, 295–302.

Chen C.-S. and Tan A.-H. (2007), Imperfection sensitivity in the nonlinear vibration of initially stresses functionally graded plates, *Composite Structures*, **78**, 529–536.

Chen C.-S., Chen T.J., and Chien R.-D. (2006), Nonlinear vibration of initially stressed functionally graded plates, *Thin-Walled Structures*, **44**, 844–851.

Cheng Z.-Q. and Reddy J.N. (2003), Frequency of functionally graded plates with three-dimensional asymptotic approach, *Journal of Engineering Mechanics ASCE*, **129**, 896–900.

Efraim E. and Eisenberger M. (2007), Exact vibration analysis of variable thickness thick annular isotropic and FGM plates, *Journal of Sound and Vibration*, **299**, 720–738.

Elishakoff I., Gentilini C., and Viola E. (2005), Forced vibrations of functionally graded plates in the three-dimensional setting, *AIAA Journal*, **43**, 2000–2007.

Fung C.-P. and Chen C.-S. (2006), Imperfection sensitivity in the nonlinear vibration of functionally graded plates, *European Journal of Mechanics—A/Solids*, **25**, 425–436.

He X.Q., Ng T.Y., Sivashanker S., and Liew K.M. (2001), Active control of FGM plates with integrated piezoelectric sensors and actuators, *International Journal of Solids and Structures*, **38**, 1641–1655.

Huang X.-L. and Shen H.-S. (2004), Nonlinear vibration and dynamic response of functionally graded plates in thermal environments, *International Journal of Solids and Structures*, **41**, 2403–2427.

Huang X.-L. and Shen H.-S. (2006), Vibration and dynamic response of functionally graded plates with piezoelectric actuators in thermal environments, *Journal of Sound and Vibration*, **289**, 25–53.

Kim Y.-W. (2005), Temperature dependent vibration analysis of functionally graded rectangular plates, *Journal of Sound and Vibration*, **284**, 531–549.

Kitipornchai S., Yang J., and Liew K.M. (2004), Semi-analytical solution for nonlinear vibration of laminated FGM plates with geometric imperfections, *International Journal of Solids and Structures*, **41**, 2235–2257.

Li Q., Iu V.P., and Kou K.P. (2008), Three-dimensional vibration analysis of functionally graded material sandwich plates, *Journal of Sound and Vibration*, **311**, 498–515.

Park J.S. and Kim J.-H. (2006), Thermal postbuckling and vibration analyses of functionally graded plates, *Journal of Sound and Vibration*, **289**, 77–93.

Pearson C.E. (1986), *Numerical Methods in Engineering and Science*, Van Nostrand Reinhold Company, Inc., New York, NY.

Prakash T. and Ganapathi M. (2006), Asymmetric flexural vibration and thermoelastic stability of FGM circular plates using finite element method, *Composites Part B*, **37**, 642–649.

Praveen G.N. and Reddy J.N. (1998), Nonlinear transient thermoelastic analysis of functionally graded ceramic-metal plates, *International Journal of Solids and Structures*, **35**, 4457–4476.

Reddy J.N. (2000), Analysis of functionally graded plates, *International Journal for Numerical Methods in Engineering*, **47**, 663–684.

Shen H.-S. (2007), Nonlinear thermal bending response of FGM plates due to heat conduction, *Composites Part B*, **38**, 201–215.

Sundararajan N., Prakash T., and Ganapathi M. (2005), Nonlinear free flexural vibrations of functionally graded rectangular and skew plates under thermal environments, *Finite Element in Analysis and Design*, **42**, 152–168.

Vel S.S. and Batra R.C. (2004), Three dimensional exact solution for the vibration of functionally graded rectangular plates, *Journal of Sound and Vibration*, **272**, 703–730.

Wu L., Wang H., and Wang D. (2007), Dynamic stability analysis of FGM plates by the moving least squares differential quadrature method, *Composite Structures*, **77**, 383–394.

Xia X.-K. and Shen H.-S. (2008), Vibration of post-buckled sandwich plates with FGM face sheets in a thermal environment, *Journal of Sound and Vibration*, **314**, 254–274.

Yang T.Y. and Han A.D. (1983), Buckled plate vibrations and large amplitude vibrations using high-order triangular elements, *AIAA Journal*, **21**, 758–766.

Yang J. and Huang, X.-L. (2007), Nonlinear transient response of functionally graded plates with general imperfections in thermal environments, *Computer Methods in Applied Mechanics and Engineering*, **196**, 2619–2630.

Yang J. and Shen H.-S. (2001), Dynamic response of initially stressed functionally graded rectangular thin plates, *Composite Structures*, **54**, 497–508.

Yang J. and Shen H.-S. (2002), Vibration characteristics and transient response of shear deformable functionally graded plates in thermal environment, *Journal of Sound and Vibration*, **255**, 579–602.

Yang J., Kitipornchai S. and Liew K.M. (2003), Large amplitude vibration of thermo-electric-mechanically stressed FGM laminated plates, *Computational Methods in Applied Mechanics and Engineering*, **192**, 3861–3885.

5

Postbuckling of Shear Deformable FGM Shells

5.1 Introduction

Buckling of circular cylindrical shells has posed baffling problems to engineering for many years. This is due to the fact that large discrepancies between theoretical prediction and experimental results had been the focus of long debate in the case of compressive buckling of cylindrical shells. As von Kármán and Tsien (1941) argued at the time, geometrical nonlinearity must play an important part in the phenomenon of thin shell buckling. Donnell and Wan (1950) reported that the initial geometric imperfection has a significant effect on the buckling and postbuckling behavior of cylindrical shells subjected to axial compression. In their analysis, however, the membrane prebuckling state was assumed, like von Kármán and Tsien (1941) did and, therefore, the boundary conditions cannot be incorporated accurately. The importance of the nonlinear prebuckling deformations and its role in the buckling analysis of cylindrical shells has been discussed by Stein (1962, 1964). On the other hand, Koiter (1945) provided a general theory of the initial postbuckling behavior of elastic bodies under static conservative load. Following the pioneer works of Kármán and Tsien (1941), Donnell and Wan (1950), Stein (1962, 1964), and Koiter (1945, 1963), numerous researches have been made on this topic. The problem may be considered to be solved completely in the domain of homogeneous, isotropic elastic materials.

In the design of an FGM shell as well as a homogeneous, isotropic shell, it is of technical importance to examine its resistance to buckling under expected loading conditions. For that purpose, the determination of the buckling load alone is not sufficient in general, but it is further required to clarify the postbuckling behavior, that is, the behavior of the shell after passing through the buckling load. One of the reasons is to estimate the effect of practically unavoidable imperfections on the buckling load and the second is to evaluate the ultimate strength to exploit the load-carrying capacity of the shell structure.

Shen (2002, 2003) provided, respectively, the postbuckling solutions of FGM cylindrical shells under axial compression and external pressure in thermal environments. In his studies, a boundary layer theory for the shell buckling suggested by Shen and Chen (1988, 1990) was adopted. However, because the shells were considered as being relatively thin and therefore the transverse shear deformation was not accounted for. These works were then extended to the case of FGM hybrid cylindrical shells under axial compression, external pressure, or their combination in thermal environments by Shen (2005), and Shen and Noda (2005, 2007) based on a higher order shear deformation shell theory. Furthermore, Shahsiah and Eslami (2003a,b) performed the thermal buckling of FGM cylindrical shells under three types of thermal loading as uniform temperature rise, linear and nonlinear temperature variation through the thickness, based on the first-order shear deformation shell theory. Similar work was then done by Wu et al. (2005) based on classical shell theory of Donnell-type. Subsequently, the effect of initial geometric imperfections and applied actuator voltage on thermal buckling of FGM cylindrical shells was discussed by Mirzavand et al. (2005) and Mirzavand and Eslami (2006, 2007). As we all know, the imperfect cylindrical shell only has limit point load, which could be obtained by solving nonlinear governing equations as initial postbuckling or full postbuckling analysis. Based on a higher order shear deformation shell theory, Woo et al. (2005) presented Fourier series solutions for the thermomechanical postbuckling of FGM plates and shallow shells, from which the results for an initially heated cylindrical shell were obtained as a limiting case. Sofiyev (2007) studied linear free vibration and buckling of FGM laminated cylindrical shells by using Galerkin method. Sheng and Wang (2008) performed the linear vibration, buckling and dynamic stability of FGM cylindrical shells embedded in an elastic medium and subjected to mechanical and thermal loads based on the first-order shear deformation shell theory. In the above studies, however, the materials properties were virtually assumed to be temperature-independent (T-ID).

Moreover, Shen (2004, 2007) provided a thermal postbuckling analysis for FGM cylindrical shells under uniform temperature field or heat conduction based on classical shell theory and higher order shear deformation shell theory, respectively. In his study, the material properties were considered to be temperature-dependent (T-D) and the effect of imperfections on the thermal postbuckling response was reported. Recently, Kadoli and Ganesan (2006) presented linear thermal buckling and free vibration analysis for FGM cylindrical shells with clamped boundary conditions. In their analysis, the material properties were assumed to be temperature-dependent, and finite element equations based on the first-order shear deformation shell theory were formulated. On the other hand, due to the temperature gradient the shell is subjected to additional moments along with the membrane forces and the problem cannot be posed as an eigenvalue problem, i.e., no buckling temperature is evident, when the edges of the shell are simply supported. Therefore, the existing solutions for simply supported FGM shells subjected to transverse temperature variation, i.e., linear and/or nonlinear gradient through the thickness, may be physically incorrect.

5.2   Boundary Layer Theory for the Buckling of FGM Cylindrical Shells

It has been shown in Shen and Chen (1988, 1990) that in shell buckling, there exists a boundary layer phenomenon where prebuckling and buckling displacements vary rapidly. Consider a circular cylindrical shell with mean radius R, length L, and thickness h. The shell is referred to a coordinate system (X, Y, Z) in which X and Y are in the axial and circumferential directions of the shell and Z is in the direction of the inward normal to the middle surface (see Figure 5.1). The corresponding displacements are designated by $\overline{U}$, $\overline{V}$, and $\overline{W}$. $\overline{\Psi}_x$ and $\overline{\Psi}_y$ are the rotations of normals to the middle surface with respect to the Y- and X-axes, respectively. The origin of the coordinate system is located at the end of the shell on the middle plane. The shell is assumed to be relatively thick, geometrically imperfect. Denoting the initial geometric imperfection by $\overline{W}^*(X, Y)$, let $\overline{W}(X, Y)$ be the additional deflection and $\overline{F}(X, Y)$ be the stress function for the stress resultants defined by $\overline{N}_x = \overline{F}_{,yy}$, $\overline{N}_y = \overline{F}_{,xx}$, and $\overline{N}_{xy} = -\overline{F}_{,xy}$, where a comma denotes partial differentiation with respect to the corresponding coordinates.

5.2.1   Donnell Theory

In 1933, Donnell established his nonlinear theory of circular cylindrical shells, in connection with the analysis of torsional buckling of thin-walled tubes. Owing to its relative simplicity and practical accuracy, this theory has been the most widely used for analyzing both buckling and postbuckling problems, despite criticism concerning its applicability.

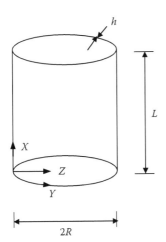

FIGURE 5.1
Geometry and coordinate system of a cylindrical shell.

The Donnell theory is based on the following assumptions (see Yamaki 1984):

1. The shell is sufficiently thin, i.e., $h/R \leq 1$ and $h/L \leq 1$.
2. The strains ε are sufficiently small, i.e., $\varepsilon \leq 1$, and, therefore, Hooke's law holds.
3. Straight lines normal to the undeformed middle surface remain straight and normal to the deformed middle surface with their length unchanged.
4. The normal stress acting in the direction normal to the middle surface may be neglected in comparison with the stresses acting in the direction parallel to the middle surface.
5. In-plane displacements $\overline{U}$ and $\overline{V}$ are infinitesimal, while normal displacement $\overline{W}$ is of the same order as the shell thickness, i.e., $|\overline{U}| \leq h$, $|\overline{V}| \leq h$, and $|\overline{W}| = O(h)$.
6. The derivatives of $\overline{W}$ are small, but their squares and products are of the same order as the strain ε considered. As a result

$$\left(\left| \frac{\partial \overline{W}}{\partial X} \right|, \left| \frac{\partial \overline{W}}{\partial Y} \right| \right) \leq 1, \quad \left[\left(\frac{\partial \overline{W}}{\partial X} \right)^2, \left(\frac{\partial \overline{W}}{\partial X} \frac{\partial \overline{W}}{\partial Y} \right), \left(\frac{\partial \overline{W}}{\partial Y} \right)^2 \right] = O(\varepsilon).$$

7. Curvature changes are small and the influence of $\overline{U}$ and $\overline{V}$ are negligible so that they can be represented by linear function of $\overline{W}$ only.

The assumptions (3) and (4) constitute the so-called Kirchhoff–Love hypotheses while those from (5) to (7) correspond to the shallow shell approximations applicable for deformations dominated by the normal displacement $\overline{W}$.

5.2.2 Generalized Kármán–Donnell-Type Nonlinear Equations

The classical nonlinear shell theory suggested by Donnell (1933) does not include the transverse shear and normal stresses and strains, which will play an important role in the analysis of composite shell structures. Reddy and Liu (1985) developed a simple higher order shear deformation shell theory, in which the transverse shear strains are assumed to be parabolically distributed across the shell thickness and which contains the same number of dependent unknowns as in the first-order shear deformation theory, and no shear correction factors are required. The governing equations of motion were derived by means of the principle of virtual work and can be expressed as

$$\frac{\partial \overline{N}_1}{\partial X} + \frac{\partial \overline{N}_6}{\partial Y} = I_1 \frac{\partial^2 \overline{U}}{\partial t^2} + \overline{I}_2 \frac{\partial^2 \overline{\Psi}_x}{\partial t^2} - c_1 I_4 \frac{\partial^3 \overline{W}}{\partial X \partial t^2}$$

$$\frac{\partial \overline{N}_6}{\partial X} + \frac{\partial \overline{N}_2}{\partial Y} = I_1 \frac{\partial^2 \overline{V}}{\partial t^2} + \overline{I}_2 \frac{\partial^2 \overline{\Psi}_y}{\partial t^2} - c_1 I_4 \frac{\partial^3 \overline{W}}{\partial Y \partial t^2}$$

$$\frac{\partial \overline{Q}_1}{\partial X} + \frac{\partial \overline{Q}_2}{\partial Y} + \frac{\partial}{\partial X}\left(\overline{N}_1 \frac{\partial \overline{W}}{\partial X} + \overline{N}_6 \frac{\partial \overline{W}}{\partial Y}\right) + \frac{\partial}{\partial Y}\left(\overline{N}_6 \frac{\partial \overline{W}}{\partial X} + \overline{N}_2 \frac{\partial \overline{W}}{\partial Y}\right) - \frac{\overline{N}_2}{R}$$

$$+ q - c_2\left(\frac{\partial \overline{R}_1}{\partial X} + \frac{\partial \overline{R}_2}{\partial Y}\right) + c_1\left(\frac{\partial^2 \overline{P}_1}{\partial X^2} + 2\frac{\partial^2 \overline{P}_6}{\partial X \partial Y} + \frac{\partial^2 \overline{P}_2}{\partial Y^2}\right)$$

$$= I_1 \frac{\partial^2 \overline{W}}{\partial t^2} - c_1^2 I_7 \frac{\partial^2}{\partial t^2}\left(\frac{\partial^2 \overline{W}}{\partial X^2} + \frac{\partial^2 \overline{W}}{\partial Y^2}\right) + c_1 I_4 \frac{\partial^2}{\partial t^2}\left(\frac{\partial \overline{U}}{\partial X} + \frac{\partial \overline{V}}{\partial Y}\right) + c_1 \overline{I}_5 \frac{\partial^2}{\partial t^2}\left(\frac{\partial \overline{\Psi}_x}{\partial X} + \frac{\partial \overline{\Psi}_y}{\partial Y}\right)$$

$$\frac{\partial \overline{M}_1}{\partial X} + \frac{\partial \overline{M}_6}{\partial Y} - \overline{Q}_1 + c_2 \overline{R}_1 - C_1\left(\frac{\partial \overline{P}_1}{\partial X} + \frac{\partial \overline{P}_6}{\partial Y}\right) = \overline{I}_2 \frac{\partial^2 \overline{U}}{\partial t^2} + \overline{I}_3 \frac{\partial^2 \overline{\Psi}_x}{\partial t^2} - c_1 \overline{I}_5 \frac{\partial^3 \overline{W}}{\partial X \partial t^2}$$

$$\frac{\partial \overline{M}_6}{\partial X} + \frac{\partial \overline{M}_2}{\partial Y} - \overline{Q}_2 + c_2 \overline{R}_2 - c_1\left(\frac{\partial \overline{P}_6}{\partial X} + \frac{\partial \overline{P}_2}{\partial Y}\right) = \overline{I}_2 \frac{\partial^2 \overline{V}}{\partial t^2} + \overline{I}_3 \frac{\partial^2 \overline{\Psi}_y}{\partial t^2} - c_1 \overline{I}_5 \frac{\partial^3 \overline{W}}{\partial Y \partial t^2} \qquad (5.1)$$

All symbols used in Equation 5.1 are defined as in Equation 1.26.

Introducing the reduced stiffness matrices $[A_{ij}^*]$, $[B_{ij}^*]$, $[D_{ij}^*]$, $[E_{ij}^*]$, $[F_{ij}^*]$, and $[H_{ij}^*]$ ($i, j = 1, 2, 6$) and by using the same manner described in Section 1.4, we derive the generalized Kármán–Donnell-type nonlinear equations of motion, which can be expressed in terms of a stress function $\overline{F}$, two rotations $\overline{\Psi}_x$ and $\overline{\Psi}_y$, and a transverse displacement $\overline{W}$, along with the initial geometric imperfection $\overline{W}^*$. These equations are then extended to the case of shear deformable FGM cylindrical shells including thermal effects.

$$\tilde{L}_{11}(\overline{W}) - \tilde{L}_{12}(\overline{\Psi}_x) - \tilde{L}_{13}(\overline{\Psi}_y) + \tilde{L}_{14}(\overline{F}) - \tilde{L}_{15}(\overline{N}^T) - \tilde{L}_{16}(\overline{M}^T) - \frac{1}{R}\overline{F}_{,xx}$$

$$= \tilde{L}(\overline{W} + \overline{W}^*, \overline{F}) + \tilde{L}_{17}(\ddot{\overline{W}}) - \overline{I}_8(\ddot{\overline{\Psi}}_{x,x} + \ddot{\overline{\Psi}}_{y,y}) + q \qquad (5.2)$$

$$\tilde{L}_{21}(\overline{F}) + \tilde{L}_{22}(\overline{\Psi}_x) + \tilde{L}_{23}(\overline{\Psi}_y) - \tilde{L}_{24}(\overline{W}) - \tilde{L}_{25}(\overline{N}^T) + \frac{1}{R}\overline{W}_{,xx} = -\frac{1}{2}\tilde{L}(\overline{W} + 2\overline{W}^*, \overline{W}) \quad (5.3)$$

$$\tilde{L}_{31}(\overline{W}) + \tilde{L}_{32}(\overline{\Psi}_x) - \tilde{L}_{33}(\overline{\Psi}_y) + \tilde{L}_{34}(\overline{F}) - \tilde{L}_{35}(\overline{N}^T) - \tilde{L}_{36}(\overline{S}^T) = \overline{I}_5\ddot{\overline{W}}_{,x} - \overline{I}_3\ddot{\overline{\Psi}}_x \qquad (5.4)$$

$$\tilde{L}_{41}(\overline{W}) - \tilde{L}_{42}(\overline{\Psi}_x) + \tilde{L}_{43}(\overline{\Psi}_y) + \tilde{L}_{44}(\overline{F}) - \tilde{L}_{45}(\overline{N}^T) - \tilde{L}_{46}(\overline{S}^T) = \overline{I}_5\ddot{\overline{W}}_{,y} - \overline{I}_3\ddot{\overline{\Psi}}_y \qquad (5.5)$$

Note that the geometric nonlinearity in the von Kármán sense is given in terms of $\tilde{L}()$ in Equations 5.2 and 5.3, and the other linear operators $\tilde{L}_{ij}()$ are defined by Equation 1.37, and the forces, moments and higher order moments caused by elevated temperature are defined by Equation 1.28.

5.2.3   Boundary Layer-Type Equations

Introducing the following dimensionless quantities

$$x = \pi X/L, \quad y = Y/R, \quad \beta = L/\pi R, \quad \overline{Z} = L^2/Rh,$$

$$\varepsilon = (\pi^2 R/L^2)[D_{11}^* D_{22}^* A_{11}^* A_{22}^*]^{1/4}$$

$$(W, W^*) = \varepsilon(\overline{W}, \overline{W}^*)/[D_{11}^* D_{22}^* A_{11}^* A_{22}^*]^{1/4}, \quad F = \varepsilon^2 \overline{F}/[D_{11}^* D_{22}^*]^{1/2}$$

$$(\Psi_x, \Psi_y) = \varepsilon^2 (\overline{\Psi}_x, \overline{\Psi}_y)(L/\pi)/[D_{11}^* D_{22}^* A_{11}^* A_{22}^*]^{1/4}$$

$$\gamma_{14} = [D_{22}^*/D_{11}^*]^{1/2}, \quad \gamma_{24} = [A_{11}^*/A_{22}^*]^{1/2}, \quad \gamma_5 = -A_{12}^*/A_{22}^*$$

$$N_x = \varepsilon^2 \overline{N}_k L^2/\pi^2 D_{11}^*, \quad (M_x, P_x) = \varepsilon^2 (\overline{M}_x, 4\overline{P}_x/3h^2)L^2/\pi^2 D_{11}^*[D_{11}^* D_{22}^* A_{11}^* A_{22}^*]^{1/4}$$

$$(\gamma_{T1}, \gamma_{T2}) = \left(A_x^T, A_y^T\right) R[A_{11}^* A_{22}^*/D_{11}^* D_{22}^*]^{1/4}$$

$$\tau = \frac{\pi t}{L}\sqrt{\frac{E_0}{\rho_0}}, \quad \gamma_{170} = -\frac{I_1 E_0 L^2}{\pi^2 \rho_0 D_{11}^*}, \quad \gamma_{171} = \frac{c_1 E_0 (I_5 I_1 - I_4 I_2)}{\rho_0 I_1 D_{11}^*}$$

$$(\gamma_{80}, \gamma_{90}, \gamma_{10}) = (-\bar{I}_8, \bar{I}_5, -\bar{I}_3)\frac{E_0}{\rho_0 D_{11}^*},$$

$$\lambda_q = q(3)^{3/4} LR^{3/2}[A_{11}^* A_{22}^*]^{1/8}/4\pi[D_{11}^* D_{22}^*]^{3/8} \tag{5.6}$$

in which E_0 and ρ_0 are the reference values. The nonlinear equations (Equations 5.11 through 5.14) may then be written in dimensionless form as

$$\varepsilon^2 L_{11}(W) - \varepsilon L_{12}(\Psi_x) - \varepsilon L_{13}(\Psi_y) + \varepsilon\gamma_{14}L_{14}(F) - \varepsilon L_{15}(N^T) - \varepsilon L_{16}(M^T) - \gamma_{14}F_{,xx}$$

$$= \gamma_{14}\beta^2 L(W + W^*, F) + \varepsilon^2 L_{17}(\ddot{W}) + \varepsilon\gamma_{80}\left(\frac{\partial\ddot{\Psi}_x}{\partial x} + \beta\frac{\partial\ddot{\Psi}_y}{\partial y}\right) + \gamma_{14}\frac{4}{3}(3)^{1/4}\lambda_q\varepsilon^{3/2} \tag{5.7}$$

$$L_{21}(F) + \gamma_{24}L_{22}(\Psi_x) + \gamma_{24}L_{23}(\Psi_y) - \varepsilon\gamma_{24}L_{24}(W) - L_{25}(N^T) + \gamma_{24}W_{,xx}$$

$$= -\frac{1}{2}\gamma_{24}\beta^2 L(W + 2W^*, W) \tag{5.8}$$

$$\varepsilon L_{31}(W) + L_{32}(\Psi_x) - L_{33}(\Psi_y) + \gamma_{14}L_{34}(F) - L_{35}(N^T) - L_{36}(S^T)$$

$$= \varepsilon\gamma_{90}\frac{\partial\ddot{W}}{\partial x} + \gamma_{10}\ddot{\Psi}_x \tag{5.9}$$

$$\varepsilon L_{41}(W) - L_{42}(\Psi_x) + L_{43}(\Psi_y) + \gamma_{14}L_{44}(F) - L_{45}(N^T) - L_{46}(S^T)$$

$$= \varepsilon\gamma_{90}\beta\frac{\partial\ddot{W}}{\partial y} + \gamma_{10}\ddot{\Psi}_y \tag{5.10}$$

where all the dimensionless operators $L_{ij}()$ and $L()$ are defined by Equations 2.14 and 4.12.

In Equation 5.6, we introduce an important parameter ε. For most FGMs $[D_{11}^* D_{22}^* A_{11}^* A_{22}^*]^{1/4} \cong 0.3h$. Furthermore, when $\overline{Z} = (L^2/Rh) > 2.96$, then from Equation 5.6 $\varepsilon < 1$. In particular, for homogeneous isotropic cylindrical shells,

$\varepsilon = \pi^2/\overline{Z}_B\sqrt{12}$, where $\overline{Z}_B = (L^2/Rh)[1 - \nu^2]^{1/2}$ is the Batdorf shell parameter, which should be greater than 2.85 in the case of classical linear buckling analysis (Batdorf 1947). In practice, the shell structure will have $\overline{Z} > 10$, so that we always have $\varepsilon \ll 1$. When $\varepsilon < 1$, Equations 5.7 through 5.10 are of the boundary layer type. We will show that, in Section 5.3, (1) the nonlinear prebuckling deformations, large deflections in the postbuckling range and initial geometric imperfections of the shell could be considered simultaneously; (2) the full postbuckling and imperfection sensitivity analysis could be performed; and (3) it is no longer to guess the forms of solutions which will be obtained step by step, and such solutions satisfy both governing equations and boundary conditions accurately in the asymptotic sense.

5.3 Postbuckling Behavior of FGM Cylindrical Shells under Axial Compression

The buckling and postbuckling behavior of axially compressed FGM cylindrical shells is a major concern for structural instability and it represents one of the best known examples of the very complicated stability behavior of shell structures. It is assumed that the effective Young's modulus E_f, thermal expansion coefficient α_f, and thermal conductivity κ_f are assumed to be functions of temperature, so that E_f, α_f, and κ_f are both temperature- and position-dependent. The Poisson ratio ν_f depends weakly on temperature change and is assumed to be a constant. We assume the volume fraction V_m follows a simple power law. According to rule of mixture, we have

$$E_f(Z, T) = [E_m(T) - E_c(T)]\left(\frac{Z - t_1}{t_2 - t_1}\right)^N + E_c(T) \tag{5.11a}$$

$$\alpha_f(Z, T) = [\alpha_m(T) - \alpha_c(T)]\left(\frac{Z - t_1}{t_2 - t_1}\right)^N + \alpha_c(T) \tag{5.11b}$$

$$\kappa_f(Z, T) = [\kappa_m(T) - \kappa_c(T)]\left(\frac{Z - t_1}{t_2 - t_1}\right)^N + \kappa_c(T) \tag{5.11c}$$

It is evident that when $Z = t_1$, $E_f = E_c$, $\alpha_f = \alpha_c$, and $\kappa_f = \kappa_c$, and when $Z = t_2$, $E_f = E_m$, $\alpha_f = \alpha_m$, and $\kappa_f = \kappa_m$.

We assume that the temperature variation occurs in the thickness direction only and one-dimensional temperature field is assumed to be constant in the XY plane of the shell. In such a case, the temperature distribution along the thickness can be obtained by solving a steady-state heat transfer equation

$$-\frac{d}{dZ}\left[\kappa\frac{dT}{dZ}\right] = 0 \tag{5.12}$$

Equation 5.12 is solved by imposing the boundary conditions $T = T_U$ at $Z = t_1$ and $T = T_L$ at $Z = t_2$, and the solution, by means of polynomial series, is

$$T_f = T_U + (T_L - T_U)\eta(Z) \tag{5.13}$$

in which

$$\eta(Z) = \frac{1}{C}\left[\left(\frac{Z - t_1}{t_2 - t_1}\right) - \frac{\kappa_{mc}}{(N+1)\kappa_c}\left(\frac{Z - t_1}{t_2 - t_1}\right)^{N+1} + \frac{\kappa_{mc}^2}{(2N+1)\kappa_c^2}\left(\frac{Z - t_1}{t_2 - t_1}\right)^{2N+1} \right.$$
$$- \frac{\kappa_{mc}^3}{(3N+1)\kappa_c^3}\left(\frac{Z - t_1}{t_2 - t_1}\right)^{3N+1} + \frac{\kappa_{mc}^4}{(4N+1)\kappa_c^4}\left(\frac{Z - t_1}{t_2 - t_1}\right)^{4N+1}$$
$$\left. - \frac{\kappa_{mc}^5}{(5N+1)\kappa_c^5}\left(\frac{Z - t_1}{t_2 - t_1}\right)^{5N+1} \right] \tag{5.14a}$$

$$C = 1 - \frac{\kappa_{mc}}{(N+1)\kappa_c} + \frac{\kappa_{mc}^2}{(2N+1)\kappa_c^2} - \frac{\kappa_{mc}^3}{(3N+1)\kappa_c^3} + \frac{\kappa_{mc}^4}{(4N+1)\kappa_c^4}$$
$$- \frac{\kappa_{mc}^5}{(5N+1)\kappa_c^5} \tag{5.14b}$$

where $\kappa_{mc} = \kappa_m - \kappa_c$.

In the case of axial compression, since there is no transverse dynamic load applied, the nonlinear governing Equations 5.2 through 5.5 can be written in simple forms as

$$\tilde{L}_{11}(\overline{W}) - \tilde{L}_{12}(\overline{\Psi}_x) - \tilde{L}_{13}(\overline{\Psi}_y) + \tilde{L}_{14}(\overline{F}) - \tilde{L}_{15}(\overline{N}^T) - \tilde{L}_{16}(\overline{M}^T) - \frac{1}{R}\overline{F}_{,xx}$$
$$= \tilde{L}(\overline{W} + \overline{W}^*, \overline{F}) \tag{5.15}$$

$$\tilde{L}_{21}(\overline{F}) + \tilde{L}_{22}(\overline{\Psi}_x) + \tilde{L}_{23}(\overline{\Psi}_y) - \tilde{L}_{24}(\overline{W}) - \tilde{L}_{25}(\overline{N}^T) + \frac{1}{R}\overline{W}_{,xx}$$
$$= -\frac{1}{2}\tilde{L}(\overline{W} + 2\overline{W}^*, \overline{W}) \tag{5.16}$$

$$\tilde{L}_{31}(\overline{W}) + \tilde{L}_{32}(\overline{\Psi}_x) - \tilde{L}_{33}(\overline{\Psi}_y) + \tilde{L}_{34}(\overline{F}) - \tilde{L}_{35}(\overline{N}^T) - \tilde{L}_{36}(\overline{S}^T) = 0 \tag{5.17}$$

$$\tilde{L}_{41}(\overline{W}) - \tilde{L}_{42}(\overline{\Psi}_x) + \tilde{L}_{43}(\overline{\Psi}_y) + \tilde{L}_{44}(\overline{F}) - \tilde{L}_{45}(\overline{N}^T) - \tilde{L}_{46}(\overline{S}^T) = 0 \tag{5.18}$$

The two end edges of the shell are assumed to be simply supported or clamped, so that the boundary conditions are

$X = 0, L$:

$$\overline{W} = \overline{\Psi}_y = 0, \quad \overline{M}_x = \overline{P}_x = 0 \text{ (simply supported)} \tag{5.19a}$$

$$\overline{W} = \overline{\Psi}_x = \overline{\Psi}_y = 0 \text{ (clamped)} \tag{5.19b}$$

$$\int_0^{2\pi R} \overline{N}_x dY + 2\pi R h \sigma_x = 0 \tag{5.19c}$$

where
  σ_x is the average axial compressive stress
  $\overline{M}_x$ is the bending moment
  $\overline{P}_x$ is the higher order moment as defined by Equation 1.32

Also, we have the closed (or periodicity) condition

$$\int_0^{2\pi R} \frac{\partial \overline{V}}{\partial Y} dY = 0 \tag{5.20a}$$

or

$$\int_0^{2\pi R} \left[\left(A_{22}^* \frac{\partial^2 \overline{F}}{\partial X^2} + A_{12}^* \frac{\partial^2 \overline{F}}{\partial Y^2} \right) + \left(B_{21}^* - \frac{4}{3h^2} E_{21}^* \right) \frac{\partial \overline{\Psi}_x}{\partial X} + \left(B_{22}^* - \frac{4}{3h^2} E_{22}^* \right) \frac{\partial \overline{\Psi}_y}{\partial Y} \right.$$
$$- \frac{4}{3h^2} \left(E_{21}^* \frac{\partial^2 \overline{W}}{\partial X^2} + E_{22}^* \frac{\partial^2 \overline{W}}{\partial Y^2} \right) + \frac{\overline{W}}{R} - \frac{1}{2} \left(\frac{\partial \overline{W}}{\partial Y} \right)^2 - \frac{\partial \overline{W}}{\partial Y} \frac{\partial \overline{W}^*}{\partial Y}$$
$$\left. - \left(A_{12}^* \overline{N}_x^T + A_{22}^* \overline{N}_y^T \right) \right] dY = 0 \tag{5.20b}$$

Because of Equation 5.20, the in-plane boundary condition $\overline{V} = 0$ (at $X = 0, L$) is not needed in Equation 5.19.
  The average end-shortening relationship is defined by

$$\frac{\Delta_x}{L} = -\frac{1}{2\pi RL} \int_0^{2\pi R} \int_0^L \frac{\partial \overline{U}}{\partial X} dXdY = -\frac{1}{2\pi RL} \int_0^{2\pi R} \int_0^L \left[\left(A_{11}^* \frac{\partial^2 \overline{F}}{\partial Y^2} + A_{12}^* \frac{\partial^2 \overline{F}}{\partial X^2} \right) \right.$$
$$+ \left(B_{11}^* - \frac{4}{3h^2} E_{11}^* \right) \frac{\partial \overline{\Psi}_x}{\partial X} + \left(B_{12}^* - \frac{4}{3h^2} E_{12}^* \right) \frac{\partial \overline{\Psi}_y}{\partial Y} - \frac{4}{3h^2} \left(E_{11}^* \frac{\partial^2 \overline{W}}{\partial X^2} + E_{12}^* \frac{\partial^2 \overline{W}}{\partial Y^2} \right)$$
$$\left. - \frac{1}{2} \left(\frac{\partial \overline{W}}{\partial X} \right)^2 - \frac{\partial \overline{W}}{\partial X} \frac{\partial \overline{W}^*}{\partial X} - \left(A_{11}^* \overline{N}_x^T + A_{12}^* \overline{N}_y^T \right) \right] dXdY \tag{5.21}$$

where Δ_x is the shell end-shortening displacements in the X-direction.
  Introducing dimensionless quantities of Equation 5.6, and

$$\lambda_p = \sigma_x / (2/Rh)[D_{11}^* D_{22}^* / A_{11}^* A_{22}^*]^{1/4},$$
$$\delta_p = (\Delta_x/L) / (2/R)[D_{11}^* D_{22}^* A_{11}^* A_{22}^*]^{1/4} \tag{5.22}$$

Also let

$$\begin{bmatrix} A_x^T \\ A_y^T \end{bmatrix} \Delta T = - \int_{t_1}^{t_2} \begin{bmatrix} A_x \\ A_y \end{bmatrix} \Delta T(Z) \mathrm{d}Z \tag{5.23}$$

where $\Delta T = (T_U + T_L - 2T_0)/2$. When $\Delta T = 0, A_x^T (= A_y^T)$ can be found in Appendix B, and if $\Delta T \neq 0$, then A_x^T can be found in Appendix D.

The nonlinear equations (Equations 5.15 through 5.18) may then be written in dimensionless form (boundary layer type) as

$$\varepsilon^2 L_{11}(W) - \varepsilon L_{12}(\Psi_x) - \varepsilon L_{13}(\Psi_y) + \varepsilon \gamma_{14} L_{14}(F) - \gamma_{14} F_{,xx} = \gamma_{14} \beta^2 L(W + W^*, F) \tag{5.24}$$

$$L_{21}(F) + \gamma_{24} L_{22}(\Psi_x) + \gamma_{24} L_{23}(\Psi_y) - \varepsilon \gamma_{24} L_{24}(W) + \gamma_{24} W_{,xx}$$

$$= -\frac{1}{2} \gamma_{24} \beta^2 L(W + 2W^*, W) \tag{5.25}$$

$$\varepsilon L_{31}(W) + L_{32}(\Psi_x) - L_{33}(\Psi_y) + \gamma_{14} L_{34}(F) = 0 \tag{5.26}$$

$$\varepsilon L_{41}(W) - L_{42}(\Psi_x) + L_{43}(\Psi_y) + \gamma_{14} L_{44}(F) = 0 \tag{5.27}$$

As mentioned in Section 5.2.3, all the dimensionless operators $L_{ij}()$ and $L()$ are defined by Equation 2.14.

The boundary conditions expressed by Equation 5.19 become

$x = 0, \pi$:

$$W = \Psi_y = 0, \quad M_x = P_x = 0 \text{ (simply supported)} \tag{5.28a}$$

$$W = \Psi_x = \Psi_y = 0 \text{ (clamped)} \tag{5.28b}$$

$$\frac{1}{2\pi} \int_0^{2\pi} \beta^2 \frac{\partial^2 F}{\partial y^2} \mathrm{d}y + 2\lambda_p \varepsilon = 0 \tag{5.28c}$$

and the closed condition becomes

$$\int_0^{2\pi} \left[\left(\frac{\partial^2 F}{\partial x^2} - \gamma_5 \beta^2 \frac{\partial^2 F}{\partial y^2} \right) + \gamma_{24} \left(\gamma_{220} \frac{\partial \Psi_x}{\partial x} + \gamma_{522} \beta \frac{\partial \Psi_y}{\partial y} \right) \right.$$

$$- \varepsilon \gamma_{24} \left(\gamma_{240} \frac{\partial^2 W}{\partial x^2} + \gamma_{622} \beta^2 \frac{\partial^2 W}{\partial y^2} \right) + \gamma_{24} W - \frac{1}{2} \gamma_{24} \beta^2 \left(\frac{\partial W}{\partial y} \right)^2$$

$$\left. - \gamma_{24} \beta^2 \frac{\partial W}{\partial y} \frac{\partial W^*}{\partial y} + \varepsilon (\gamma_{T2} - \gamma_5 \gamma_{T1}) \Delta T \right] \mathrm{d}y = 0 \tag{5.29}$$

It has been shown (Shen 2008a) that the effect of the boundary layer on the solution of a shell in compression is of the order ε^1, hence the unit end-shortening relationship may be written in dimensionless form as

$$
\delta_p = -\frac{1}{4\pi^2 \gamma_{24}} \varepsilon^{-1} \int_0^{2\pi} \int_0^\pi \left[\left(\gamma_{24}^2 \beta^2 \frac{\partial^2 F}{\partial y^2} - \gamma_5 \frac{\partial^2 F}{\partial x^2} \right) \right.
$$

$$
+ \gamma_{24} \left(\gamma_{511} \frac{\partial \Psi_x}{\partial x} + \gamma_{233} \beta \frac{\partial \Psi_y}{\partial y} \right) - \varepsilon \gamma_{24} \left(\gamma_{611} \frac{\partial^2 W}{\partial x^2} + \gamma_{244} \beta^2 \frac{\partial^2 W}{\partial y^2} \right)
$$

$$
\left. -\frac{1}{2} \gamma_{24} \left(\frac{\partial W}{\partial x} \right)^2 - \gamma_{24} \frac{\partial W}{\partial x} \frac{\partial W^*}{\partial x} + \varepsilon (\gamma_{24}^2 \gamma_{T1} - \gamma_5 \gamma_{T2}) \Delta T \right] dx\, dy \qquad (5.30)
$$

By virtue of the fact that $\Delta T[= (T_U + T_L - 2T_0)/2 = (\Delta T_U + \Delta T_L)/2]$ is assumed to be uniform, the thermal effects in Equations 5.15 through 5.18 vanish, but terms in ΔT intervene in Equations 5.29 and 5.30.

From Equations 5.15 through 5.30, one can determine the postbuckling behavior of perfect and imperfect FGM cylindrical shells subjected to axial compression in thermal environments by means of a singular perturbation technique. The essence of this procedure, in the present case, is to assume that

$$
W = w(x,y,\varepsilon) + \tilde{W}(x,\xi,y,\varepsilon) + \hat{W}(x,\zeta,y,\varepsilon)
$$

$$
F = f(x,y,\varepsilon) + \tilde{F}(x,\xi,y,\varepsilon) + \hat{F}(x,\zeta,y,\varepsilon)
$$

$$
\Psi_x = \psi_x(x,y,\varepsilon) + \tilde{\Psi}_x(x,\xi,y,\varepsilon) + \hat{\Psi}_x(x,\zeta,y,\varepsilon) \qquad (5.31)
$$

$$
\Psi_y = \psi_y(x,y,\varepsilon) + \tilde{\Psi}_y(x,\xi,y,\varepsilon) + \hat{\Psi}_y(x,\zeta,y,\varepsilon)
$$

where ε is a small perturbation parameter (provided $\overline{Z} > 2.96$) as defined in Equation 5.6 and $w(x,y,\varepsilon)$, $f(x,y,\varepsilon)$, $\psi_x(x,y,\varepsilon)$, $\psi_y(x,y,\varepsilon)$ are called outer solutions or regular solutions of the shell, $\tilde{W}(x,\xi,y,\varepsilon)$, $\tilde{F}(x,\xi,y,\varepsilon)$, $\tilde{\Psi}_x(x,\xi,y,\varepsilon)$, $\tilde{\Psi}_y(x,\xi,y,\varepsilon)$, and $\hat{W}(x,\zeta,y,\varepsilon)$, $\hat{F}(x,\zeta,y,\varepsilon)$, $\hat{\Psi}_x(x,\zeta,y,\varepsilon)$, $\hat{\Psi}_y(x,\zeta,y,\varepsilon)$ are the boundary layer solutions near the $x = 0$ and $x = \pi$ edges, respectively, and ξ and ζ are the boundary layer variables, defined by

$$
\xi = x/\sqrt{\varepsilon}, \quad \zeta = (\pi - x)/\sqrt{\varepsilon} \qquad (5.32)
$$

This means for homogeneous isotropic cylindrical shells the width of the boundary layers is of order $\sqrt{Rt}$. In Equation 5.31, the regular and boundary layer solutions are expressed in the form of perturbation expansions as

$$
w(x,y,\varepsilon) = \sum_{j=1} \varepsilon^j w_j(x,y), \quad f(x,y,\varepsilon) = \sum_{j=0} \varepsilon^j f_j(x,y)
$$

$$
\psi_x(x,y,\varepsilon) = \sum_{j=1} \varepsilon^j (\psi_x)_j(x,y), \quad \psi_y(x,y,\varepsilon) = \sum_{j=1} \varepsilon^j (\psi_y)_j(x,y) \qquad (5.33a)
$$

$$\tilde{W}(x,\xi,y,\varepsilon) = \sum_{j=0} \varepsilon^{j+1} \tilde{W}_{j+1}(x,\xi,y), \quad \tilde{F}(x,\xi,y,\varepsilon) = \sum_{j=0} \varepsilon^{j+2} \tilde{F}_{j+2}(x,\xi,y)$$

$$\tilde{\Psi}_x(x,\xi,y,\varepsilon) = \sum_{j=0} \varepsilon^{j+3/2} (\tilde{\Psi}_x)_{j+3/2}(x,\xi,y), \quad \tilde{\Psi}_y(x,\xi,y,\varepsilon) = \sum_{j=0} \varepsilon^{j+2} (\tilde{\Psi}_y)_{j+2}(x,\xi,y)$$

$$(5.33b)$$

$$\hat{W}(x,\zeta,y,\varepsilon) = \sum_{j=0} \varepsilon^{j+1} \hat{W}_{j+1}(x,\zeta,y), \quad \hat{F}(x,\zeta,y,\varepsilon) = \sum_{j=0} \varepsilon^{j+2} \hat{F}_{j+2}(x,\zeta,y)$$

$$\hat{\Psi}_x(x,\zeta,y,\varepsilon) = \sum_{j=0} \varepsilon^{j+3/2} (\hat{\Psi}_x)_{j+3/2}(x,\zeta,y), \quad \hat{\Psi}_y(x,\zeta,y,\varepsilon) = \sum_{j=0} \varepsilon^{j+2} (\hat{\Psi}_y)_{j+2}(x,\zeta,y)$$

$$(5.33c)$$

Substituting Equation 5.31 into Equations 5.24 through 5.27 and collecting the terms of the same order of ε, we can obtain three sets of perturbation equations for the regular and boundary layer solutions, respectively, e.g., the regular solutions $w(x, y)$, $f(x, y)$ $\psi_x(x, y)$, and $\psi_y(x, y)$ should satisfy

$$O(\varepsilon^0): -\gamma_{14}(f_0)_{,xx} = \gamma_{14}\beta^2 L(w_0, f_0) \tag{5.34a}$$

$$L_{21}(f_0) + \gamma_{24}L_{22}(\psi_{x0}) + \gamma_{24}L_{23}(\psi_{y0}) + \gamma_{24}(w_0)_{,xx} = -\frac{1}{2}\gamma_{24}\beta^2 L(w_0, w_0) \tag{5.34b}$$

$$L_{32}(\psi_{x0}) - L_{33}(\psi_{y0}) + \gamma_{14}L_{34}(f_0) = 0 \tag{5.34c}$$

$$-L_{42}(\psi_{x0}) + L_{43}(\psi_{y0}) + \gamma_{14}L_{44}(f_0) = 0 \tag{5.34d}$$

$$O(\varepsilon^1): -L_{12}(\psi_{x0}) - L_{13}(\psi_{y0}) + \gamma_{14}L_{14}(f_0) - \gamma_{14}(f_1)_{,xx}$$

$$= \gamma_{14}\beta^2 [L(w_1, f_0) + L(w_0, f_1)] \tag{5.35a}$$

$$L_{21}(f_1) + \gamma_{24}L_{22}(\psi_{x1}) + \gamma_{24}L_{23}(\psi_{y1}) - \gamma_{24}L_{24}(w_0) + \gamma_{24}(w_1)_{,xx}$$

$$= -\frac{1}{2}\gamma_{24}\beta^2 L(w_0, w_1) \tag{5.35b}$$

$$L_{31}(w_0) + L_{32}(\psi_{x1}) - L_{33}(\psi_{y1}) + \gamma_{14}L_{34}(f_1) = 0 \tag{5.35c}$$

$$L_{41}(w_0) - L_{42}(\psi_{x1}) + L_{43}(\psi_{y1}) + \gamma_{14}L_{44}(f_1) = 0 \tag{5.35d}$$

$$O(\varepsilon^2): L_{11}(w_0) - L_{12}(\psi_{x1}) - L_{13}(\psi_{y1}) + \gamma_{14}L_{12}(f_1) - \gamma_{14}(f_2)_{,xx}$$

$$= \gamma_{14}\beta^2 [L(w_2 + W^*, f_0) + L(w_1, f_1) + L(w_0, f_2)] \tag{5.36a}$$

$$L_{21}(f_2) + \gamma_{24}L_{22}(\psi_{x2}) + \gamma_{24}L_{23}(\psi_{y2}) - \gamma_{24}L_{24}(w_1) + \gamma_{24}(w_2)_{,xx}$$

$$= -\frac{1}{2}\gamma_{24}\beta^2 [L(w_1, w_1) + L(w_2 + 2W^*, w_0)] \tag{5.36b}$$

$$L_{31}(w_1) + L_{32}(\psi_{x2}) - L_{33}(\psi_{y2}) + \gamma_{14}L_{34}(f_2) = 0 \tag{5.36c}$$

$$L_{41}(w_1) - L_{42}(\psi_{x2}) + L_{43}(\psi_{y2}) + \gamma_{14}L_{44}(f_2) = 0 \tag{5.36c}$$

Because of the definition of W given in Equation 5.6, we assume that $w_0 = 0$ and $w_1 = A_{00}^{(1)}$, along with $\psi_{x0} = \psi_{y0} = \psi_{x1} = \psi_{y1} = 0$, $f_0 = -B_{00}^{(0)} y^2/2$, and $f_1 = -B_{00}^{(1)} y^2/2$. The initial buckling mode is assumed to have the form

$$w_2(x, y) = A_{11}^{(2)} \sin mx \sin ny + A_{02}^{(2)} \cos 2ny \tag{5.37}$$

and the initial geometric imperfection is assumed to have the form

$$W^*(x, y, \varepsilon) = \varepsilon^2 a_{11}^* \sin mx \sin ny = \varepsilon^2 \mu A_{11}^{(2)} \sin mx \sin ny \tag{5.38}$$

where $\mu = a_{11}^*/A_{11}^{(2)}$ is the imperfection parameter. From Equation 5.36 one has

$$f_2 = -B_{00}^{(2)} \frac{y^2}{2} + B_{11}^{(2)} \sin mx \sin ny$$

$$\psi_{x2} = C_{11}^{(2)} \cos mx \sin ny \tag{5.39}$$

$$\psi_{y2} = D_{11}^{(2)} \sin mx \cos ny$$

and

$$B_{11}^{(2)} = \frac{\gamma_{24} m^2}{g_{06}} A_{11}^{(2)}, \quad C_{11}^{(2)} = \gamma_{14} m \frac{g_{01}}{g_{00}} B_{11}^{(2)}, \quad D_{11}^{(2)} = \gamma_{14} n \beta \frac{g_{02}}{g_{00}} B_{11}^{(2)},$$

$$\beta^2 B_{00}^{(0)} = \frac{\gamma_{24} m^2}{(1 + \mu) g_{06}} \tag{5.40}$$

where g_{ij} are given in detail in Appendix L.

Now we turn our attention to the boundary layer solutions $\tilde{W}(x, \xi, y)$ and $\tilde{F}(x, \xi, y)$ which should satisfy boundary layer equations of each order, for example

$$O(\varepsilon^2): \gamma_{110} \frac{\partial^4 \tilde{W}_1}{\partial \xi^4} - \gamma_{120} \frac{\partial^3 \tilde{\Psi}_{x(3/2)}}{\partial \xi^3} + \gamma_{14} \gamma_{140} \frac{\partial^4 \tilde{F}_2}{\partial \xi^4} - \gamma_{14} \frac{\partial^2 \tilde{F}_2}{\partial \xi^2} = 0 \tag{5.41a}$$

$$\frac{\partial^4 \tilde{F}_2}{\partial \xi^4} + \gamma_{24} \gamma_{220} \frac{\partial^3 \tilde{\Psi}_{x(3/2)}}{\partial \xi^3} - \gamma_{24} \gamma_{240} \frac{\partial^4 \tilde{W}_1}{\partial \xi^4} + \gamma_{24} \frac{\partial^2 \tilde{W}_1}{\partial \xi^2} = 0 \tag{5.41b}$$

$$\gamma_{310} \frac{\partial^3 \tilde{W}_1}{\partial \xi^3} - \gamma_{320} \frac{\partial^2 \tilde{\Psi}_{x(3/2)}}{\partial \xi^2} + \gamma_{14} \gamma_{220} \frac{\partial^3 \tilde{F}_2}{\partial \xi^3} = 0 \tag{5.41c}$$

$$\gamma_{430} \frac{\partial^2 \tilde{\Psi}_{y2}}{\partial \xi^2} = 0 \tag{5.41d}$$

which leads to

$$\frac{\partial^4 \tilde{W}_1}{\partial \xi^4} + 2c \frac{\partial^2 \tilde{W}_1}{\partial \xi^2} + b^2 \tilde{W}_1 = 0 \tag{5.42}$$

where

$$c = \frac{\gamma_{14}\gamma_{24}\gamma_{320}g_{15}}{2g_{16}}, \quad b = \left(\frac{\gamma_{14}\gamma_{24}\gamma_{320}^2}{g_{16}}\right)^{1/2} \tag{5.43}$$

where g_{15} and g_{16} are also given in detail in Appendix L. The solution of Equation 5.42 can be written as

$$\tilde{W}_1 = \left(a_{10}^{(1)} \sin \phi\xi + a_{01}^{(1)} \cos \phi\xi\right)e^{-\vartheta\xi} \tag{5.44a}$$

and

$$\tilde{F}_2 = \left(b_{10}^{(2)} \sin \phi\xi + b_{01}^{(2)} \cos \phi\xi\right)e^{-\vartheta\xi} \tag{5.44b}$$

$$\tilde{\Psi}_{x(3/2)} = \left(c_{10}^{(3/2)} \sin \phi\xi + c_{01}^{(3/2)} \cos \phi\xi\right)e^{-\vartheta\xi} \tag{5.44c}$$

$$\tilde{\Psi}_{y2} = 0 \tag{5.44d}$$

where

$$\vartheta = \left[\frac{b-c}{2}\right]^{1/2}, \quad \phi = \left[\frac{b+c}{2}\right]^{1/2} \tag{5.45}$$

Similarly, the boundary layer solutions $\hat{W}(x, \zeta, y)$ and $\hat{F}(x, \zeta, y)$ can be obtained in the same manner.

Then matching the regular solutions with the boundary layer solutions at the each end of the shell, e.g., let $(w_1 + \tilde{W}_1) = (w_1 + \tilde{W}_1)_{,x} = 0$ and $(w_1 + \hat{W}_1) = (w_1 + \hat{W}_1)_{,x} = 0$ at $x = 0$ and $x = \pi$ edges, respectively, so that the asymptotic solutions satisfying the clamped boundary conditions are constructed as

$$\begin{aligned}
W = \varepsilon &\left[A_{00}^{(1)} - A_{00}^{(1)}\left(a_{01}^{(1)}\cos\phi\frac{x}{\sqrt{\varepsilon}} + a_{10}^{(1)}\sin\phi\frac{x}{\sqrt{\varepsilon}}\right)\exp\left(-\vartheta\frac{x}{\sqrt{\varepsilon}}\right)\right.\\
&\left. - A_{00}^{(1)}\left(a_{01}^{(1)}\cos\phi\frac{\pi-x}{\sqrt{\varepsilon}} + a_{10}^{(1)}\sin\phi\frac{\pi-x}{\sqrt{\varepsilon}}\right)\exp\left(-\vartheta\frac{\pi-x}{\sqrt{\varepsilon}}\right)\right]\\
&+ \varepsilon^2\left[A_{11}^{(2)}\sin mx \sin ny + A_{02}^{(2)}\cos 2ny \quad - \left(A_{02}^{(2)}\cos 2ny\right)\right.\\
&\times \left(a_{01}^{(1)}\cos\phi\frac{x}{\sqrt{\varepsilon}} + a_{10}^{(1)}\sin\phi\frac{x}{\sqrt{\varepsilon}}\right)\exp\left(-\vartheta\frac{x}{\sqrt{\varepsilon}}\right)\\
&\left. - \left(A_{02}^{(2)}\cos 2ny\right)\left(a_{01}^{(1)}\cos\phi\frac{\pi-x}{\sqrt{\varepsilon}} + a_{10}^{(1)}\sin\phi\frac{\pi-x}{\sqrt{\varepsilon}}\right)\exp\left(-\vartheta\frac{\pi-x}{\sqrt{\varepsilon}}\right)\right]\\
&+ \varepsilon^3\left[A_{11}^{(3)}\sin mx \sin ny + A_{02}^{(3)}\cos 2ny\right] + \varepsilon^4\left[A_{00}^{(4)} + A_{11}^{(4)}\sin mx \sin ny\right.\\
&\left. + A_{20}^{(4)}\cos 2mx + A_{02}^{(4)}\cos 2ny + A_{13}^{(4)}\sin mx \sin 3ny + A_{04}^{(4)}\cos 4ny\right] + O(\varepsilon^5) \tag{5.46}
\end{aligned}$$

$$F = -B_{00}^{(0)}\frac{y^2}{2} + \varepsilon\left[-B_{00}^{(1)}\frac{y^2}{2}\right] + \varepsilon^2\left[-B_{00}^{(2)}\frac{y^2}{2} + B_{11}^{(2)}\sin mx\sin ny\right.$$

$$+ A_{00}^{(1)}\left(b_{01}^{(2)}\cos\phi\,\frac{x}{\sqrt{\varepsilon}} + b_{10}^{(2)}\sin\phi\,\frac{x}{\sqrt{\varepsilon}}\right)\exp\left(-\vartheta\,\frac{x}{\sqrt{\varepsilon}}\right)$$

$$\left.+ A_{00}^{(1)}\left(b_{01}^{(2)}\cos\phi\,\frac{\pi-x}{\sqrt{\varepsilon}} + b_{10}^{(2)}\sin\phi\,\frac{\pi-x}{\sqrt{\varepsilon}}\right)\exp\left(-\vartheta\,\frac{\pi-x}{\sqrt{\varepsilon}}\right)\right]$$

$$+ \varepsilon^3\left[-B_{00}^{(3)}\frac{y^2}{2} + B_{02}^{(3)}\cos 2ny + \left(A_{02}^{(2)}\cos 2ny\right)\left(b_{01}^{(3)}\cos\phi\,\frac{x}{\sqrt{\varepsilon}} + b_{10}^{(3)}\sin\phi\,\frac{x}{\sqrt{\varepsilon}}\right)\exp\left(-\vartheta\,\frac{x}{\sqrt{\varepsilon}}\right)\right.$$

$$\left.+ \left(A_{02}^{(2)}\cos 2ny\right)\left(b_{01}^{(3)}\cos\phi\,\frac{\pi-x}{\sqrt{\varepsilon}} + b_{10}^{(3)}\sin\phi\,\frac{\pi-x}{\sqrt{\varepsilon}}\right)\exp\left(-\vartheta\,\frac{\pi-x}{\sqrt{\varepsilon}}\right)\right]$$

$$+ \varepsilon^4\left[-B_{00}^{(4)}\frac{y^2}{2} + B_{20}^{(4)}\cos 2mx + B_{02}^{(4)}\cos 2ny + B_{13}^{(4)}\sin mx\sin 3ny\right] + O(\varepsilon^5) \tag{5.47}$$

$$\Psi_x = \varepsilon^{3/2}\left[A_{00}^{(1)}c_{10}^{(3/2)}\sin\phi\,\frac{x}{\sqrt{\varepsilon}}\exp\left(-\vartheta\,\frac{x}{\sqrt{\varepsilon}}\right) + A_{00}^{(1)}c_{10}^{(3/2)}\sin\phi\,\frac{\pi-x}{\sqrt{\varepsilon}}\exp\left(-\vartheta\,\frac{\pi-x}{\sqrt{\varepsilon}}\right)\right]$$

$$+ \varepsilon^2\left[C_{11}^{(2)}\cos mx\sin ny\right] + \varepsilon^{5/2}\left[\left(A_{02}^{(2)}\cos 2ny\right)c_{10}^{(5/2)}\sin\phi\,\frac{x}{\sqrt{\varepsilon}}\exp\left(-\vartheta\,\frac{x}{\sqrt{\varepsilon}}\right)\right.$$

$$\left.+ \left(A_{02}^{(2)}\cos 2ny\right)c_{10}^{(5/2)}\sin\phi\,\frac{\pi-x}{\sqrt{\varepsilon}}\exp\left(-\vartheta\,\frac{\pi-x}{\sqrt{\varepsilon}}\right)\right] + \varepsilon^3\left[C_{11}^{(3)}\cos mx\sin ny\right]$$

$$+ \varepsilon^4\left[C_{11}^{(4)}\cos mx\sin ny + C_{20}^{(4)}\sin mx + C_{13}^{(4)}\cos mx\sin 3ny\right] + O(\varepsilon^5) \tag{5.48}$$

$$\Psi_y = \varepsilon^2\left[D_{11}^{(2)}\sin mx\cos ny\right] + \varepsilon^3\left[D_{11}^{(3)}\sin mx\cos ny + D_{02}^{(3)}\sin 2ny\right]$$

$$+ \varepsilon^4\left[D_{11}^{(4)}\sin mx\cos ny + D_{02}^{(4)}\sin 2ny + D_{13}^{(4)}\sin mx\cos 3ny\right] + O(\varepsilon^5) \tag{5.49}$$

Note that all of the coefficients in Equations 5.46 through 5.49 are related and can be expressed in terms of $A_{11}^{(2)}$, as shown in Equation 5.40 for $B_{11}^{(2)}$, except for $A_{00}^{(j)}$ ($j=1-4$) which may be determined by substituting Equations 5.46 through 5.49 into closed condition (Equation 5.29), and then we obtain

$$A_{00}^{(1)} = -2\frac{\gamma_5}{\gamma_{24}}\lambda_p - \frac{1}{\gamma_{24}}[\gamma_{T2} - \gamma_5\gamma_{T1}]\Delta T$$

$$A_{00}^{(2)} = A_{00}^{(3)} = 0 \tag{5.50}$$

$$A_{00}^{(4)} = \frac{1}{8}n^2\beta^2(1+2\mu)\left(A_{11}^{(2)}\right)^2 + n^2\beta^2\left(A_{02}^{(2)}\right)^2$$

Next, upon substitution of Equations 5.46 through 5.49 into the boundary condition (Equation 5.28c) and into Equation 5.30, the postbuckling equilibrium paths can be written as

$$\lambda_p = \lambda_p^{(0)} - \lambda_p^{(2)}\left(A_{11}^{(2)}\varepsilon\right)^2 + \lambda_p^{(4)}\left(A_{11}^{(2)}\varepsilon\right)^4 + \cdots \tag{5.51}$$

and

$$\delta_p = \delta_x^{(0)} - \delta_x^{(T)} + \delta_x^{(2)}\left(A_{11}^{(2)}\varepsilon\right)^2 + \delta_x^{(4)}\left(A_{11}^{(2)}\varepsilon\right)^4 + \cdots \tag{5.52}$$

In Equations 5.51 and 5.52, $\left(A_{11}^{(2)}\varepsilon\right)$ is taken as the second perturbation parameter relating to the dimensionless maximum deflection. From Equation 5.46, by taking $(x, y) = (\pi/2m, \pi/2n)$, one has

$$A_{11}^{(2)}\varepsilon = W_m - \Theta_2 W_m^2 + \cdots \qquad (5.53)$$

where W_m is the dimensionless form of maximum deflection of the shell that can be written as

$$W_m = \frac{1}{C_3}\left[\frac{h}{[D_{11}^*D_{22}^*A_{11}^*A_{22}^*]^{1/4}}\frac{\overline{W}}{h} + \Theta_1\right] \qquad (5.54)$$

All symbols used in Equations 5.51 through 5.54 are also described in detail in Appendix L. The perturbation scheme described here is quite different from that of traditional one (Koiter 1963, Dym and Hoff 1968). In the present analysis, ε is definitely a small perturbation parameter, but in the second step $(A_{11}^{(2)}\varepsilon)$ may be large. In contrast, the perturbation parameter ε is defined as a characteristic amplitude nondimensionalized with respect to the shell thickness (Dym and Hoff 1968), which is no longer a small perturbation parameter in the deep postbuckling range when the shell deflection is suffi-ciently large, e.g., $\overline{W}/h > 1$, and in such a case the solution is questionable.

Equations 5.51 through 5.54 can be employed to obtain numerical results for full nonlinear postbuckling load–shortening or load–deflection curves of FGM cylindrical shells subjected to axial compression. The initial buckling load of a perfect shell can readily be obtained numerically, by setting $\overline{W}^*/h = 0$ (or $\mu = 0$), while taking $\overline{W}/h = 0$ (note that $W_m \neq 0$). In this case, the minimum buckling load is determined by considering Equation 5.51 for various values of the buckling mode (m, n), which determine the number of half-waves in the X-direction and of full waves in the Y-direction. Note that because of Equation 5.46, the prebuckling deformation of the shell is nonlinear.

For numerical illustrations, two sets of material mixture are considered for an FGM cylindrical shell. One is silicon nitride and stainless steel (referred to as $Si_3N_4/SUS304$) and the other is zirconium oxide and titanium alloy (referred to as ZrO_2/Ti-6Al-4V). The material properties are assumed to be nonlinear function of temperature of Equation 1.4, and typical values, in the present case, can be found in Tables 1.1 through 1.3. Poisson's ratio is assumed to be a constant, that is, $\nu_f = 0.28$ for the $Si_3N_4/SUS304$ cylindrical shell, and $\nu_f = 0.29$ for the ZrO_2/Ti-6Al-4V one. For these examples, the shell geometric parameter $R/h = 30$, $h = 10$ mm, and $T_0 = 300$ K (room tempera-ture). It should be appreciated that in all figures $\overline{W}^*/h = 0.1$ denotes the dimensionless maximum initial geometric imperfection of the shell.

We first examine the postbuckling load–shortening curves for an isotropic cylindrical shell with a local geometric imperfection subjected to axial com-pression. The results are compared in Figure 5.2 with the experimental

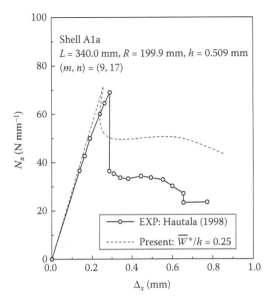

FIGURE 5.2

Comparisons of postbuckling load–shortening curves for an isotropic cylindrical shell with a local geometric imperfection under axial compression.

results of Hautala (1998). The computing data adopted here are $L = 340$ mm, $R = 199.9$ mm, $h = 0.509$ mm, $E = 190$ GPa, and $\nu = 0.3$. The results calculated show that when an initial geometric imperfection was present ($\overline{W}^*/h = 0.25$), the limit point load $N_x = 71.63$ N mm^{-1}, then the present results are in reasonable agreement with the experimental results.

Tables 5.1 and 5.2 present buckling load P (in kN) for perfect, $Si_3N_4/SUS304$ and ZrO_2/Ti-6Al-4V cylindrical shells with different values

TABLE 5.1

Comparisons of Buckling Loads P_{cr} (kN) for Perfect, $Si_3N_4/SUS304$ Cylindrical Shells Subjected to Axial Compression and under Thermal Environments ($R/h = 30$, $h = 10$ mm, $T_0 = 300$ K)

	$\overline{Z}$	$N = 0$	$N = 0.2$	$N = 1.0$	$N = 2.0$	$N = 5.0$
T-ID						
$T_U = 300$ K	300	74942.67[a]	82424.49[a]	94842.38[a]	100394.40[a]	106562.10[a]
$T_L = 300$ K	500	70484.06[b]	77472.75[b]	89258.90[b]	94590.73[b]	100443.90[b]
	800	68067.06[c]	74876.27[c]	86125.56[c]	91151.51[c]	96739.29[c]
$T_U = 500$ K	300	73175.55[a]	80711.32[a]	93379.08[a]	99057.63[a]	105298.20[a]
$T_L = 300$ K	500	67156.96[b]	74226.40[b]	86428.66[b]	91970.67[b]	97961.20[b]
	800	68185.81[c]	74912.29[c]	85843.08[c]	90738.77[c]	96262.37[c]
$T_U = 700$ K	300	71132.21[a]	78753.39[a]	91745.87[a]	97575.29[a]	103913.40[a]
$T_L = 300$ K	500	63881.80[b]	71002.34[b]	83574.78[b]	89317.30[b]	95442.02[b]
	800	66541.07[d]	74116.27[d]	86708.57[d]	91453.34[c]	96743.75[c]

(*continued*)

TABLE 5.1 (continued)

Comparisons of Buckling Loads P_{cr} (kN) for Perfect, Si_3N_4/SUS304 Cylindrical Shells Subjected to Axial Compression and under Thermal Environments ($R/h = 30$, $h = 10$ mm, $T_0 = 300$ K)

	$\overline{Z}$	$N=0$	$N=0.2$	$N=1.0$	$N=2.0$	$N=5.0$
T-D						
($T_U = 500$ K,	300	73175.55[a]	79927.86[a]	91263.13[a]	96301.30[a]	101806.70[a]
$T_L = 300$ K)	500	67156.96[b]	73504.35[b]	84416.09[b]	89316.01[b]	94572.83[b]
	800	68185.81[c]	74188.52[c]	83967.88[c]	88318.46[c]	93208.09[c]
($T_U = 700$ K,	300	71132.21[a]	77422.60[a]	87873.71[a]	92449.94[a]	97348.63[a]
$T_L = 300$ K)	500	63881.80[b]	69838.80[b]	79904.80[b]	84338.26[b]	88957.43[b]
	800	66541.07[d]	72870.63[d]	82973.04[d]	87234.89[d]	91594.48[c]

[a] Buckling mode $(m, n) = (4, 5)$.
[b] Buckling mode $(m, n) = (6, 5)$.
[c] Buckling mode $(m, n) = (8, 5)$.
[d] Buckling mode $(m, n) = (9, 6)$.

TABLE 5.2

Comparisons of Buckling Loads P_{cr} (kN) for Perfect, ZrO_2/Ti-6Al-4V Cylindrical Shells Subjected to Axial Compression and under Thermal Environments ($R/h = 30$, $h = 10$ mm, $T_0 = 300$ K)

	$\overline{Z}$	$N=0$	$N=0.2$	$N=1.0$	$N=2.0$	$N=5.0$
T-ID						
($T_U = 300$ K,	300	38137.02[a]	42208.14[a]	48959.04[a]	51989.80[a]	55358.61[a]
$T_L = 300$ K)	500	35788.11[b]	39583.18[b]	45977.06[b]	48877.66[b]	52068.91[b]
	800	34956.13[c]	38702.66[c]	44856.76[c]	47607.31[c]	50681.81[c]
($T_U = 500$ K,	300	37625.46[a]	41193.16[a]	46802.43[a]	49126.52[a]	50900.87[e]
$T_L = 300$ K)	500	34831.12[b]	37753.29[b]	42313.59[b]	44165.85[b]	45465.31[b]
	800	35032.66[c]	39161.86[c]	44039.69[d]	45793.84[d]	47233.32[d]
($T_U = 700$ K,	300	37064.95[a]	40030.25[a]	44018.71[e]	44823.28[e]	44092.91[e]
$T_L = 300$ K)	500	33875.69[b]	35995.33[b]	39245.23[b]	40646.90[b]	42102.80[b]
	800	35325.80[d]	37744.23[d]	41347.57[d]	42934.16[d]	45384.85[d]
T-D						
($T_U = 500$ K,	300	37625.46[a]	39069.95[a]	40717.95[a]	40924.73[a]	39109.79[a]
$T_L = 300$ K)	500	34831.12[b]	35575.15[b]	36249.34[b]	36340.39[b]	36158.93[b]
	800	35032.66[c]	37173.96[d]	38060.16[d]	38349.19[d]	39316.14[d]
($T_U = 700$ K,	300	37064.95[a]	35303.03[a]	29385.48[a]	26235.67[a]	41797.12[a]
$T_L = 300$ K)	500	33875.69[b]	31333.34[b]	34159.14[b]	32257.93[e]	14388.71[b]
	800	35325.80[d]	33277.44[d]	31713.80[b]	27780.47[b]	25829.02[b]

[a] Buckling mode $(m, n) = (4, 5)$.
[b] Buckling mode $(m, n) = (6, 5)$.
[c] Buckling mode $(m, n) = (8, 5)$.
[d] Buckling mode $(m, n) = (9, 6)$.
[e] Buckling mode $(m, n) = (5, 6)$.

of volume fraction index N ($=0.0$, 0.2, 1.0, 2.0, and 5.0) subjected to axial compression. Three thermal environmental conditions, referred to as I, II, and III, are considered. For case I, $T_U = T_L = 300\,\mathrm{K}$, for case II, $T_U = 500\,\mathrm{K}$, $T_L = 300\,\mathrm{K}$, and for case III, $T_U = 700\,\mathrm{K}$, $T_L = 300\,\mathrm{K}$. Here, T-D represents material properties for FGMs are temperature-dependent. T-ID represents material properties for FGMs are temperature-independent, i.e., in a fixed temperature $T_0 = 300\,\mathrm{K}$, as previously used in Loy et al. (1999), Pradhan et al. (2000), and Ng et al. (2001). It can be seen that, for the Si_3N_4/SUS304 cylindrical shell, a fully metallic shell ($N = 0$) has lowest buckling load and that the buckling load increases as the volume fraction index N increases. It can be seen that an increase is about $+39\%$, $+41\%$, and $+37\%$, respectively, for the Si_3N_4/SUS304 cylindrical shell with $\overline{Z} = 300$, 500, and 800 from $N = 0$ to $N = 5$ at the same thermal environmental condition II under T-D case. In contrast, for the ZrO_2/Ti-6Al-4V cylindrical shell, the buckling load is lower than that of the Si_3N_4/SUS304 shell at the same loading conditions and erratic behavior can be observed under T-D case. It can also be seen that the temperature reduces the buckling load when the temperature dependency is put into consideration. The percentage decrease is about -7.9%, -10.8%, and -4.3% for the Si_3N_4/SUS304 cylindrical shell and about -49%, -34%, and -42% for the ZrO_2/Ti-6Al-4V cylindrical shell with $\overline{Z} = 300$, 500, and 800 from thermal environmental condition case I to case III under the same volume fraction distribution $N = 2$. In the following parametric study only T-D case is considered except for Figure 5.3. Typical results are shown in Figures 5.3 through 5.5 for the Si_3N_4/SUS304 cylindrical shell with $\overline{z} = 500$.

Figure 5.3 gives the postbuckling load–shortening and load–deflection curves for a Si_3N_4/SUS304 cylindrical shell with volume fraction index $N = 2.0$ subjected to axial compression under thermal environmental condition II and two cases of thermoelastic properties T-ID and T-D. It can be seen that the postbuckling equilibrium path becomes lower when the temperature-dependent properties are taken into account.

Figure 5.4 presents the postbuckling load–shortening and load–deflection curves for the same cylindrical shell subjected to axial compression and under three thermal environmental conditions I, II, and III. It is found that an initial extension occurs as the temperature increases and the buckling loads are reduced with increases in temperature, and postbuckling path becomes lower.

Figure 5.5 shows the effect of the volume fraction index N ($= 0.2$, 1.0, and 2.0) on the postbuckling behavior of Si_3N_4/SUS304 cylindrical shell subjected to axial compression and under thermal environmental condition II. The results show that the slope of the postbuckling load–shortening curve for the shell with $N = 2.0$ are larger than others, and the shell has considerable postbuckling strength.

From Figures 5.3 through 5.5, the well-known snap-through phenomenon could be found. The elastic limit loads for imperfect shells can be achieved and imperfection sensitivity can be predicted. The postbuckling

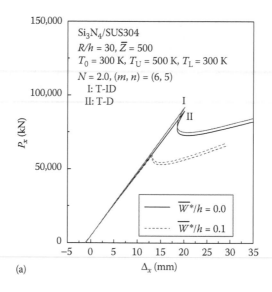

(a)

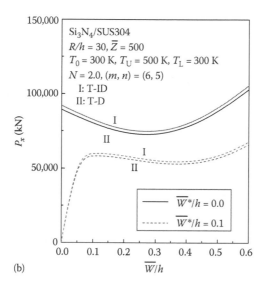

(b)

FIGURE 5.3

Effect of temperature dependency on the postbuckling behavior of a $Si_3N_4/SUS304$ cylindrical shell subjected to axial compression: (a) load shortening; (b) load deflection.

load–shortening and load–deflection curves for imperfect $Si_3N_4/SUS304$ cylindrical shells have been plotted, along with the perfect shell results, in Figures 5.3 through 5.5. Table 5.3 gives imperfection sensitivity λ^* for the imperfect $Si_3N_4/SUS304$ cylindrical shells with $\bar{Z} = 300$, 500, and 800,

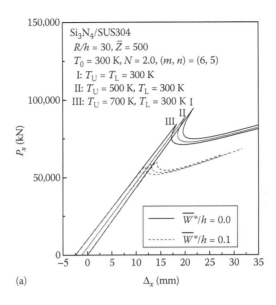

(a)

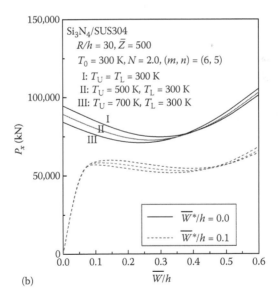

(b)

FIGURE 5.4
Effect of temperature field on the postbuckling behavior of a Si_3N_4/SUS304 cylindrical shell subjected to axial compression: (a) load shortening; (b) load deflection.

and with different values of volume fraction index N ($=0.2$ and 2.0) subjected to axial compression and under thermal environmental conditions I and III. Here, λ^* is the maximum value of P_x for the imperfect shell, made

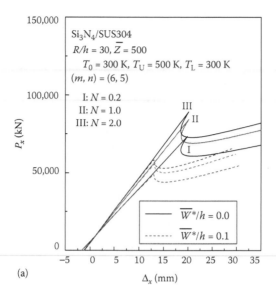

(a)

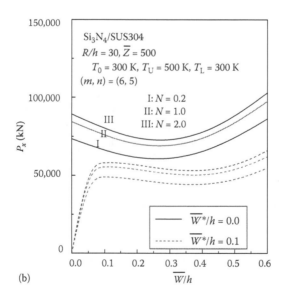

(b)

FIGURE 5.5
Effect of volume fraction index N on the postbuckling behavior of Si_3N_4/SUS304 cylindrical shells subjected to axial compression: (a) load shortening; (b) load deflection.

dimensionless by dividing by the critical value of P_x for the perfect shell as shown in Table 5.1. These results show that the imperfection sensitivity becomes weak when temperature is increased and fiber volume fraction

TABLE 5.3

Imperfection Sensitivity λ^* for $Si_3N_4/SUS304$ Cylindrical Shells Subjected to Axial Compression in Thermal Environments ($R/h = 30$, $h = 10$ mm, $T_0 = 300$ K)

Thermal Environmental Conditions	N	$\overline{Z}$	$\overline{W}^*/h$				
			0.0	0.05	0.1	0.15	0.20
($T_U = 300$ K,	0.2	300	1.0	0.776	0.653	0.564	0.495
$T_L = 300$ K)		500	1.0	0.774	0.641	0.548	0.479
		800	1.0	0.777	0.641	0.548	—
	2.0	300	1.0	0.773	0.649	0.560	0.490
		500	1.0	0.770	0.636	0.543	0.475
		800	1.0	0.775	0.638	0.545	—
($T_U = 700$ K,	0.2	300	1.0	0.812	0.683	0.590	—
$T_L = 300$ K)		500	1.0	0.842	0.698	0.597	—
		800	1.0	0.838	0.692	0.584	—
	2.0	300	1.0	0.799	0.671	—	—
		500	1.0	0.822	0.679	—	—
		800	1.0	0.823	0.672	—	—

only has small effect on the imperfection sensitivity. Note that for larger imperfection amplitudes, e.g., $\overline{W}^*/h > 0.2$, the postbuckling equilibrium path becomes stable, so that no imperfection sensitivity can be predicted.

5.4 Postbuckling Behavior of FGM Cylindrical Shells under External Pressure

Buckling under external pressure is another basic problem for cylindrical shells. We consider an FGM cylindrical shell subjected to external pressure q. Now the nonlinear differential equations in the Donnell sense are

$$\tilde{L}_{11}(\overline{W}) - \tilde{L}_{12}(\overline{\Psi}_x) - \tilde{L}_{13}(\overline{\Psi}_y) + \tilde{L}_{14}(\overline{F}) - \tilde{L}_{15}(\overline{N}^T) - \tilde{L}_{16}(\overline{M}^T) - \frac{1}{R}\overline{F}_{,xx}$$

$$= \tilde{L}(\overline{W} + \overline{W}^*, \overline{F}) + q \tag{5.55}$$

$$\tilde{L}_{21}(\overline{F}) + \tilde{L}_{22}(\overline{\Psi}_x) + \tilde{L}_{23}(\overline{\Psi}_y) - \tilde{L}_{24}(\overline{W}) - \tilde{L}_{25}(\overline{N}^T) + \frac{1}{R}\overline{W}_{,xx}$$

$$= -\frac{1}{2}\tilde{L}(\overline{W} + 2\overline{W}^*, \overline{W}) \tag{5.56}$$

$$\tilde{L}_{31}(\overline{W}) + \tilde{L}_{32}(\overline{\Psi}_x) - \tilde{L}_{33}(\overline{\Psi}_y) + \tilde{L}_{34}(\overline{F}) - \tilde{L}_{35}(\overline{N}^T) - \tilde{L}_{36}(\overline{S}^T) = 0 \qquad (5.57)$$

$$\tilde{L}_{41}(\overline{W}) - \tilde{L}_{42}(\overline{\Psi}_x) + \tilde{L}_{43}(\overline{\Psi}_y) + \tilde{L}_{44}(\overline{F}) - \tilde{L}_{45}(\overline{N}^T) - \tilde{L}_{46}(\overline{S}^T) = 0 \qquad (5.58)$$

Note that only Equation 5.55 has a small change with respect to Equation 5.15, whereas Equations 5.56 through 5.58 are identical in forms to those of Equations 5.16 through 5.18.

The two end edges of the shell are assumed to be simply supported or clamped, in the present case, the boundary conditions are

$X = 0, L$:

$$\overline{W} = \overline{\Psi}_y = 0, \quad \overline{M}_x = \overline{P}_x = 0 \quad \text{(simply supported)} \qquad (5.59a)$$

$$\overline{W} = \overline{\Psi}_x = \overline{\Psi}_y = 0 \quad \text{(clamped)} \qquad (5.59b)$$

$$\int_0^{2\pi R} \overline{N}_x dY + \pi R^2 qa = 0 \qquad (5.59c)$$

where $a = 0$ and $a = 1$ for lateral and hydrostatic pressure loading case, respectively. Also the closed (or periodicity) condition is expressed by Equation 5.20, and the average end-shortening relationship is defined by Equation 5.21.

Introducing dimensionless quantities of Equation 5.6, and

$$\delta_q = (\Delta_x/L)(3)^{3/4} L R^{1/2} / 4\pi [D_{11}^* D_{22}^* A_{11}^* A_{22}^*]^{3/8} \qquad (5.60)$$

The nonlinear Equations 5.55 through 5.58 may then be written in dimensionless form (boundary layer type) as

$$\varepsilon^2 L_{11}(W) - \varepsilon L_{12}(\Psi_x) - \varepsilon L_{13}(\Psi_y) + \varepsilon \gamma_{14} L_{14}(F) - \gamma_{14} F_{,xx}$$
$$= \gamma_{14} \beta^2 L(W + W^*, F) + \gamma_{14} \frac{4}{3} (3)^{1/4} \lambda_q \varepsilon^{3/2} \qquad (5.61)$$

$$L_{21}(F) + \gamma_{24} L_{22}(\Psi_x) + \gamma_{24} L_{23}(\Psi_y) - \varepsilon \gamma_{24} L_{24}(W) + \gamma_{24} W_{,xx}$$
$$= -\frac{1}{2} \gamma_{24} \beta^2 L(W + 2W^*, W) \qquad (5.62)$$

$$\varepsilon L_{31}(W) + L_{32}(\Psi_x) - L_{33}(\Psi_y) + \gamma_{14} L_{34}(F) = 0 \qquad (5.63)$$

$$\varepsilon L_{41}(W) - L_{42}(\Psi_x) + L_{43}(\Psi_y) + \gamma_{14} L_{44}(F) = 0 \qquad (5.64)$$

The boundary conditions expressed by Equation 5.59 become

$x = 0, \pi$:

$$W = \Psi_y = 0, \quad M_x = P_x = 0 \quad \text{(simply supported)} \qquad (5.65a)$$

$$W = \Psi_x = \Psi_y = 0 \quad \text{(clamped)} \tag{5.65b}$$

$$\frac{1}{2\pi} \int_0^{2\pi} \beta^2 \frac{\partial^2 F}{\partial y^2} \, dy + \frac{2}{3} (3)^{1/4} \lambda_q \varepsilon^{3/2} a = 0 \tag{5.65c}$$

and the closed condition is expressed by Equation 5.29.

It has been shown (Shen 2008b) that the effect of the boundary layer on the solution of a shell under external pressure is of the order $\varepsilon^{3/2}$, hence the unit end-shortening relationship may be written in dimensionless form as

$$
\begin{aligned}
\delta_q = {}&-\frac{(3)^{3/4}}{8\pi^2 \gamma_{24}} \varepsilon^{-3/2} \int_0^{2\pi} \int_0^\pi \left[\left(\gamma_{24}^2 \beta^2 \frac{\partial^2 F}{\partial y^2} - \gamma_5 \frac{\partial^2 F}{\partial x^2} \right) + \gamma_{24} \left(\gamma_{511} \frac{\partial \Psi_x}{\partial x} + \gamma_{233} \beta \frac{\partial \Psi_y}{\partial y} \right) \right. \\
&- \varepsilon \gamma_{24} \left(\gamma_{611} \frac{\partial^2 W}{\partial x^2} + \gamma_{244} \beta^2 \frac{\partial^2 W}{\partial y^2} \right) - \frac{1}{2} \gamma_{24} \left(\frac{\partial W}{\partial x} \right)^2 \\
&\left. - \gamma_{24} \frac{\partial W}{\partial x} \frac{\partial W^*}{\partial x} + \varepsilon (\gamma_{24}^2 \gamma_{T1} - \gamma_5 \gamma_{T2}) \Delta T \right] dx \, dy
\end{aligned}
\tag{5.66}
$$

Having developed the theory, we are now in a position to solve Equations 5.61 through 5.64 with boundary condition (Equation 5.65) by means of a singular perturbation technique. The essence of this procedure, in the present case, is to assume that

$$
\begin{aligned}
W &= w(x, y, \varepsilon) + \tilde{W}(x, \xi, y, \varepsilon) + \hat{W}(x, \zeta, y, \varepsilon) \\
F &= f(x, y, \varepsilon) + \tilde{F}(x, \xi, y, \varepsilon) + \hat{F}(x, \zeta, y, \varepsilon) \\
\Psi_x &= \psi_x(x, y, \varepsilon) + \tilde{\Psi}_x(x, \xi, y, \varepsilon) + \hat{\Psi}_x(x, \zeta, y, \varepsilon) \\
\Psi_y &= \psi_y(x, y, \varepsilon) + \tilde{\Psi}_y(x, \xi, y, \varepsilon) + \hat{\Psi}_y(x, \zeta, y, \varepsilon)
\end{aligned}
\tag{5.67}
$$

in which the regular and boundary layer solutions are expressed in the forms of perturbation expansions as

$$w(x, y, \varepsilon) = \sum_{j=1} \varepsilon^{j/2} w_{j/2}(x, y), \quad f(x, y, \varepsilon) = \sum_{j=0} \varepsilon^{j/2} f_{j/2}(x, y)$$

$$\psi_x(x, y, \varepsilon) = \sum_{j=1} \varepsilon^{j/2} (\psi_x)_{j/2}(x, y), \quad \psi_y(x, y, \varepsilon) = \sum_{j=1} \varepsilon^{j/2} (\psi_y)_{j/2}(x, y) \tag{5.68a}$$

$$\tilde{W}(x, \xi, y, \varepsilon) = \sum_{j=0} \varepsilon^{j/2+1} \tilde{W}_{j/2+1}(x, \xi, y), \quad \tilde{F}(x, \xi, y, \varepsilon) = \sum_{j=0} \varepsilon^{j/2+2} \tilde{F}_{j/2+2}(x, \xi, y)$$

$$\tilde{\Psi}_x(x, \xi, y, \varepsilon) = \sum_{j=0} \varepsilon^{(j+3)/2} (\tilde{\Psi}_x)_{(j+3)/2}(x, \xi, y), \quad \tilde{\Psi}_y(x, \xi, y, \varepsilon) = \sum_{j=0} \varepsilon^{j/2+2} (\tilde{\Psi}_y)_{j/2+2}(x, \xi, y)$$

$$\tag{5.68b}$$

$$\hat{W}(x,\zeta,y,\varepsilon) = \sum_{j=0} \varepsilon^{j/2+1}\hat{W}_{j/2+1}(x,\zeta,y), \quad \hat{F}(x,\zeta,y,\varepsilon) = \sum_{j=0} \varepsilon^{j/2+2}\hat{F}_{j/2+2}(x,\zeta,y)$$

$$\hat{\Psi}_x(x,\zeta,y,\varepsilon) = \sum_{j=0} \varepsilon^{(j+3)/2}(\hat{\Psi}_x)_{(j+3)/2}(x,\zeta,y), \quad \hat{\Psi}_y(x,\zeta,y,\varepsilon) = \sum_{j=0} \varepsilon^{j/2+2}(\hat{\Psi}_y)_{j/2+2}(x,\zeta,y)$$

$$(5.68c)$$

We also let

$$\frac{4}{3}(3)^{1/4}\lambda_q\varepsilon^{3/2} = \sum_{j=0} \varepsilon^j(k_y)_j \tag{5.69}$$

Substituting Equation 5.67 into Equations 5.61 through 5.64 and collecting the terms of the same order of ε, we can obtain three sets of perturbation equations for the regular and boundary layer solutions, respectively, e.g., the regular solutions $w(x, y)$ and $f(x, y)$ should satisfy

$$O(\varepsilon^0): -\gamma_{14}(f_0)_{,xx} = \gamma_{14}\beta^2 L(w_0, f_0) + k_{y0} \tag{5.70a}$$

$$L_{21}(f_0) + \gamma_{24}L_{22}(\psi_{x0}) + \gamma_{24}L_{23}(\psi_{y0}) + \gamma_{24}(w_0)_{,xx} = -\frac{1}{2}\gamma_{24}\beta^2 L(w_0, w_0) \tag{5.70b}$$

$$L_{32}(\psi_{x0}) - L_{33}(\psi_{y0}) + \gamma_{14}L_{34}(f_0) = 0 \tag{5.70c}$$

$$-L_{42}(\psi_{x0}) + L_{43}(\psi_{y0}) + \gamma_{14}L_{44}(f_0) = 0 \tag{5.70d}$$

$$O(\varepsilon^{1/2}): -\gamma_{14}(f_{1/2})_{,xx} = \gamma_{14}\beta^2\left[L(w_{1/2}, f_0) + L(w_0, f_{1/2})\right] \tag{5.71a}$$

$$L_{21}(f_{1/2}) + \gamma_{24}L_{22}(\psi_{x(1/2)}) + \gamma_{24}L_{23}(\psi_{y(1/2)}) + \gamma_{24}(w_{1/2})_{,xx}$$
$$= -\frac{1}{2}\gamma_{24}\beta^2 L(w_0, w_{1/2}) \tag{5.71b}$$

$$L_{32}(\psi_{x(1/2)}) - L_{33}(\psi_{y(1/2)}) + \gamma_{14}L_{34}(f_{1/2}) = 0 \tag{5.71c}$$

$$-L_{42}(\psi_{x(1/2)}) + L_{43}(\psi_{y(1/2)}) + \gamma_{14}L_{44}(f_{1/2}) = 0 \tag{5.71d}$$

$$O(\varepsilon^1): -L_{12}(\psi_{x0}) - L_{13}(\psi_{y0}) + \gamma_{14}L_{14}(f_0) - \gamma_{14}(f_1)_{,xx}$$
$$= \gamma_{14}\beta^2[L(w_1, f_0) + L(w_0, f_1)] + k_{y1} \tag{5.72a}$$

$$L_{21}(f_1) + \gamma_{24}L_{22}(\psi_{x1}) + \gamma_{24}L_{23}(\psi_{y1}) - \gamma_{24}L_{24}(w_0) + \gamma_{24}(w_1)_{,xx}$$
$$= -\frac{1}{2}\gamma_{24}\beta^2 L(w_0, w_1) \tag{5.72b}$$

$$L_{31}(w_0) + L_{32}(\psi_{x1}) - L_{33}(\psi_{y1}) + \gamma_{14}L_{34}(f_1) = 0 \tag{5.72c}$$

$$L_{41}(w_0) - L_{42}(\psi_{x1}) + L_{43}(\psi_{y1}) + \gamma_{14}L_{44}(f_1) = 0 \tag{5.72d}$$

$$O(\varepsilon^{3/2}): -L_{12}(\psi_{x(1/2)}) - L_{13}(\psi_{y(1/2)}) + \gamma_{14}L_{12}(f_{1/2}) - \gamma_{14}(f_{3/2})_{,xx}$$
$$= \gamma_{14}\beta^2[L(w_{3/2}, f_0) + L(w_1, f_{1/2}) + L(w_{1/2}, f_1) + L(w_0, f_{3/2})] \tag{5.73a}$$

$$L_{21}(f_{3/2}) + \gamma_{24}L_{22}(\psi_{x(3/2)}) + \gamma_{24}L_{23}(\psi_{y(3/2)}) - \gamma_{24}L_{22}(w_{1/2}) + \gamma_{24}(w_{3/2})_{,xx}$$
$$= -\frac{1}{2}\gamma_{24}\beta^2[L(w_{1/2}, w_1) + L(w_{3/2}, w_0)] \tag{5.73b}$$

$$L_{31}(w_{1/2}) + L_{32}(\psi_{x(3/2)}) - L_{33}(\psi_{y(3/2)}) + \gamma_{14}L_{34}(f_{3/2}) = 0 \tag{5.73c}$$

$$L_{41}(w_{1/2}) - L_{42}(\psi_{x(3/2)}) + L_{43}(\psi_{y(3/2)}) + \gamma_{14}L_{44}(f_{3/2}) = 0 \tag{5.73d}$$

$$O(\varepsilon^2): L_{11}(w_0) - L_{12}(\psi_{x1}) - L_{13}(\psi_{y1}) + \gamma_{14}L_{12}(f_1) - \gamma_{14}(f_2)_{,xx}$$
$$= \gamma_{14}\beta^2[L(w_2 + W^*, f_0) + L(w_{3/2}, f_{1/2}) + L(w_1, f_1)$$
$$+ L(w_{1/2}, f_{3/2}) + L(w_0, f_2)] + k_{y2} \tag{5.74a}$$

$$L_{21}(f_2) + \gamma_{24}L_{22}(\psi_{x2}) + \gamma_{24}L_{23}(\psi_{y2}) - \gamma_{24}L_{24}(w_1) + \gamma_{24}(w_2)_{,xx}$$
$$= -\frac{1}{2}\gamma_{24}\beta^2[L(w_1, w_1) + L(w_{3/2}, w_{1/2}) + L(w_2 + 2W^*, w_0)] \tag{5.74b}$$

$$L_{31}(w_1) + L_{32}(\psi_{x2}) - L_{33}(\psi_{y2}) + \gamma_{14}L_{34}(f_2) = 0 \tag{5.74c}$$

$$L_{41}(w_1) - L_{42}(\psi_{x2}) + L_{43}(\psi_{y2}) + \gamma_{14}L_{44}(f_2) = 0 \tag{5.74d}$$

From Equation 5.68a, we assume that $w_0 = w_{1/2} = w_1 = 0$ and $w_{3/2} = A_{00}^{(3/2)}$, along with $\psi_{x0} = \psi_{x(1/2)} = \psi_{x1} = \psi_{x(3/2)} = 0$, $\psi_{y0} = \psi_{y(1/2)} = \psi_{y1} = \psi_{y(3/2)} = 0$, $f_{1/2} = f_{3/2} = 0$, $f_0 = -B_{00}^{(0)}(\beta^2 x^2 + ay^2/2)$, and $f_1 = -B_{00}^{(1)}(\beta^2 x^2 + ay^2/2)$, from Equations 5.70a and 5.72a we have $k_{y0} = \beta^2 B_{00}^{(0)}$ and $k_{y1} = \beta^2 B_{00}^{(1)}$.

The initial buckling mode is assumed to have the form

$$w_2(x, y) = A_{11}^{(2)} \sin mx \sin ny \tag{5.75}$$

and the initial geometric imperfection is assumed to have the similar form

$$W^*(x, y, \varepsilon) = \varepsilon^2 a_{11}^* \sin mx \sin ny = \varepsilon^2 \mu A_{11}^{(2)} \sin mx \sin ny \tag{5.76}$$

where $\mu = a_{11}^*/A_{11}^{(2)}$ is the imperfection parameter. From Equation 5.74 one has

$$f_2 = -B_{00}^{(2)}\left(\beta^2 x^2 + \frac{1}{2} ay^2\right) + B_{11}^{(2)} \sin mx \sin ny$$
$$\psi_{x2} = C_{11}^{(2)} \cos mx \sin ny \tag{5.77}$$
$$\psi_{y2} = D_{11}^{(2)} \sin mx \cos ny$$

and

$$k_{y2} = \beta^2 B_{00}^{(2)}, \quad B_{11}^{(2)} = \frac{\gamma_{24}m^2}{g_{06}}A_{11}^{(2)}, \quad C_{11}^{(2)} = \gamma_{14}m\frac{g_{01}}{g_{00}}B_{11}^{(2)},$$

$$D_{11}^{(2)} = \gamma_{14}n\beta\frac{g_{02}}{g_{00}}B_{11}^{(2)}, \quad \beta^2 B_{00}^{(0)} = \frac{\gamma_{24}m^4}{(1+\mu)(n^2\beta^2 + (1/2)\,am^2)g_{06}} \tag{5.78}$$

For boundary layer solutions, $\tilde{W}(x, \xi, y)$ and $\tilde{F}(x, \xi, y)$ should satisfy boundary layer equations of each order, for example

$$O(\varepsilon^{5/2}): \gamma_{110}\frac{\partial^4 \tilde{W}_{3/2}}{\partial \xi^4} - \gamma_{120}\frac{\partial^3 \tilde{\Psi}_{x2}}{\partial \xi^3} + \gamma_{14}\gamma_{140}\frac{\partial^4 \tilde{F}_{5/2}}{\partial \xi^4} - \gamma_{14}\frac{\partial^2 \tilde{F}_{5/2}}{\partial \xi^2} = 0 \quad (5.79a)$$

$$\frac{\partial^4 \tilde{F}_{5/2}}{\partial \xi^4} + \gamma_{24}\gamma_{220}\frac{\partial^3 \tilde{\Psi}_{x2}}{\partial \xi^3} - \gamma_{24}\gamma_{240}\frac{\partial^4 \tilde{W}_{3/2}}{\partial \xi^4} + \gamma_{24}\frac{\partial^2 \tilde{W}_{3/2}}{\partial \xi^2} = 0 \quad (5.79b)$$

$$\gamma_{310}\frac{\partial^3 \tilde{W}_{3/2}}{\partial \xi^3} - \gamma_{320}\frac{\partial^2 \tilde{\Psi}_{x2}}{\partial \xi^2} + \gamma_{14}\gamma_{220}\frac{\partial^3 \tilde{F}_{5/2}}{\partial \xi^3} = 0 \quad (5.79c)$$

$$\gamma_{430}\frac{\partial^2 \tilde{\Psi}_{y(5/2)}}{\partial \xi^2} = 0 \quad (5.79d)$$

which leads to

$$\frac{\partial^4 \tilde{W}_{3/2}}{\partial \xi^4} + 2c\frac{\partial^2 \tilde{W}_{3/2}}{\partial \xi^2} + b^2 \tilde{W}_{3/2} = 0 \quad (5.80)$$

where c and b are defined by Equation 5.43. The solution of Equation 5.80 can be written as

$$\tilde{W}_{3/2} = \left(a_{10}^{(3/2)} \sin \phi\xi + a_{01}^{(3/2)} \cos \phi\xi\right)e^{-\vartheta\xi} \quad (5.81a)$$

and

$$\tilde{F}_{5/2} = \left(b_{10}^{(5/2)} \sin \phi\xi + b_{01}^{(5/2)} \cos \phi\xi\right)e^{-\vartheta\xi} \quad (5.81b)$$

$$\tilde{\Psi}_{x2} = \left(c_{10}^{(2)} \sin \phi\xi + c_{01}^{(2)} \cos \phi\xi\right)e^{-\vartheta\xi} \quad (5.81c)$$

$$\tilde{\Psi}_{y(5/2)} = 0 \quad (5.81d)$$

where ϑ and ϕ are defined by Equation 5.45.

Similarly, the boundary layer solutions $\hat{W}(x, \zeta, y)$ and $\hat{F}(x, \zeta, y)$ can be obtained in the same manner.

Then matching the regular solutions with the boundary layer solutions at the each end of the shell, e.g., let $(w_{3/2} + \tilde{W}_{3/2}) = (w_{3/2} + \tilde{W}_{3/2})_{,x} = 0$ and $(w_{3/2} + \hat{W}_{3/2}) = (w_{3/2} + \hat{W}_{3/2})_{,x} = 0$ at $x = 0$ and $x = \pi$ edges, respectively, so that the asymptotic solutions satisfying the clamped boundary conditions are constructed as

$$W = \varepsilon^{3/2}\left[A_{00}^{(3/2)} - A_{00}^{(3/2)}\left(a_{01}^{(3/2)} \cos \phi\frac{x}{\sqrt{\varepsilon}} + a_{10}^{(3/2)} \sin \phi\frac{x}{\sqrt{\varepsilon}}\right)\exp\left(-\vartheta\frac{x}{\sqrt{\varepsilon}}\right)\right.$$
$$\left. - A_{00}^{(3/2)}\left(a_{01}^{(3/2)} \cos \phi\frac{\pi - x}{\sqrt{\varepsilon}} + a_{10}^{(3/2)} \sin \phi\frac{\pi - x}{\sqrt{\varepsilon}}\right)\exp\left(-\vartheta\frac{\pi - x}{\sqrt{\varepsilon}}\right)\right]$$
$$+ \varepsilon^2\left[A_{11}^{(2)} \sin mx \sin ny\right] + \varepsilon^3\left[A_{11}^{(3)} \sin mx \sin ny\right]$$
$$+ \varepsilon^4\left[A_{00}^{(4)} + A_{11}^{(4)} \sin mx \sin ny + A_{20}^{(4)} \cos 2mx + A_{02}^{(4)} \cos 2ny\right] + O(\varepsilon^5) \quad (5.82)$$

$$F = -\frac{1}{2}B_{00}^{(0)}\left(\beta^2 x^2 + a\frac{y^2}{2}\right) + \varepsilon\left[-\frac{1}{2}B_{00}^{(1)}\left(\beta^2 x^2 + a\frac{y^2}{2}\right)\right]$$

$$+ \varepsilon^2\left[-\frac{1}{2}B_{00}^{(2)}\left(\beta^2 x^2 + a\frac{y^2}{2}\right) + B_{11}^{(2)}\sin mx \sin ny\right]$$

$$+ \varepsilon^{5/2}\left[A_{00}^{(3/2)}\left(b_{01}^{(5/2)}\cos\phi\frac{x}{\sqrt{\varepsilon}} + b_{10}^{(5/2)}\sin\phi\frac{x}{\sqrt{\varepsilon}}\right)\exp\left(-\vartheta\frac{x}{\sqrt{\varepsilon}}\right)\right.$$

$$\left. + A_{00}^{(3/2)}\left(b_{01}^{(5/2)}\cos\phi\frac{\pi - x}{\sqrt{\varepsilon}} + b_{10}^{(5/2)}\sin\phi\frac{\pi - x}{\sqrt{\varepsilon}}\right)\exp\left(-\vartheta\frac{\pi - x}{\sqrt{\varepsilon}}\right)\right]$$

$$+ \varepsilon^3\left[-\frac{1}{2}B_{00}^{(3)}\left(\beta^2 x^2 + a\frac{y^2}{2}\right)\right] + \varepsilon^4\left[-\frac{1}{2}B_{00}^{(4)}\left(\beta^2 x^2 + a\frac{y^2}{2}\right)\right.$$

$$\left. + B_{20}^{(4)}\cos 2mx + B_{02}^{(4)}\cos 2ny\right] + O(\varepsilon^5) \tag{5.83}$$

$$\Psi_x = \varepsilon^2\left[C_{11}^{(2)}\cos mx \sin ny + \left(c_{01}^{(2)}\cos\phi\frac{x}{\sqrt{\varepsilon}} + c_{10}^{(2)}\sin\phi\frac{x}{\sqrt{\varepsilon}}\right)\exp\left(-\vartheta\frac{x}{\sqrt{\varepsilon}}\right)\right.$$

$$\left. + \left(c_{01}^{(2)}\cos\phi\frac{\pi - x}{\sqrt{\varepsilon}} + c_{10}^{(2)}\sin\phi\frac{\pi - x}{\sqrt{\varepsilon}}\right)\exp\left(-\vartheta\frac{\pi - x}{\sqrt{\varepsilon}}\right)\right]$$

$$+ \varepsilon^3\left[C_{11}^{(3)}\cos mx \sin ny\right] + \varepsilon^4\left[C_{11}^{(4)}\cos mx \sin ny + C_{20}^{(4)}\sin 2mx\right] + O(\varepsilon^5) \tag{5.84}$$

$$\Psi_y = \varepsilon^2\left[D_{11}^{(2)}\sin mx \cos ny\right] + \varepsilon^3\left[D_{11}^{(3)}\sin mx \cos ny\right]$$

$$+ \varepsilon^4\left[D_{11}^{(4)}\sin mx \cos ny + D_{02}^{(4)}\sin 2ny\right] + O(\varepsilon^5) \tag{5.85}$$

It can be seen that the solutions expressed by Equations 5.82 through 5.85 are quite different from those of the shell subjected to axial compression. As in the case of axial compression, all of the coefficients in Equations 5.82 through 5.85 are related and can be expressed in terms of $A_{11}^{(2)}$, as shown in Equation 5.78 for $B_{11}^{(2)}$, except for $A_{00}^{(j)}$ ($j = 3/2, 2, 3, 4$) which may be determined by substituting Equations 5.82 through 5.85 into closed condition, and can be written as

$$A_{00}^{(3/2)} = \frac{1}{\gamma_{24}}\left(1 - \frac{1}{2}a\gamma_5\right)\frac{4}{3}(3)^{1/4}\lambda_q - \varepsilon^{-1/2}\frac{1}{\gamma_{24}}[\gamma_{T2} - \gamma_5\gamma_{T1}]\Delta T,$$

$$A_{00}^{(2)} = A_{00}^{(3)} = 0, \quad A_{00}^{(4)} = \frac{1}{8}n^2\beta^2(1 + 2\mu)\left(A_{11}^{(2)}\right)^2 \tag{5.86}$$

Next, upon substitution of Equations 5.82 through 5.85 into the boundary condition (Equation 5.65c) and into Equation 5.66, the postbuckling equilibrium paths can be written as

$$\lambda_q = \frac{1}{4}(3)^{3/4}\varepsilon^{-3/2}\left[\lambda_q^{(0)} + \lambda_q^{(2)}\left(A_{11}^{(2)}\varepsilon^2\right)^2 + \cdots\right] \tag{5.87}$$

and

$$\delta_q = \delta_x^{(0)} - \delta_x^{(T)} + \delta_x^{(2)}\left(A_{11}^{(2)}\varepsilon^2\right)^2 + \cdots \tag{5.88}$$

In Equations 5.87 and 5.88, $\left(A_{11}^{(2)}\varepsilon^2\right)$ is taken as the second perturbation parameter relating to the dimensionless maximum deflection. From Equation 5.82, by taking $(x,y) = (\pi/2m, \pi/2n)$, one has

$$A_{11}^{(2)}\varepsilon^2 = W_m - \Theta_4 W_m^2 + \cdots \tag{5.89}$$

where W_m is the dimensionless form of maximum deflection of the shell that can be written as

$$W_m = \frac{1}{C_3}\left[\varepsilon\frac{h}{[D_{11}^* D_{22}^* A_{11}^* A_{22}^*]^{1/4}}\frac{\overline{W}}{h} + \Theta_3\right] \tag{5.90}$$

All symbols used in Equations 5.87 through 5.90 are described in detail in Appendix M. The perturbation scheme described here is quite different from that of traditional one (Budiansky and Amazigo 1968, Amazigo 1974). As argued before, the traditional asymptotic solutions are only suitable for initial postbuckling analysis and cannot be extended to the deep postbuckling range.

Equations 5.87 through 5.90 can be employed to obtain numerical results for full nonlinear postbuckling load–shortening and load–deflection curves of FGM cylindrical shells subjected to external pressure. For numerical illustrations, two sets of material mixture, as shown in Section 5.3, for FGM cylindrical shells are considered.

We first examine the buckling loads for simply supported, isotropic cylindrical shells under hydrostatic pressure. The results are calculated and compared in Table 5.4 with classical shell theory results of Hutchinson and Amazigo (1967), and finite element results obtained by Kasagi and Sridharan (1993). The material properties used in this case, as given in Hutchinson and Amazigo (1967), are $E = 10 \times 10^6$ psi, and $\nu = 0.33$. It can be seen that, for most cases the present results agree well with existing results. In contrast, for very short cylinders ($\overline{Z}_B = 10$), the present results are lower than those of Hutchinson and Amazigo (1967) and Kasagi and Sridharan (1993). This is because in the present analysis the nonlinear prebuckling deformations are considered, and the effect of the boundary layer is great for very short cylinders. In addition, the postbuckling load–deflection curves for an isotropic interring short cylindrical shell (model no. 7) subjected to lateral pressure are compared in Figure 5.6 with the experimental results of Seleim and Roorda (1987). The computing data adopted here are $L = 3.5$ in., $R = 5.04$ in., $h = 0.08$ in., and $\nu = 0.3$. Then the postbuckling load–deflection curves for an isotropic long cylindrical shell subjected to lateral pressure are compared in Figure 5.7 with the experimental and FEM results of Djerroud et al. (1991). The computing data adopted here are $L = 150$ mm, $R = 75$ mm, $h = 0.16$ mm, $E = 166.473$ GPa, and $\nu = 0.34$. The results calculated show that when an initial geometric imperfection was present, e.g., $\overline{W}^*/h = 0.01$ for Figure 5.6 and $\overline{W}^*/h = 0.1$ for Figure 5.7, the present results are in reasonable agreement with the experimental results.

TABLE 5.4

Comparisons of Buckling Loads q_{cr} (psi) for Perfect Isotropic Cylindrical Shells under Hydrostatic Pressure ($E = 10 \times 10^6$ psi, $\nu = 0.33$)

R/h	$\overline{Z}_B$	Present HSDT	Hutchinson and Amazigo (1967)	Kasagi and Sridharan FEM (1993)
50	10	1383.623 (1, 9)[a]	1425.0 (1, 9)	1390.0 (1, 9)
	50	566.085 (1, 7)	570.2 (1, 7)	560.0 (1, 7)
	100	389.620 (1, 6)	391.8 (1, 6)	385.6 (1, 6)
	500	166.770 (1, 4)	167.3 (1, 4)	165.0 (1, 4)
	1,000	124.984 (1, 3)	125.1 (1, 3)	123.5 (1, 3)
	5,000	56.566 (1, 2)	56.6 (1, 2)	55.9 (1, 2)
	10,000	37.022 (1, 2)	37.1 (1, 2)	36.65 (1, 2)
200	10	87.077 (1, 17)	89.07 (1, 18)	88.65 (1, 18)
	50	35.167 (1, 13)	35.25 (1, 13)	35.09 (1, 13)
	100	24.305 (1, 11)	24.35 (1, 11)	24.26 (1, 11)
	500	10.436 (1, 8)	10.45 (1, 8)	10.42 (1, 8)
	1,000	7.398 (1, 7)	7.412 (1, 7)	7.388 (1, 7)
	5,000	3.416 (1, 5)	3.423 (1, 5)	3.412 (1, 5)
	10,000	2.315 (1, 4)	2.319 (1, 4)	2.312 (1, 4)

[a] The numbers in brackets indicate the buckling mode (m, n).

Tables 5.5 and 5.6 present buckling pressure q_{cr} (in kPa) for perfect, $Si_3N_4/SUS304$ and ZrO_2/Ti-6Al-4V cylindrical shells with different values of volume fraction index N ($= 0.0, 0.2, 1.0, 2.0,$ and 5.0) subjected to lateral pressure. As in the case of axial compression, three thermal

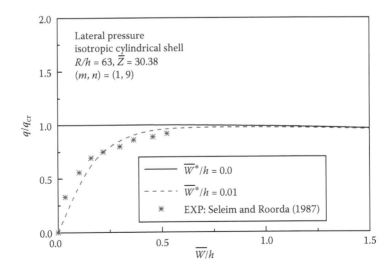

FIGURE 5.6

Comparisons of postbuckling load–deflection curves for an isotropic short cylindrical shell under lateral pressure.

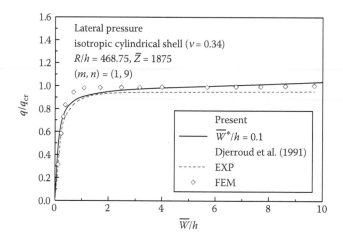

FIGURE 5.7

Comparisons of postbuckling load–deflection curves for an isotropic long cylindrical shell under lateral pressure.

TABLE 5.5

Comparisons of Buckling Loads q_{cr} (kPa) for Perfect, Si_3N_4/SUS304 Cylindrical Shells Subjected to Lateral Pressure and under Thermal Environments ($R/h = 30$, $h = 10\,mm$, $T_0 = 300\,K$)

	$\bar{Z}$	$N=0$	$N=0.2$	$N=1.0$	$N=2.0$	$N=5.0$
T-ID						
($T_U = 300\,K$,	300	13945.96[a]	15425.75[a]	17541.49[a]	18383.41[a]	19426.02[a]
$T_L = 300\,K$)	500	10026.28[b]	11060.52[b]	12648.07[b]	13318.71[b]	14103.74[b]
	800	7969.70[b]	8809.92[b]	10031.34[b]	10524.54[b]	11126.91[b]
($T_U = 500\,K$,	300	13945.88[a]	15425.66[a]	17541.41[a]	18383.33[a]	19425.95[a]
$T_L = 300\,K$)	500	10026.25[b]	11060.49[b]	12648.04[b]	13318.68[b]	14103.71[b]
	800	7969.69[b]	8809.91[b]	10031.33[b]	10524.53[b]	11126.90[b]
($T_U = 700\,K$,	300	13945.90[a]	15425.66[a]	17541.38[a]	18383.30[a]	19425.91[a]
$T_L = 300\,K$)	500	10026.28[b]	11060.51[b]	12648.04[b]	13318.67[b]	14103.70[b]
	800	7969.72[b]	8809.93[b]	10031.34[b]	10524.53[b]	11126.90[b]
T-D						
($T_U = 500\,K$,	300	13945.88[a]	15270.61[a]	17170.87[a]	17919.19[a]	18839.65[a]
$T_L = 300\,K$)	500	10026.25[b]	10951.44[b]	12374.80[b]	12971.41[b]	13666.14[b]
	800	7969.69[b]	8721.74[b]	9818.30[b]	10256.76[b]	10788.86[b]
($T_U = 700\,K$,	300	13945.90[a]	15143.17[a]	16865.60[a]	17538.02[a]	18360.08[a]
$T_L = 300\,K$)	500	10026.28[b]	10861.94[b]	12150.18[b]	12686.65[b]	13308.43[b]
	800	7969.72[b]	8649.31[b]	9642.90[b]	10036.95[b]	10512.42[b]

[a] Buckling mode $(m, n) = (1, 4)$.
[b] Buckling mode $(m, n) = (1, 3)$.

TABLE 5.6

Comparisons of Buckling Loads q_{cr} (kPa) for Perfect, ZrO_2/Ti-6Al-4V Cylindrical Shells Subjected to Lateral Pressure and under Thermal Environments ($R/h = 30$, $h = 10$ mm, $T_0 = 300$ K)

	$\overline{Z}$	$N = 0$	$N = 0.2$	$N = 1.0$	$N = 2.0$	$N = 5.0$
T-ID						
($T_U = 300$ K,	300	7133.13[a]	7941.64[a]	9095.10[a]	9557.11[a]	10131.67[a]
$T_L = 300$ K)	500	5122.51[b]	5687.28[b]	6552.62[b]	6919.87[b]	7351.27[b]
	800	4075.35[b]	4534.36[b]	5200.24[b]	5470.73[b]	5802.46[b]
($T_U = 500$ K,	300	7133.10[a]	7941.59[a]	9095.05[a]	9557.07[a]	10131.70[a]
$T_L = 300$ K)	500	5122.50[b]	5687.27[b]	6552.62[b]	6919.88[b]	7351.34[b]
	800	4075.34[b]	4534.36[b]	5200.24[b]	5470.74[b]	5802.50[b]
($T_U = 700$ K,	300	7133.09[a]	7941.60[a]	9095.20[a]	9557.36[a]	10132.36[a]
$T_L = 300$ K)	500	5122.49[b]	5687.29[b]	6552.74[b]	6920.09[b]	7351.80[b]
	800	4075.34[b]	4534.37[b]	5200.30[b]	5470.85[b]	5802.72[b]
T-D						
($T_U = 500$ K,	300	7133.10[a]	7582.31[a]	8233.86[a]	8482.65[a]	8781.91[a]
$T_L = 300$ K)	500	5122.50[b]	5435.24[b]	5920.03[b]	6118.53[b]	6345.85[b]
	800	4075.34[b]	4330.18[b]	4705.55[b]	4851.32[b]	5024.47[b]
($T_U = 700$ K,	300	7133.09[a]	7381.57[a]	7747.77[a]	7885.16[a]	8057.13[a]
$T_L = 300$ K)	500	5122.49[b]	5295.18[b]	5566.09[b]	5675.56[b]	5805.49[b]
	800	4075.34[b]	4216.22[b]	4426.50[b]	4506.44[b]	4602.22[b]

[a] Buckling mode $(m, n) = (1, 4)$.
[b] Buckling mode $(m, n) = (1, 3)$.

environmental conditions, referred to as I, II, and III, are considered. It can be found that the buckling pressure of ZrO_2/Ti-6Al-4V cylindrical shell is lower than that of Si_3N_4/SUS304 cylindrical shell at the same loading conditions. It can be seen that a fully metallic shell ($N = 0$) has lowest buckling load and that the buckling pressure increases as the volume fraction index N increases. It can be seen that an increase is about +36% for the Si_3N_4/SUS304 cylindrical shell and about +24% for the ZrO_2/Ti-6Al-4V cylindrical shell from $N = 0$ to $N = 5$ at the same thermal environmental condition II under T-D case. It can also be seen that the temperature reduces the buckling pressure when the temperature dependency is put into consideration. The percentage decrease is about −4.7% for the Si_3N_4/SUS304 cylindrical shell and about −18% for the ZrO_2/Ti-6Al-4V cylindrical shell from thermal environmental condition case I to case III under the same volume fraction distribution $N = 2$.

Figures 5.8 through 5.10 are postbuckling results for the same Si_3N_4/SUS304 cylindrical shell analogous to the compressive postbuckling results of Figures 5.3 through 5.5, which are for the loading case of hydrostatic pressure under environmental conditions. It can be seen that an increase in pressure is usually required to obtain an increase in deformation, and the postbuckling path is stable for both perfect and imperfect shells, and the shell structure is virtually imperfection-insensitive. This conclusion

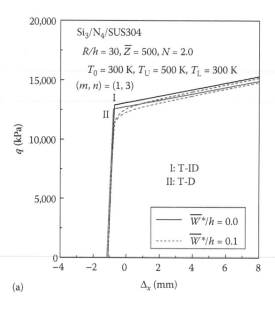

(a)

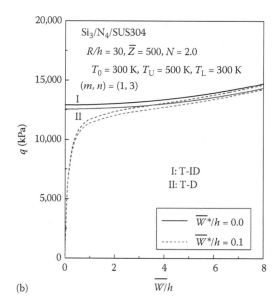

(b)

FIGURE 5.8

Effect of temperature dependency on the postbuckling behavior of a Si_3N_4/SUS304 cylindrical shell subjected to hydrostatic pressure: (a) load shortening; (b) load deflection.

was validated recently in molecular dynamics simulation tests for carbon nanotubes subjected to external pressure (Zhang and Shen 2006). To compare Figures 5.5 and 5.10, it can be seen that now the slope of the post-buckling load–shortening curve for the shell with these three values of

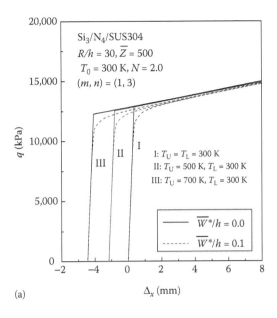

(a)

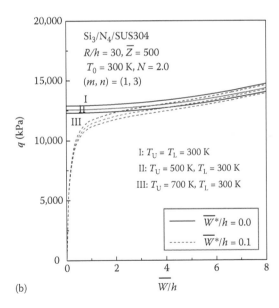

(b)

FIGURE 5.9
Effect of temperature field on the postbuckling behavior of a $Si_3N_4/SUS304$ cylindrical shell subjected to hydrostatic pressure: (a) load shortening; (b) load deflection.

volume fraction index N ($= 0.2$, 1.0, and 2.0) is almost the same, and the shell with $N = 2$ has higher buckling pressure and considerable postbuckling strength. Otherwise, they lead to broadly the same conclusions as do Figures 5.3 and 5.4.

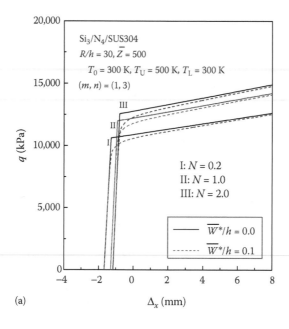

(a)

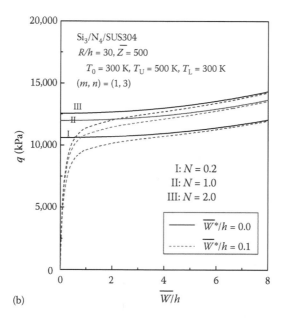

(b)

FIGURE 5.10
Effect of volume fraction index N on the postbuckling behavior of Si_3N_4/SUS304 cylindrical shells subjected to hydrostatic pressure: (a) load shortening; (b) load deflection.

5.5 Postbuckling Behavior of FGM Cylindrical Shells under Torsion

Torsional buckling is an unsolved basic problem for cylindrical shells. This is due to the fact that the solution becomes more complicated in the case of cylindrical shells under torsion. The key issue is how to model the buckling mode of the shell under torsion. Loo (1954) and Nash (1957) suggested solutions formed as

$$\overline{W} = W_1 \sin(\pi X/L) \sin(nY/R + knX/R) \tag{5.91}$$

$$\overline{W} = W_1[1 - \cos(2\pi X/L)] \sin(nY/R + knX/R) \tag{5.92}$$

where the parameter k is determined by minimizing the strain energy. Equation 5.91 was also adopted by Sofiyev and Schnak (2004) for dynamic stability analysis of FGM cylindrical shells under torsional loading. It is worthy to note that both Equations 5.91 and 5.92 cannot satisfy boundary conditions such as simply supported or clamped at the end of the cylindrical shell and can be as approximate solutions. Yamaki and Matsuda (1975) attempted to give more accurate solutions as

$$\overline{W} = \sum_{m=1} \sum_{n=0} W_{mn}(\Psi_{m-1,n} + \Psi_{m+1,n}) \tag{5.93a}$$

$$\Psi_{mn} = \cos(mx - ny) + (-1)^m \cos(mx - ny) \tag{5.93b}$$

Since sufficient numbers of unknown parameters are retained, the solutions of Equation 5.93 could satisfy both compatibility and boundary conditions, but they do not satisfy equilibrium equation, and therefore, the Galerkin method had to be performed. Therefore, we need a clear understanding of the mechanisms of cylindrical shells subjected to torsion.

We assume an FGM cylindrical shell subjected to a torque uniformly applied along the edges. Now the nonlinear differential equations in the Donnell sense are the same as in the case of axial compression, i.e., Equations 5.15 through 5.18, where all linear operators $\tilde{L}_{ij}()$ and nonlinear operator $L_{ij}()$ are defined by Equation 1.37, and the forces, moments, and higher order moments caused by elevated temperature are defined by Equation 1.28.

The two end edges of the shell are assumed to be simply supported or clamped, in the present case the boundary conditions are

$X = 0, L$:

$$\overline{W} = \overline{\Psi}_y = 0, \quad \overline{M}_x = \overline{P}_x = 0 \quad \text{(simply supported)} \tag{5.94a}$$

$$\overline{W} = \overline{\Psi}_x = \overline{\Psi}_y = 0 \quad \text{(clamped)} \tag{5.94b}$$

$$R \int\limits_{0}^{2\pi R} \overline{N}_{xy} dY - M_S = 0 \tag{5.94c}$$

where $M_S = 2\pi R^2 h \tau_s$ and τ_s is the shear stress. Also the closed (or periodicity) condition is expressed by Equation 5.20, and the average end-shortening relationship is defined by Equation 5.21.

The angle of twist is defined as

$$\Gamma = \frac{1}{L} \int\limits_{0}^{L} \left(\frac{\partial U}{\partial Y} + \frac{\partial V}{\partial X} \right) dX = -\frac{1}{L} \int\limits_{0}^{L} \left[A_{66}^* \frac{\partial^2 \overline{F}}{\partial X \partial Y} - \left(B_{66}^* - \frac{4}{3h^2} E_{66}^* \right) \left(\frac{\partial \overline{\Psi}_x}{\partial Y} + \frac{\partial \overline{\Psi}_y}{\partial X} \right) \right.$$

$$\left. + \frac{4}{3h^2} \left(2E_{66}^* \frac{\partial^2 \overline{W}}{\partial X \partial Y} \right) + \frac{\partial \overline{W}}{\partial X} \frac{\partial \overline{W}}{\partial Y} + \frac{\partial \overline{W}}{\partial X} \frac{\partial \overline{W}^*}{\partial Y} + \frac{\partial \overline{W}}{\partial Y} \frac{\partial \overline{W}^*}{\partial X} \right] dX \tag{5.95}$$

Introducing dimensionless quantities of Equation 5.6, and

$$\gamma_{566} = \left[B_{66}^* - (4/3h^2)E_{66}^* \right] / \left[D_{11}^* D_{22}^* A_{11}^* A_{22}^* \right]^{1/4}$$

$$\gamma_{666} = (4/3h^2)E_{66}^* / \left[D_{11}^* D_{22}^* A_{11}^* A_{22}^* \right]^{1/4}$$

$$\lambda_s = \tau_s L^{1/2} R^{3/4} h \left[D_{11}^* D_{22}^* A_{11}^* A_{22}^* \right]^{3/16} / \pi^{1/2} \left[D_{11}^* D_{22}^* \right]^{1/2} \tag{5.96}$$

$$(\delta_s, \gamma) = (\Delta_x/L, \Gamma) L^{1/2} R^{3/4} / \pi^{1/2} \left[D_{11}^* D_{22}^* A_{11}^* A_{22}^* \right]^{5/16}$$

The boundary layer-type equations may then be written in the same forms as in the case of axial compression, i.e.,

$$\varepsilon^2 L_{11}(W) - \varepsilon L_{12}(\Psi_x) - \varepsilon L_{13}(\Psi_y) + \varepsilon \gamma_{14} L_{14}(F) - \gamma_{14} F_{,xx} = \gamma_{14} \beta^2 L(W + W^*, F) \tag{5.97}$$

$$L_{21}(F) + \gamma_{24} L_{22}(\Psi_x) + \gamma_{24} L_{23}(\Psi_y) - \varepsilon \gamma_{24} L_{24}(W) + \gamma_{24} W_{,xx}$$

$$= -\frac{1}{2} \gamma_{24} \beta^2 L(W + 2W^*, W) \tag{5.98}$$

$$\varepsilon L_{31}(W) + L_{32}(\Psi_x) - L_{33}(\Psi_y) + \gamma_{14} L_{34}(F) = 0 \tag{5.99}$$

$$\varepsilon L_{41}(W) - L_{42}(\Psi_x) + L_{43}(\Psi_y) + \gamma_{14} L_{44}(F) = 0 \tag{5.100}$$

The boundary conditions expressed by Equation 5.91 become

$x = 0, \pi$:

$$W = \Psi_y = 0, \quad M_x = P_x = 0 \quad \text{(simply supported)} \tag{5.101a}$$

$$W = \Psi_x = \Psi_y = 0 \quad \text{(clamped)} \tag{5.101b}$$

$$\frac{1}{2\pi} \int\limits_0^{2\pi} \beta \frac{\partial^2 F}{\partial x \partial y} \, dy + \lambda_s \varepsilon^{5/4} = 0 \tag{5.101c}$$

and the closed condition is expressed by Equation 5.29.

It has been shown (Shen 2008c) that the effect of the boundary layer on the solution of a shell under torsion is of the order $\varepsilon^{5/4}$, the unit end-shortening relationship may be written in dimensionless form as

$$
\begin{aligned}
\delta_s = &-\frac{1}{2\pi^2 \gamma_{24}} \varepsilon^{-5/4} \int\limits_0^{2\pi} \int\limits_0^{\pi} \left[\left(\gamma_{24}^2 \beta^2 \frac{\partial^2 F}{\partial y^2} - \gamma_5 \frac{\partial^2 F}{\partial x^2} \right) + \gamma_{24} \left(\gamma_{511} \frac{\partial \Psi_x}{\partial x} + \gamma_{233} \beta \frac{\partial \Psi_y}{\partial y} \right) \right. \\
&- \varepsilon \gamma_{24} \left(\gamma_{611} \frac{\partial^2 W}{\partial x^2} + \gamma_{244} \beta^2 \frac{\partial^2 W}{\partial y^2} \right) - \frac{1}{2} \gamma_{24} \left(\frac{\partial W}{\partial x} \right)^2 - \gamma_{24} \frac{\partial W}{\partial x} \frac{\partial W^*}{\partial x} \\
&\left. + \varepsilon (\gamma_{24}^2 \gamma_{T1} - \gamma_5 \gamma_{T2}) \Delta T \right] dx \, dy \tag{5.102}
\end{aligned}
$$

and the angle of twist may be written in dimensionless form as

$$
\begin{aligned}
\gamma = &-\frac{1}{\pi \gamma_{24}} \varepsilon^{-5/4} \int\limits_0^{\pi} \left[\left(\gamma_{266} \beta \frac{\partial^2 F}{\partial x \partial y} \right) - \gamma_{24} \gamma_{566} \left(\beta \frac{\partial \Psi_x}{\partial y} + \frac{\partial \Psi_y}{\partial x} \right) \right. \\
&\left. + \varepsilon \gamma_{24} \left(2\gamma_{666} \beta \frac{\partial^2 W}{\partial x \partial y} \right) + \gamma_{24} \beta \left(\frac{\partial W}{\partial x} \frac{\partial W}{\partial y} + \frac{\partial W}{\partial y} \frac{\partial W^*}{\partial x} + \frac{\partial W}{\partial x} \frac{\partial W^*}{\partial y} \right) \right] dx \tag{5.103}
\end{aligned}
$$

Note that $\overline{N}_{xy}^T$ is zero-valued for an FGM cylindrical shell, so that no thermal effect remains in Equation 5.103.

Having developed the theory, we are now in a position to solve Equations 5.97 through 5.100 with boundary condition (Equation 5.101) by means of a singular perturbation technique. The essence of this procedure, in the present case, is to assume that

$$
\begin{aligned}
W &= w(x,y,\varepsilon) + \tilde{W}(x,\xi,y,\varepsilon) + \hat{W}(x,\zeta,y,\varepsilon) \\
F &= f(x,y,\varepsilon) + \tilde{F}(x,\xi,y,\varepsilon) + \hat{F}(x,\zeta,y,\varepsilon) \\
\Psi_x &= \psi_x(x,y,\varepsilon) + \tilde{\Psi}_x(x,\xi,y,\varepsilon) + \hat{\Psi}_x(x,\zeta,y,\varepsilon) \\
\Psi_y &= \psi_y(x,y,\varepsilon) + \tilde{\Psi}_y(x,\xi,y,\varepsilon) + \hat{\Psi}_y(x,\zeta,y,\varepsilon)
\end{aligned} \tag{5.104}
$$

in which the regular and boundary layer solutions are expressed in the forms of perturbation expansions as

$$w(x,y,\varepsilon) = \sum_{j=1} \varepsilon^{j/4+1} w_{j/4+1}(x,y), \quad f(x,y,\varepsilon) = \sum_{j=0} \varepsilon^{j/4} f_{j/4}(x,y)$$

$$\psi_x(x,y,\varepsilon) = \sum_{j=1} \varepsilon^{j/4+1} (\psi_x)_{j/4+1}(x,y), \quad \psi_y(x,y,\varepsilon) = \sum_{j=1} \varepsilon^{j/4+1} (\psi_y)_{j/4+1}(x,y)$$

$$(5.105a)$$

$$\tilde{W}(x,\xi,y,\varepsilon) = \sum_{j=1} \varepsilon^{j/4+1} \tilde{W}_{j/4+1}(x,\xi,y), \quad \tilde{F}(x,\xi,y,\varepsilon) = \sum_{j=1} \varepsilon^{j/4+2} \tilde{F}_{j/4+2}(x,\xi,y)$$

$$\tilde{\Psi}_x(x,\xi,y,\varepsilon) = \sum_{j=1} \varepsilon^{j/4+3/2} (\tilde{\Psi}_x)_{j/4+3/2}(x,\xi,y), \quad \tilde{\Psi}_y(x,\xi,y,\varepsilon) = \sum_{j=1} \varepsilon^{j/4+2} (\tilde{\Psi}_y)_{j/4+2}(x,\xi,y)$$

$$(5.105b)$$

$$\hat{W}(x,\zeta,y,\varepsilon) = \sum_{j=1} \varepsilon^{j/4+1} \hat{W}_{j/4+1}(x,\zeta,y), \quad \hat{F}(x,\zeta,y,\varepsilon) = \sum_{j=1} \varepsilon^{j/4+2} \hat{F}_{j/4+2}(x,\zeta,y)$$

$$\hat{\Psi}_x(x,\zeta,y,\varepsilon) = \sum_{j=1} \varepsilon^{j/4+3/2} (\hat{\Psi}_x)_{j/4+3/2}(x,\zeta,y), \quad \hat{\Psi}_y(x,\zeta,y,\varepsilon) = \sum_{j=1} \varepsilon^{j/4+2} (\hat{\Psi}_y)_{j/4+2}(x,\zeta,y)$$

$$(5.105c)$$

Substituting Equation 5.104 into Equations 5.97 and 5.100 and collecting the terms of the same order of ε, we can obtain three sets of perturbation equations for the regular and boundary layer solutions, respectively. From Equation 5.105a, we assume that $w_0 = w_{1/4} = w_{1/2} = w_{3/4} = w_1 = 0$, $\psi_{x0} = \psi_{x(1/2)} = \psi_{x(1/2)} = \psi_{x(3/4)} = \psi_{x1} = 0$, $\psi_{y0} = \psi_{y(1/4)} = \psi_{y(1/2)} = \psi_{y(3/4)} = \psi_{y1} = 0$, then the regular solutions $w(x, y)$ and $f(x, y)$ should satisfy

$$O(\varepsilon^0): -\gamma_{14}(f_0)_{,xx} = 0 \tag{5.106a}$$

$$L_{21}(f_0) = 0 \tag{5.106b}$$

$$\gamma_{14} L_{34}(f_0) = 0 \tag{5.106c}$$

$$\gamma_{14} L_{44}(f_0) = 0 \tag{5.106d}$$

$$O(\varepsilon^{5/4}): -\gamma_{14}(f_{5/4})_{,xx} = \gamma_{14}\beta^2[L(w_{5/4}, f_0)] \tag{5.107a}$$

$$L_{21}(f_{5/4}) + \gamma_{24} L_{22}(\psi_{x(5/4)}) + \gamma_{24} L_{23}(\psi_{y(5/4)}) + \gamma_{24}(w_{5/4})_{,xx} = 0 \tag{5.107b}$$

$$L_{32}(\psi_{x(5/4)}) - L_{33}(\psi_{y(5/4)}) + \gamma_{14} L_{34}(f_{5/4}) = 0 \tag{5.107c}$$

$$-L_{42}(\psi_{x(5/4)}) + L_{43}(\psi_{y(5/4)}) + \gamma_{14} L_{44}(f_{5/4}) = 0 \tag{5.107d}$$

$$O(\varepsilon^2): \gamma_{14} L_{12}(f_1) - \gamma_{14}(f_2)_{,xx} = \gamma_{14}\beta^2[L(w_2 + W^*, f_0) + L(w_{5/4}, f_{3/4})] \tag{5.108a}$$

$$L_{21}(f_2) + \gamma_{24} L_{22}(\psi_{x2}) + \gamma_{24} L_{23}(\psi_{y2}) + \gamma_{24}(w_2)_{,xx} = 0 \tag{5.108b}$$

$$L_{32}(\psi_{x2}) - L_{33}(\psi_{y2}) + \gamma_{14} L_{34}(f_2) = 0 \tag{5.108c}$$

$$-L_{42}(\psi_{x2}) + L_{43}(\psi_{y2}) + \gamma_{14} L_{44}(f_2) = 0 \tag{5.108d}$$

Furthermore, we assume that $w_{5/4} = A_{00}^{(5/4)}$, along with $\psi_{x(5/4)} = \psi_{y(5/4)} = 0$, $f_{1/4} = f_{1/2} = f_{3/4} = 0, f_0 = -B_{00}^{(0)} y^2/2 - b_{00}^{(0)} xy$, and $f_1 = -B_{00}^{(1)} y^2/2 - b_{00}^{(1)} xy$. As argued in Shen (2008c), $(\sin mx \sin ny)$ is no longer the solution of Equation 5.108, hence we assume the initial buckling mode to have the form

$$w_2(x,y) = A_{00}^{(2)} + A_{11}^{(2)} \sin(mx - ky) \sin ny + a_{11}^{(2)} \cos(mx - ky) \cos ny$$
$$+ A_{02}^{(2)} \cos 2ny \qquad (5.109)$$

in which k is a continuous variable and can be determined later, and the initial geometric imperfection is assumed to have the form

$$W^*(x,y,\varepsilon) = \varepsilon^2 \mu \left[A_{11}^{(2)} \sin(mx - ky) \sin ny + a_{11}^{(2)} \cos(mx - ky) \cos ny \right] \quad (5.110)$$

where μ is the imperfection parameter.
From Equation 5.108 one has

$$f_2 = -B_{00}^{(2)} \frac{y^2}{2} - b_{00}^{(2)} xy + B_{11}^{(2)} \sin(mx - ky) \sin ny$$

$$\psi_{x2} = C_{11}^{(2)} \cos(mx - ky) \sin ny + c_{11}^{(2)} \sin(mx - ky) \cos ny \qquad (5.111)$$

$$\psi_{y2} = D_{11}^{(2)} \sin(mx - ky) \cos ny + d_{11}^{(2)} \cos(mx - ky) \sin ny$$

and

$$B_{11}^{(2)} = \frac{\gamma_{24} m^2}{g_{210}} A_{11}^{(2)}, \quad a_{11}^{(2)} = \frac{g_{220}}{g_{210}} A_{11}^{(2)}, \quad C_{11}^{(2)} = \gamma_{14} g_{211} B_{11}^{(2)}$$

$$c_{11}^{(2)} = \gamma_{14} g_{213} B_{11}^{(2)}, \quad D_{11}^{(2)} = \gamma_{14} g_{212} B_{11}^{(2)}, \quad d_{11}^{(2)} = \gamma_{14} g_{214} B_{11}^{(2)} \qquad (5.112)$$

$$\beta^2 B_{00}^{(0)} = \frac{\gamma_{24} m^2 (g_{210} n\beta + g_{220} k\beta)}{n\beta (g_{210}^2 - g_{220}^2)(1 + \mu)}, \quad \beta b_{00}^{(0)} = -\frac{\gamma_{24} m^3 g_{220}}{2n\beta (g_{210}^2 - g_{220}^2)(1 + \mu)}$$

where g_{ijk} are given in detail in Appendix N.
For the boundary layer solutions, $\tilde{W}(x, \xi, y)$ and $\tilde{F}(x, \xi, y)$ should satisfy boundary layer equations of each order, for example,

$$O(\varepsilon^{9/4}): \gamma_{110} \frac{\partial^4 \tilde{W}_{5/4}}{\partial \xi^4} - \gamma_{120} \frac{\partial^3 \tilde{\Psi}_{x(7/4)}}{\partial \xi^3} + \gamma_{14} \gamma_{140} \frac{\partial^4 \tilde{F}_{9/4}}{\partial \xi^4} - \gamma_{14} \frac{\partial^2 \tilde{F}_{9/4}}{\partial \xi^2} = 0 \quad (5.113a)$$

$$\frac{\partial^4 \tilde{F}_{9/4}}{\partial \xi^4} + \gamma_{24} \gamma_{220} \frac{\partial^3 \tilde{\Psi}_{x(7/4)}}{\partial \xi^3} - \gamma_{24} \gamma_{240} \frac{\partial^4 \tilde{W}_{5/4}}{\partial \xi^4} + \gamma_{24} \frac{\partial^2 \tilde{W}_{5/4}}{\partial \xi^2} = 0 \qquad (5.113b)$$

$$\gamma_{310} \frac{\partial^3 \tilde{W}_{5/4}}{\partial \xi^3} - \gamma_{320} \frac{\partial^2 \tilde{\Psi}_{x(7/4)}}{\partial \xi^2} + \gamma_{14} \gamma_{220} \frac{\partial^3 \tilde{F}_{9/4}}{\partial \xi^3} = 0 \qquad (5.113c)$$

$$\gamma_{430} \frac{\partial^2 \tilde{\Psi}_{y(9/4)}}{\partial \xi^2} = 0 \qquad (5.113d)$$

which leads to

$$\frac{\partial^4 \tilde{W}_{5/4}}{\partial \xi^4} + 2c\frac{\partial^2 \tilde{W}_{5/4}}{\partial \xi^2} + b^2 \tilde{W}_{5/4} = 0 \tag{5.114}$$

where c and b are defined by Equation 5.43. The solution of Equation 5.114 can be written as

$$\tilde{W}_{5/4} = \left(a_{10}^{(5/4)} \sin\phi\xi + a_{01}^{(5/4)} \cos\phi\xi\right)e^{-\vartheta\xi} \tag{5.115a}$$

and

$$\tilde{F}_{9/4} = \left(b_{10}^{(9/4)} \sin\phi\xi + b_{01}^{(9/4)} \cos\phi\xi\right)e^{-\vartheta\xi} \tag{5.115b}$$

$$\tilde{\Psi}_{x(7/4)} = \left(c_{10}^{(7/4)} \sin\phi\xi + c_{01}^{(7/4)} \cos\phi\xi\right)e^{-\vartheta\xi} \tag{5.115c}$$

$$\tilde{\Psi}_{y(9/4)} = 0 \tag{5.115d}$$

where ϑ and ϕ are defined by Equation 5.45.

Similarly, the boundary layer solutions $\hat{W}(x, \zeta, y)$ and $\hat{F}(x, \zeta, y)$ can be obtained in the same manner.

Then matching the regular solutions with the boundary layer solutions at the each end of the shell, e.g., let $(w_{5/4} + \tilde{W}_{5/4}) = (w_{5/4} + \tilde{W}_{5/4})_{,x} = 0$ and $(w_{5/4} + \hat{W}_{5/4}) = (w_{5/4} + \hat{W}_{5/4})_{,x} = 0$ at $x=0$ and $x=\pi$ edges, respectively, so that the asymptotic solutions satisfying the clamped boundary conditions are constructed as

$$
\begin{aligned}
W = \varepsilon^{5/4}&\left[A_{00}^{(5/4)} - A_{00}^{(5/4)}\left(a_{01}^{(5/4)}\cos\phi\frac{x}{\sqrt{\varepsilon}} + a_{10}^{(5/4)}\sin\phi\frac{x}{\sqrt{\varepsilon}}\right)\exp\left(-\vartheta\frac{x}{\sqrt{\varepsilon}}\right)\right.\\
&\left. - A_{00}^{(5/4)}\left(a_{01}^{(5/4)}\cos\phi\frac{\pi-x}{\sqrt{\varepsilon}} + a_{10}^{(5/4)}\sin\phi\frac{\pi-x}{\sqrt{\varepsilon}}\right)\exp\left(-\vartheta\frac{\pi-x}{\sqrt{\varepsilon}}\right)\right]\\
+ \varepsilon^2&\left[A_{00}^{(2)} + A_{11}^{(2)}\sin(mx-ky)\sin ny + a_{11}^{(2)}\cos(mx-ky)\cos ny + A_{02}^{(2)}\cos 2ny\right.\\
&- \left(A_{00}^{(2)} - A_{11}^{(2)}\sin ky\sin ny + a_{11}^{(2)}\cos ky\cos ny + A_{02}^{(2)}\cos 2ny\right)\\
&\times \left(a_{01}^{(2)}\cos\phi\frac{x}{\sqrt{\varepsilon}} + a_{10}^{(2)}\sin\phi\frac{x}{\sqrt{\varepsilon}}\right)\exp\left(-\vartheta\frac{x}{\sqrt{\varepsilon}}\right) - \left(A_{00}^{(2)} + (-1)^{m-1}A_{11}^{(2)}\sin ky\sin ny\right.\\
&\left. + (-1)^m a_{11}^{(2)}\cos ky\cos ny + A_{02}^{(2)}\cos 2ny\right)\left(a_{01}^{(2)}\cos\phi\frac{\pi-x}{\sqrt{\varepsilon}}\right.\\
&\left.\left. + a_{10}^{(2)}\sin\phi\frac{\pi-x}{\sqrt{\varepsilon}}\right)\exp\left(-\vartheta\frac{\pi-x}{\sqrt{\varepsilon}}\right)\right] + \varepsilon^3\left[A_{00}^{(3)} + A_{11}^{(3)}\sin(mx-ky)\sin ny\right.\\
&\left. + a_{11}^{(3)}\cos(mx-ky)\cos ny + A_{02}^{(3)}\cos 2ny\right] + \varepsilon^4\left[A_{00}^{(4)} + A_{11}^{(4)}\sin(mx-ky)\sin ny\right.\\
&+ a_{11}^{(4)}\cos(mx-ky)\cos ny + A_{20}^{(4)}\cos 2(mx-ky) + A_{02}^{(4)}\cos 2ny + A_{13}^{(4)}\sin(mx-ky)\\
&\left.\times \sin 3ny + a_{13}^{(4)}\cos(mx-ky)\cos 3ny + A_{04}^{(4)}\cos 4ny\right] + O(\varepsilon^5)
\end{aligned}
\tag{5.116}
$$

$$F = -B_{00}^{(0)}\frac{y^2}{2} - b_{00}^{(0)}xy + \varepsilon\left[-B_{00}^{(1)}\frac{y^2}{2} - b_{00}^{(1)}xy\right] + \varepsilon^2\left[-B_{00}^{(2)}\frac{y^2}{2} - b_{00}^{(2)}xy + B_{11}^{(2)}\sin(mx - ky)\sin ny\right]$$

$$+ \varepsilon^{9/4}\left[A_{00}^{(5/4)}\left(b_{01}^{(9/4)}\cos\phi\frac{x}{\sqrt{\varepsilon}} + b_{10}^{(9/4)}\sin\phi\frac{x}{\sqrt{\varepsilon}}\right)\exp\left(-\vartheta\frac{x}{\sqrt{\varepsilon}}\right)\right.$$

$$+ A_{00}^{(5/4)}\left(b_{01}^{(9/4)}\cos\phi\frac{\pi - x}{\sqrt{\varepsilon}} + b_{10}^{(9/4)}\sin\phi\frac{\pi - x}{\sqrt{\varepsilon}}\right)\exp\left(-\alpha\frac{\pi - x}{\sqrt{\varepsilon}}\right)\right] + \varepsilon^3\left[-B_{00}^{(3)}\frac{y^2}{2} - b_{00}^{(3)}xy + B_{02}^{(3)}\cos 2ny\right.$$

$$+ \left(A_{00}^{(2)} - A_{11}^{(2)}\sin ky\sin ny + a_{11}^{(2)}\cos ky\cos ny + A_{02}^{(2)}\cos 2ny\right)$$

$$\times \left(b_{01}^{(3)}\cos\phi\frac{x}{\sqrt{\varepsilon}} + b_{10}^{(3)}\sin\phi\frac{x}{\sqrt{\varepsilon}}\right)\exp\left(-\alpha\frac{x}{\sqrt{\varepsilon}}\right) + \left(A_{00}^{(2)} + (-1)^{m-1}A_{11}^{(2)}\sin ky\sin ny\right.$$

$$+ (-1)^m a_{11}^{(2)}\cos ky\cos ny + A_{02}^{(2)}\cos 2ny\right)\left(b_{01}^{(3)}\cos\phi\frac{\pi - x}{\sqrt{\varepsilon}} + b_{10}^{(3)}\sin\phi\frac{\pi - x}{\sqrt{\varepsilon}}\right)\exp\left(-\alpha\frac{\pi - x}{\sqrt{\varepsilon}}\right)\right]$$

$$+ \varepsilon^4\left[-B_{00}^{(4)}\frac{y^2}{2} - b_{00}^{(4)}xy + B_{20}^{(4)}\cos 2(mx - ky) + B_{02}^{(4)}\cos 2ny + B_{13}^{(4)}\sin(mx - ky)\sin 3ny\right.$$

$$+ b_{13}^{(4)}\cos(mx - ky)\cos 3ny\right] + O(\varepsilon^5) \tag{5.117}$$

$$\Psi_x = \varepsilon^{7/4}\left[A_{00}^{(5/4)}c_{10}^{(7/4)}\sin\phi\frac{x}{\sqrt{\varepsilon}}\exp\left(-\vartheta\frac{x}{\sqrt{\varepsilon}}\right) + A_{00}^{(5/4)}c_{10}^{(7/4)}\sin\phi\frac{\pi - x}{\sqrt{\varepsilon}}\exp\left(-\vartheta\frac{\pi - x}{\sqrt{\varepsilon}}\right)\right]$$

$$+ \varepsilon^2\left[C_{11}^{(2)}\cos(mx - ky)\sin ny + c_{11}^{(2)}\sin(mx - ky)\cos ny\right]$$

$$+ \varepsilon^{5/2}\left[\left(A_{00}^{(2)} - A_{11}^{(2)}\sin ky\sin ny + a_{11}^{(2)}\cos\cos ny + A_{02}^{(2)}\cos 2ny\right)c_{10}^{(5/2)}\right.$$

$$\times \sin\phi\frac{x}{\sqrt{\varepsilon}}\exp\left(-\vartheta\frac{x}{\sqrt{\varepsilon}}\right) + \left(A_{00}^{(2)} + (-1)^{m-1}A_{11}^{(2)}\sin ky\sin ny\right.$$

$$+ (-1)^m a_{11}^{(2)}\cos ky\cos ny + A_{02}^{(2)}\cos 2ny\right)c_{10}^{(5/2)}\sin\phi\frac{\pi - x}{\sqrt{\varepsilon}}\exp\left(-\vartheta\frac{\pi - x}{\sqrt{\varepsilon}}\right)\right]$$

$$+ \varepsilon^3\left[C_{11}^{(3)}\cos(mx - ky)\sin ny + c_{11}^{(3)}\sin(mx - ky)\cos ny\right]$$

$$+ \varepsilon^4\left[C_{11}^{(4)}\cos(mx - ky)\sin ny + c_{11}^{(4)}\sin(mx - ky)\cos ny + C_{20}^{(4)}\sin 2(mx - ky)\right.$$

$$+ C_{13}^{(4)}\cos(mx - ky)\sin 3ny + c_{13}^{(4)}\sin(mx - ky)\cos 3ny\right] + O(\varepsilon^5) \tag{5.118}$$

$$\Psi_y = \varepsilon^2\left[D_{11}^{(2)}\sin(mx - ky)\cos ny + d_{11}^{(2)}\cos(mx - ky)\sin ny\right]$$

$$+ \varepsilon^3\left[D_{11}^{(3)}\sin(mx - ky)\cos ny + d_{11}^{(3)}\cos(mx - ky)\sin ny + D_{02}^{(3)}\sin 2ny\right]$$

$$+ \varepsilon^4\left[D_{11}^{(4)}\sin(mx - ky)\cos ny + d_{11}^{(4)}\cos(mx - ky)\sin ny + D_{02}^{(4)}\sin 2ny\right.$$

$$+ D_{13}^{(4)}\sin(mx - ky)\cos 3ny + d_{13}^{(4)}\cos(mx - ky)\sin 3ny\right] + O(\varepsilon^5) \tag{5.119}$$

It can be seen that the solutions expressed by Equations 5.116 through 5.119 are more complicated than those of the shell subjected to axial compression. As mentioned before, all of the coefficients in Equations 5.116 through 5.119 are related and can be expressed in terms of $A_{11}^{(2)}$, as shown in Equation 5.112 for $B_{11}^{(2)}$, except for $A_{00}^{(j)}$ ($j = 5/4$, 2, 3, 4) which may be determined by substituting Equations 5.116 through 5.119 into closed condition, and can be written as

$$A_{00}^{(5/4)} = -2\frac{\gamma_5}{\gamma_{24}}\lambda_p - \varepsilon^{-1/4}\frac{1}{\gamma_{24}}[\gamma_{T2} - \gamma_5\gamma_{T1}]\Delta T, \quad A_{00}^{(2)} = A_{00}^{(3)} = 0,$$

$$A_{00}^{(4)} = \frac{1}{8}(n^2\beta^2 + k^2\beta^2)\frac{g_{210}^2 + g_{220}^2}{g_{210}^2}(1 + 2\mu)\left(A_{11}^{(2)}\right)^2 + n^2\beta^2\left(A_{02}^{(2)}\right)^2 \qquad (5.120)$$

where λ_p is the nondimensional compressive stress defined by

$$\lambda_p = \lambda_{xs} + \lambda_{xp} = \lambda_p^{(0)} - \lambda_p^{(2)}\left(A_{11}^{(2)}\varepsilon\right)^2 + \lambda_p^{(4)}\left(A_{11}^{(2)}\varepsilon\right)^4 + \cdots \qquad (5.121)$$

in which λ_{xs} is the compressive stress caused by twisting. Since there is no axial load applied, the second term λ_{xp} must be zero-valued, from which k may be determined.

Next, upon substitution of Equations 5.116 through 5.119 into the boundary condition (Equation 5.101c) and into Equations 5.102 and 5.103, the postbuckling equilibrium paths can be written as

$$\lambda_s = \lambda_s^{(0)} - \lambda_s^{(2)}\left(A_{11}^{(2)}\varepsilon\right)^2 + \lambda_s^{(4)}\left(A_{11}^{(2)}\varepsilon\right)^4 + \cdots \qquad (5.122)$$

and

$$\delta_s = \delta_x^{(0)} - \delta_x^{(T)} + \delta_x^{(2)}\left(A_{11}^{(2)}\varepsilon\right)^2 + \delta_x^{(4)}\left(A_{11}^{(2)}\varepsilon\right)^4 + \cdots \qquad (5.123)$$

$$\gamma = \gamma^{(0)} + \gamma^{(2)}\left(A_{11}^{(2)}\varepsilon\right)^2 + \gamma^{(4)}\left(A_{11}^{(2)}\varepsilon\right)^4 + \cdots \qquad (5.124)$$

In Equations 5.121 through 5.124, $\left(A_{11}^{(2)}\varepsilon\right)$ is taken as the second perturbation parameter relating to the dimensionless maximum deflection. From Equation 5.126, by taking $(x, y) = (\pi/2m, \pi/2n)$, one has

$$A_{11}^{(2)}\varepsilon = W_m - \Theta_6 W_m^2 + \cdots \qquad (5.125)$$

where W_m is the dimensionless form of maximum deflection of the shell that can be written as

$$W_m = \frac{1}{C_3}\left[\frac{h}{[D_{11}^*D_{22}^*A_{11}^*A_{22}^*]^{1/4}}\frac{\overline{W}}{h} + \Theta_5\right] \qquad (5.126)$$

All symbols used in Equations 5.122 through 5.126 are also described in detail in Appendix N.

Unlike the traditional perturbation scheme which only suitable for initial postbuckling analysis (Budiansky 1969, Yamaki and Kodama 1980), the present two-step perturbation technique is a powerful tool for full postbuckling analysis and can avoid the weakness of the traditional one.

Equations 5.122 through 5.126 can be employed to obtain numerical results for full nonlinear postbuckling load–shortening, load–rotation, and load–deflection curves of FGM cylindrical shells subjected to torsion. Because of Equation 5.116, the prebuckling deformation of the shell is nonlinear. It is evident that, from Equation 5.117, there exists a shear stress along with an associate compressive stress when the shell is subjected to torsion. Such a compressive stress, no matter how small it is, will affect the buckling and postbuckling behavior of the FGM cylindrical shell as shown in Equations 5.122 through 5.124, but is missing in all the previously analyses. As a result, all the results published previously need to be reexamined.

We first examine the shear stress τ_s and torque M_S for isotropic cylindrical shell subjected to torsion. The results are calculated and compared in Tables 5.7 and 5.8 with the theoretical and experimental results obtained by Nash (1959), Ekstrom (1963), and Suer and Harris (1959). It can be seen that the present results are in good agreement with experimental results, but lower than theoretical results of Nash (1959), Ekstrom (1963), and Suer and Harris (1959), because the compressive stresses are included in the present analysis. In addition, the postbuckling equilibrium paths for a moderately thick chromium-molybdenum steel tube (specimen J_1) subjected to torsion are compared in Figure 5.11 with the experimental results of Ambrose et al. (1937). The computing data adopted here are $L = 47.0$ in., $R/h = 21.283$, $h = 0.0345$ in., $E = 30.0 \times 10^6$ psi, and $\nu = 0.3$. The results calculated show that when an initial geometric imperfection was present, e.g., $\overline{W}^*/h = 0.12$, the present results are in reasonable agreement with the experimental results.

TABLE 5.7

Comparisons of Buckling Shear Stress τ_s (psi) for Isotropic Cylindrical Shells Subjected to Torsion ($\nu = 0.3$)

Sources	E (psi)	L (in.)	R (in.)	h (in.)	Experimental Results	Theoretical Results
Nash (1959)	28.0e + 6	38.0	4	0.0172	6590	7493
Present						6835
Ekstrom (1963)	29.0e + 6	19.85	3	0.0075	4800	5500
Present						4997

TABLE 5.8

Comparisons of Buckling Load M_S (in./lb.) for Isotropic Cylindrical Shells Subjected to Torsion ($\nu = 0.3$)

Sources	E (psi)	L (in.)	R (in.)	h (in.)	Experimental Results	Theoretical Results
Suer and Harris (1959)	27.0e + 6	22.5	8.75	0.0087	9048	9448
Present						9315

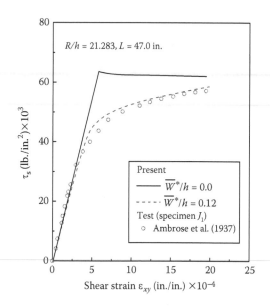

FIGURE 5.11

Comparisons of postbuckling equilibrium paths for a chromium-molybdenum steel tube under torsion.

Tables 5.9 and 5.10 give buckling shear stress τ_s (in MPa) for perfect, thin ($R/h = 100$) and moderately thick ($R/h = 30$) Si_3N_4/SUS304 cylindrical shells with shell parameter $\overline{Z} = 300$ and 500 and with different values of volume fraction index N ($= 0.0, 0.2, 0.5, 1.0, 2.0$, and 5.0) subjected to torsion. As in

TABLE 5.9

Comparisons of Buckling Shear Stress τ_s (MPa) for Perfect, Si_3N_4/SUS304 Thin Cylindrical Shells under Torsion in Thermal Environments ($R/h = 100$, $h = 1$ mm, $T_0 = 300$ K)

Temperature Field	$\overline{Z}$	$N=0$	$N=0.2$	$N=0.5$	$N=1.0$	$N=2.0$	$N=5.0$
T-ID							
($T_U = 300$ K,	300	385.331[a]	426.363[a]	458.291[a]	484.561[a]	507.646[a]	536.192[a]
$T_L = 300$ K)	500	278.354[b]	305.946[b]	329.687[b]	351.366[b]	372.192[b]	394.983[b]
($T_U = 600$ K,	300	383.483[a]	424.488[a]	456.849[a]	483.029[a]	505.997[a]	534.878[a]
$T_L = 300$ K)	500	265.161[b]	293.171[b]	319.136[b]	341.134[b]	362.272[b]	385.870[b]
($T_U = 900$ K,	300	381.509[a]	422.522[a]	455.183[a]	481.770[a]	504.820[a]	533.851[a]
$T_L = 300$ K)	500	243.525[b]	273.394[b]	301.998[b]	325.142[b]	348.629[b]	371.600[b]
T-D							
($T_U = 600$ K,	300	383.482[a]	418.583[a]	446.032[a]	468.821[a]	487.868[a]	511.325[a]
$T_L = 300$ K)	500	265.161[b]	289.493[b]	310.400[b]	330.543[b]	347.712[b]	366.852[b]
($T_U = 900$ K,	300	381.509[a]	412.466[a]	435.976[a]	455.119[a]	471.683[a]	491.636[a]
$T_L = 300$ K)	500	243.525[b]	275.867[b]	290.927[b]	307.265[b]	320.771[b]	334.817[b]

[a] Buckling mode $(m, n) = (1, 5)$.
[b] Buckling mode $(m, n) = (1, 3)$.

TABLE 5.10

Comparisons of Buckling Shear Stress τ_s (MPa) for Perfect, Si_3N_4/SUS304 Moderately Thick Cylindrical Shells under Torsion in Thermal Environments ($R/h = 30$, $h = 1$ mm, $T_0 = 300$ K)

Temperature Field	$\overline{Z}$	$N=0$	$N=0.2$	$N=0.5$	$N=1.0$	$N=2.0$	$N=5.0$
T-ID[a]							
($T_U = 300$ K,	300	1098.005	1214.271	1308.178	1386.625	1452.187	1537.305
$T_L = 300$ K)	500	1119.316	1240.542	1342.356	**1401.286**	**1466.755**	1554.256
($T_U = 600$ K,	300	1091.096	1206.660	1300.124	1378.409	1450.407	1534.401
$T_L = 300$ K)	500	1119.366	1238.680	**1326.099**	**1404.326**	**1466.855**	**1550.036**
($T_U = 900$ K,	300	1078.803	1194.088	1289.553	1371.918	1441.044	1527.409
$T_L = 300$ K)	500	1112.694	1230.576	**1326.557**	1403.768	1470.904	**1556.032**
T-D[a]							
($T_U = 600$ K,	300	1091.096	1186.372	1267.361	1333.170	1389.529	1460.884
$T_L = 300$ K)	500	1119.366	1218.748	1299.363	1359.732	1419.135	1482.154
($T_U = 900$ K,	300	1078.803	1167.604	1240.796	1292.551	1343.937	1405.144
$T_L = 300$ K)	500	1112.694	1198.552	1276.269	1329.622	1369.840	1435.896

[a] Buckling mode $(m, n) = (1, 2)$.

the case of axial compression, three thermal environmental conditions, referred to as I, II, and III, are considered. For case I, $T_U = T_L = 300$ K, for case II, $T_U = 600$ K, $T_L = 300$ K, and for case III $T_U = 900$ K, $T_L = 300$ K. It can be seen that, for the Si_3N_4/SUS304 cylindrical shell, a fully metallic shell ($N = 0$) has lowest buckling load and that the buckling load increases as the volume fraction index N increases. The increase is about $+33\%$ and $+38\%$ for thin cylindrical shells and about $+34\%$ and $+32\%$ for moderately thick cylindrical shells, with $\overline{Z} = 300$ and 500, respectively, from $N = 0$ to $N = 5$ at the same thermal environmental condition II under T-D case. It is found that, for a thin cylindrical shell under condition III, the buckling load under T-ID case is lower than that under T-D case when the shell has lower volume fraction index $N = 0.2$ (see bold numbers in Table 5.9). On the other hand, for a moderately thick shell under T-ID case, the buckling load under thermal environmental condition case II is higher than that under case I when the volume fraction index $N = 1.0$ and 2.0, and also the buckling load under case III is higher than that under case II when $N = 0.5$ and 5.0 (see bold numbers in Table 5.10). As a result, T-ID case may lead to an incorrect solution. It can also be found that the effect of temperature field on the buckling shear stress under T-ID case is small, and it becomes pronounced when the temperature dependency is put into consideration. The temperature increase reduces the buckling load, and the percentage decrease is about -3% and -7% for thin cylindrical shells and about -4% and -7% for moderately thick cylindrical shells with $\overline{Z} = 300$ from thermal environmental condition case I to case III under the same volume fraction distribution $N = 0.2$ and 2.0, respectively.

Figure 5.12 shows the postbuckling load–shortening and load–rotation curves for Si_3N_4/SUS304 thin cylindrical shell with volume fraction index $N = 2.0$ subjected to torsion under thermal environmental condition II and two cases of thermoelastic properties T-ID and T-D. It can be seen that the postbuckling equilibrium path becomes lower when the temperature-

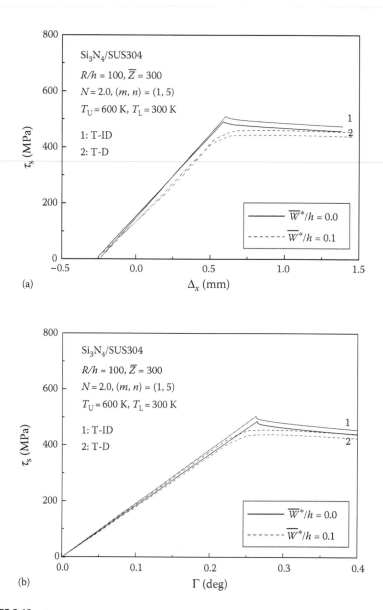

(a)

(b)

FIGURE 5.12

Effect of temperature dependency on the postbuckling behavior of a Si_3N_4/SUS304 thin cylindrical shell subjected to torsion: (a) load shortening; (b) load rotation.

dependent properties are taken into account. It can also be seen that only very weak snap-through behavior of shells occurs for perfect shells. In contrast, for imperfect shells, the postbuckling behavior is stable.

Figure 5.13 shows the effect of the volume fraction index N ($=0.2$, 2.0, and 5.0) on the postbuckling behavior of $Si_3N_4/SUS304$ thin cylindrical shell

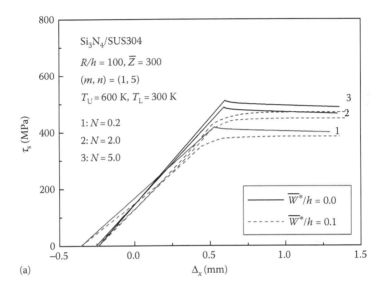

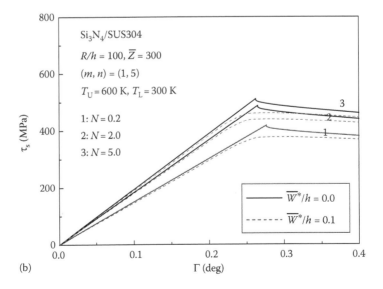

FIGURE 5.13

Effect of volume fraction index N on the postbuckling behavior of a $Si_3N_4/SUS304$ thin cylindrical shell subjected to torsion: (a) load shortening; (b) load rotation.

subjected to torsion under thermal environmental condition II. It can be seen that the initial extension is almost the same when the volume fraction index N is greater than 2. It is found that the increase in the volume fraction index N yields an increase in postbuckling strength.

Figure 5.14 gives the postbuckling load–shortening and load–rotation curves for the same cylindrical shell subjected to torsion under three thermal

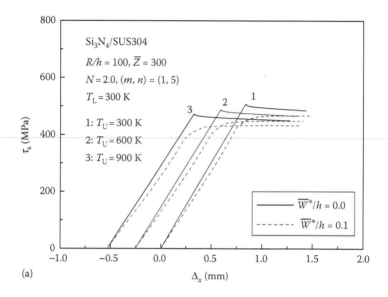

(a)

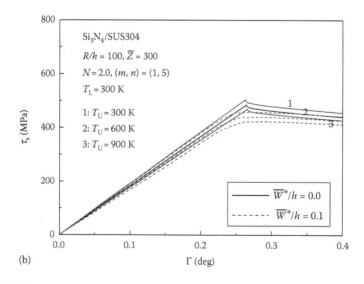

(b)

FIGURE 5.14
Effect of temperature changes on the postbuckling behavior of a Si$_3$N$_4$/SUS304 thin cylindrical shell subjected to torsion: (a) load shortening; (b) load rotation.

environmental conditions I, II, and II. It is evident that buckling loads reduce as the temperature increases, and postbuckling path becomes lower. Note that $T_U = T_L = 300$ K means uniform temperature field. The initial deflection is not zero-valued and an initial extension occurs (curves 2 and 3 in the figure) when the heat conduction is put into consideration.

Figures 5.15 through 5.17 are torsional postbuckling results for moderately thick, $Si_3N_4/SUS304$ cylindrical shells analogous to the results of Figures 5.12

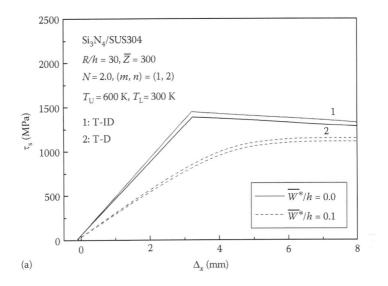

(a)

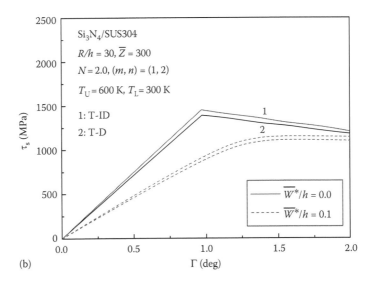

(b)

FIGURE 5.15
Effect of temperature dependency on the postbuckling behavior of a $Si_3N_4/SUS304$ moderately thick cylindrical shell subjected to torsion: (a) load shortening; (b) load rotation.

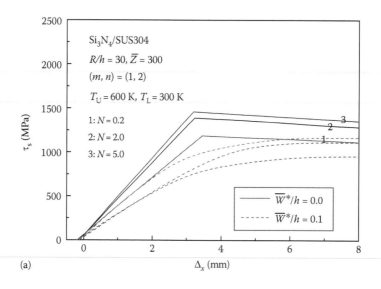

(a)

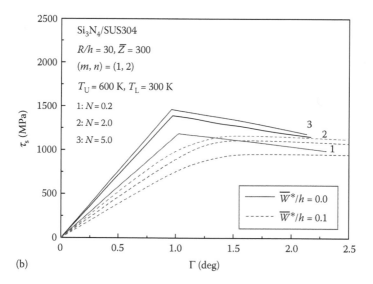

(b)

FIGURE 5.16
Effect of volume fraction index N on the postbuckling behavior of a Si_3N_4/SUS304 moderately thick cylindrical shell subjected to torsion.

through 5.14. To compare these figures, it can be seen that the initial end-shortening and rotation of a moderately thick cylindrical shell are larger than those of a thin cylindrical shell, and no "snap-through" phenomenon could be found for perfect shell. Otherwise, they lead to broadly the same conclusions as do Figures 5.12 through 5.14. In these examples, we chose the shell radius-to-thickness ratio $R/h = 100$ and 30, $\overline{Z} = 300$, and $h = 1.0$ mm.

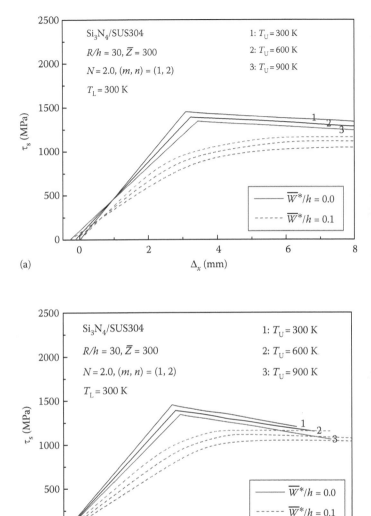

FIGURE 5.17
Effect of temperature changes on the postbuckling behavior of a Si$_3$N$_4$/SUS304 moderately thick cylindrical shell subjected to torsion: (a) load shortening; (b) load rotation.

5.6 Thermal Postbuckling Behavior of FGM Cylindrical Shells

Finally, we consider thermal buckling problem for an FGM cylindrical shell. The temperature field considered is assumed to be a uniform distribution over the shell surface and varied in the thickness direction. The material properties of FGMs are assumed to be graded in the thickness direction according to a simple power law distribution in terms of the volume fractions of the constituents, and are assumed to be temperature-dependent. In the present case, the nonlinear differential equations in the Donnell sense are the same as in the case of axial compression, i.e., Equations 5.15 through 5.18 in Section 5.3, where all linear operators $\tilde{L}_{ij}()$ and nonlinear operator $L_{ij}()$ are defined by Equation 1.37, and the forces, moments and higher order moments caused by elevated temperature are defined by Equation 1.28.

The two end edges of the shell are assumed to be simply supported or clamped, and to be restrained against expansion longitudinally while temperature is increased steadily, so that the boundary conditions are

$X = 0, L$:

$$\overline{W} = \overline{U} = \overline{\Psi}_y = 0, \quad \overline{M}_x = \overline{P}_x = 0 \quad \text{(simply supported)} \tag{5.127a}$$

$$\overline{W} = \overline{U} = \overline{\Psi}_x = \overline{\Psi}_y = 0 \quad \text{(clamped)} \tag{5.127b}$$

Also the closed (or periodicity) condition is expressed by Equation 5.20, and the average end-shortening relationship is defined by Equation 5.21.

Introducing dimensionless quantities of Equation 5.6, and

$$(\gamma_{T1}, \gamma_{T2}) = \left(A_x^T, A_y^T\right) R[A_{11}^* A_{22}^* / D_{11}^* D_{22}^*]^{1/4}$$

$$\delta_x = \left(\frac{\Delta_x}{L}\right) \frac{R}{2[D_{11}^* D_{22}^* A_{11}^* A_{22}^*]^{1/4}}, \quad \lambda_T = \alpha_0 \Delta T \tag{5.128}$$

where α_0 is an arbitrary reference value, defined by Equation 3.39. Also we let

$$\begin{bmatrix} A_x^T \\ A_y^T \end{bmatrix} \Delta T = -\int_{t_1}^{t_2} \begin{bmatrix} A_x \\ A_y \end{bmatrix} \Delta T(Z) dZ \tag{5.129}$$

where ΔT is the temperature rise for a uniform temperature field, and $\Delta T = T_U - T_L$ for the heat conduction and the details of A_x^T can be found in Appendix B (for $\Delta T = 0$) and Appendix D (for $\Delta T \neq 0$).

It is noted that from Equation 1.28 the thermal force $\overline{N}_{xy}^T$ are zero-valued, and for a uniform temperature field the thermal moments are also zero-valued. In contrast, for the heat conduction, the thermal moments are constants, hence the initial deflection of the shell with simply supported edges is not zero, but the clamped edges can prevent the transverse deflection from occurring. As a result, the uniform temperature field is considered only for simply supported shells and both uniform temperature field and heat conduction are considered for clamped shells. Then the boundary layer-type equations are in the same forms as in the case of axial compression, i.e.,

$$\varepsilon^2 L_{11}(W) - \varepsilon L_{12}(\Psi_x) - \varepsilon L_{13}(\Psi_y) + \varepsilon \gamma_{14} L_{14}(F) - \gamma_{14} F_{,xx} = \gamma_{14}\beta^2 L(W + W^*, F) \quad (5.130)$$

$$L_{21}(F) + \gamma_{24} L_{22}(\Psi_x) + \gamma_{24} L_{23}(\Psi_y) - \varepsilon \gamma_{24} L_{24}(W) + \gamma_{24} W_{,xx}$$
$$= -\frac{1}{2}\gamma_{24}\beta^2 L(W + 2W^*, W) \quad (5.131)$$

$$\varepsilon L_{31}(W) + L_{32}(\Psi_x) - L_{33}(\Psi_y) + \gamma_{14} L_{34}(F) = 0 \quad (5.132)$$

$$\varepsilon L_{41}(W) - L_{42}(\Psi_x) + L_{43}(\Psi_y) + \gamma_{14} L_{44}(F) = 0 \quad (5.133)$$

The boundary conditions of Equation 5.127 become

$x = 0, \pi$:

$$W = \delta_x = \Psi_y = 0, \quad M_x = P_x = 0 \quad \text{(simply supported)} \quad (5.134a)$$

$$W = \delta_x = \Psi_x = \Psi_y = 0 \quad \text{(clamped)} \quad (5.134b)$$

and the closed condition of Equation 5.20b becomes

$$\int_0^{2\pi} \left[\left(\frac{\partial^2 F}{\partial x^2} - \gamma_5\beta^2 \frac{\partial^2 F}{\partial y^2} \right) + \gamma_{24}\left(\gamma_{220}\frac{\partial\Psi_x}{\partial x} + \gamma_{522}\beta\frac{\partial\Psi_y}{\partial y} \right) - \varepsilon\gamma_{24}\left(\gamma_{240}\frac{\partial^2 W}{\partial x^2} + \gamma_{622}\beta^2\frac{\partial^2 W}{\partial y^2} \right) \right.$$
$$\left. + \gamma_{24}W - \frac{1}{2}\gamma_{24}\beta^2\left(\frac{\partial W}{\partial y} \right)^2 - \gamma_{24}\beta^2\frac{\partial W}{\partial y}\frac{\partial W^*}{\partial y} + \varepsilon(\gamma_{T2} - \gamma_5\gamma_{T1})\lambda_T \right] dy = 0 \quad (5.135)$$

The unit end-shortening relationship becomes

$$\delta_x = -\frac{1}{4\pi^2\gamma_{24}}\varepsilon^{-1}\int_0^{2\pi}\int_0^\pi \left[\left(\gamma_{24}^2\beta^2\frac{\partial^2 F}{\partial y^2} - \gamma_5\frac{\partial^2 F}{\partial x^2} \right) + \gamma_{24}\left(\gamma_{511}\frac{\partial\Psi_x}{\partial x} + \gamma_{233}\beta\frac{\partial\Psi_y}{\partial y} \right) \right.$$
$$- \varepsilon\gamma_{24}\left(\gamma_{611}\frac{\partial^2 W}{\partial x^2} + \gamma_{244}\beta^2\frac{\partial^2 W}{\partial y^2} \right) - \frac{1}{2}\gamma_{24}\left(\frac{\partial W}{\partial x} \right)^2 - \gamma_{24}\frac{\partial W}{\partial x}\frac{\partial W^*}{\partial x}$$
$$\left. + \varepsilon(\gamma_{24}^2\gamma_{T1} - \gamma_5\gamma_{T2})\lambda_T \right] dx\, dy \quad (5.136)$$

From Equations 5.130 through 5.136, one can determine the thermal post-buckling behavior of perfect and imperfect FGM cylindrical shells under thermal loading by means of a singular perturbation technique. In the present case, the solutions are assumed to have the same forms as in the case of axial compression, i.e., Equations 5.31 through 5.33. All the necessary steps of the solution methodology are described below, but the solutions are not repeated here for convenience.

First, the assumed solution form of Equation 5.31 is substituted into Equations 5.130 through 5.133 and collecting terms of the same order of ε, we obtain three sets of perturbation equations for the regular and boundary layer solutions, respectively.

Then, Equations 5.37 and 5.38 are used to solve these perturbation equations of each order step by step. At each step the amplitudes of the terms $w_j(x,y)$, $f_j(x,y)$, $\psi_{xj}(x,y)$, $\psi_{yj}(x,y)$ for the regular solutions, and of the terms $\tilde{W}_j(x,\xi,y)$, $\tilde{F}_j(x,\xi,y)$, $\tilde{\Psi}_{xj}(x,\xi,y)$, $\tilde{\Psi}_{yj}(x,\xi,y)$, and $\hat{W}_j(x,\zeta,y)$, $\hat{F}_j(x,\zeta,y)$, $\hat{\Psi}_{xj}(x,\zeta,y)$, $\hat{\Psi}_{yj}(x,\zeta,y)$ for the boundary layer solutions can be determined, respectively.

By matching the regular solutions with the boundary layer solutions at each end of the shell, we obtain the asymptotic solutions satisfying the clamped boundary conditions as expressed by Equations 5.46 through 5.49.

Next, upon substitution of Equations 5.46 through 5.49 into the boundary condition $\delta_x = 0$ and into closed condition (Equation 5.135), the thermal postbuckling equilibrium path can be written as

$$\lambda_T = C_{11}\left[\lambda_T^{(0)} - \lambda_T^{(2)}\left(A_{11}^{(2)}\varepsilon\right)^2 + \lambda_T^{(4)}\left(A_{11}^{(2)}\varepsilon\right)^4 + \cdots\right] \tag{5.137}$$

In Equation 5.137, $(A_{11}^{(2)}\varepsilon)$ is taken as the second perturbation parameter relating to the dimensionless maximum deflection. From Equation 5.46, by taking $(x, y) = (\pi/2m, \pi/2n)$ one has

$$A_{11}^{(2)}\varepsilon = W_m - \Theta_7 W_m^2 + \cdots \tag{5.138a}$$

where W_m is the dimensionless form of the maximum deflection of the shell that can be written as

$$W_m = \frac{1}{C_3}\left[\frac{h}{[D_{11}^* D_{22}^* A_{11}^* A_{22}^*]^{1/4}}\frac{\overline{W}}{h} + \Theta_8\right] \tag{5.138b}$$

All symbols used in Equations 5.137 and 5.138 are described in detail in Appendix O. It is noted that $\lambda_T^{(i)}$ ($i = 0, 2, \ldots$) are all functions of temperature and position.

Equations 5.137 and 5.138 can be employed to obtain numerical results for full thermal postbuckling load–deflection curves of FGM cylindrical shells. For numerical illustrations, two sets of material mixture, as shown in Section 5.3, for FGM cylindrical shells are considered.

It has been shown (Shen and Li 2002) for most moderately thick cylindrical shells that the critical value of temperature rise ΔT_{cr} is very high, and in such a case the thermal buckling will not occur. For this reason, in the present examples we chose the shell radius-to-thickness ratio $R/h = 100$ and $h = 10\,\text{mm}$.

The thermal buckling loads ΔT_{cr} (in K) for simply supported perfect FGM cylindrical shells with different values of volume fraction index N ($= 0$, 0.2, 0.5, 1.0, 2.0, and 5.0) subjected to a uniform temperature rise are calculated and compared in Table 5.11. Note that, for the thermal buckling problem, it is necessary to solve Equation 5.137 by an iterative numerical procedure, as previously shown in Section 3.3. It can be seen that, for the Si_3N_4/SUS304 cylindrical shells, a fully metallic shell ($N = 0$) has the lowest buckling temperature, and the buckling temperature increases as the volume fraction index

TABLE 5.11

Comparisons of Buckling Temperatures ΔT_{cr} (in K) for Simply Supported Perfect FGM Cylindrical Shells under Uniform Temperature Field with Temperature-Independent or Temperature-Dependent Properties ($R/h = 100$, $h = 10$ mm, $T_0 = 300$ K)

$\bar{Z}$	$N = 0$	$N = 0.2$	$N = 0.5$	$N = 1.0$	$N = 2.0$	$N = 5.0$
$Si_3N_4/SUS304$, T-ID						
300	356.4784[a]	394.3238[a]	437.9438[a]	488.1880[a]	547.4807[a]	622.0067[a]
500	351.0783[b]	389.0045[b]	431.6156[b]	479.9285[b]	536.6419[b]	608.9135[b]
800	348.2981[c]	385.7825[c]	428.1321[c]	476.3186[c]	532.9545[c]	604.9021[c]
$Si_3N_4/SUS304$, T-D						
300	298.7615[a]	325.5322[a]	355.2824[a]	387.9230[a]	424.6803[a]	468.5131[a]
500	295.1862[b]	322.1722[b]	351.3019[b]	382.6795[b]	417.8137[b]	460.4933[b]
800	293.3418[c]	320.0738[c]	349.1119[c]	380.5313[c]	415.7682[c]	458.4126[c]
$ZrO_2/Ti\text{-}6Al\text{-}4V$, T-ID						
300	616.5239[a]	519.5518[a]	442.9970[a]	385.1690[a]	341.8219[a]	310.7177[a]
500	607.6771[b]	513.0004[b]	436.9333[b]	378.8795[b]	335.2085[b]	304.3159[b]
800	602.9612[c]	508.8263[c]	433.4809[c]	376.1057[c]	332.9807[c]	302.3792[c]
$ZrO_2/Ti\text{-}6Al\text{-}4V$, T-D						
300	540.9151[a]	324.5936[a]	259.1421[a]	222.1906[a]	197.3951[a]	180.4425[a]
500	534.2328[b]	322.0919[b]	257.0812[b]	220.1058[b]	195.2136[b]	178.2761[b]
800	530.6597[c]	320.4967[c]	255.8761[c]	219.1405[c]	194.4353[c]	177.5914[c]

[a] Buckling mode $(m, n) = (2, 7)$.
[b] Buckling mode $(m, n) = (3, 8)$.
[c] Buckling mode $(m, n) = (4, 8)$.

N increases. This is expected because the metallic shell has a larger value of thermal expansion coefficient α_f than the ceramic shell does. It is found that the increase is about $+74\%$ under T-ID case, and about $+56\%$ under T-D case, from $N = 0$ to $N = 5$. In contrast, for the ZrO_2/Ti-6Al-4V cylindrical shells, the buckling temperature is decreased as the volume fraction index N increases, which confirming the finding in Shen (2004). It can also be seen that the buckling temperature reduces when the temperature dependency is put into consideration. The percentage decrease is about -22% for the $Si_3N_4/SUS304$ cylindrical shell and about -42% for the ZrO_2/Ti-6Al-4V cylindrical shell under the same volume fraction distribution $N = 2$.

Table 5.12 presents buckling loads ΔT_{cr} (in K) for clamped perfect FGM cylindrical shells with different values of volume fraction index N ($= 0$, 0.2, 0.5, 1.0, 2.0, and 5.0) under heat conduction. The temperature is now set at ambient temperature 300 K on the inner surface of the shell. It is found that, for the $Si_3N_4/SUS304$ cylindrical shell, the buckling temperature under T-D case is higher than that under T-ID case when the shell has lower volume fraction index N, and the buckling temperature is no longer increased as the volume fraction index N increases.

TABLE 5.12

Comparisons of Buckling Temperatures ΔT_{cr} (in K) for CC Perfect FGM Cylindrical Shells under Heat Conduction with Temperature-Independent or Temperature-Dependent Properties ($R/h = 100$, $h = 10\,\text{mm}$, $T_L = T_0 = 300\,\text{K}$)

$\overline{Z}$	$N=0$	$N=0.2$	$N=0.5$	$N=1.0$	$N=2.0$	$N=5.0$
$Si_3N_4/SUS304$, T-ID						
300	709.5367[a]	801.1807[a]	916.3505[a]	1046.498[b]	1184.8260[b]	1320.546[b]
500	701.0275[b]	792.2409[b]	905.2246[b]	1035.337[b]	1174.6604[b]	1310.380[b]
800	696.4399[c]	768.7820[c]	899.1711[c]	1028.960[c]	1168.1630[c]	1303.484[c]
$Si_3N_4/SUS304$, T-D						
300	709.5367[a]	809.6627[a]	911.4340[a]	971.1096[a]	1049.300[b]	950.3516[b]
500	701.0276[b]	799.7127[b]	899.4313[b]	976.5723[b]	1040.993[b]	943.4778[b]
800	696.0619[c]	793.8150[c]	892.8902[c]	981.0917[c]	1036.795[c]	939.2562[c]
ZrO_2/Ti-$6Al$-$4V$, T-ID						
300	1227.975[a]	711.9446[a]	514.2974[a]	406.1678[b]	329.1755[b]	240.9924[b]
500	1213.289[b]	704.7034[b]	508.4977[b]	401.0534[b]	325.7647[b]	238.7110[b]
800	1205.611[c]	699.9917[c]	505.2125[c]	398.6858[c]	324.0527[c]	237.5206[c]
ZrO_2/Ti-$6Al$-$4V$, T-D						
300	1227.975[a]	392.0297[a]	284.2647[a]	230.3391[b]	191.8820[b]	145.4020[b]
500	1213.289[b]	389.3818[b]	282.2638[b]	228.3273[b]	190.3960[b]	144.3751[b]
800	1205.611[c]	387.8687[c]	281.1219[c]	227.4204[c]	189.6923[c]	143.8673[c]

[a] Buckling mode $(m, n) = (2, 7)$.
[b] Buckling mode $(m, n) = (3, 9)$.
[c] Buckling mode $(m, n) = (3, 8)$.
[d] Buckling mode $(m, n) = (4, 8)$.

Figure 5.18 shows the thermal postbuckling load–deflection curves for clamped perfect and imperfect, $Si_3N_4/SUS304$ and ZrO_2/Ti-6Al-4V cylindrical shells with a volume fraction index $N = 0.2$ under heat conduction and under two cases of thermoelastic material properties, i.e., T-ID and T-D. It can be seen that, for the ZrO_2/Ti-6Al-4V shell, the thermal postbuckling equilibrium path becomes lower when the temperature-dependent properties are taken into account. It is found that, for the $Si_3N_4/SUS304$ shell, the buckling temperature and postbuckling thermal loads under T-D case are slightly higher than those under T-ID case.

Figure 5.19 shows the thermal postbuckling load–deflection curves for the same two FGM cylindrical shells with different values of the volume fraction index N ($= 0.2$, 0.5, and 2.0) under heat conduction. It can be seen that the $Si_3N_4/SUS304$ shell has lower buckling temperature and postbuckling path when it has lower volume fraction index N. As argued before, this is not a general statement. For the ZrO_2/Ti-6Al-4V cylindrical shell, the conclusion is reversed.

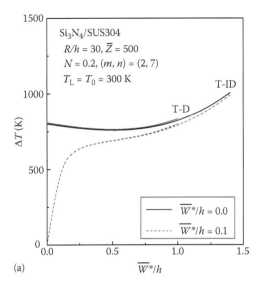

(a)

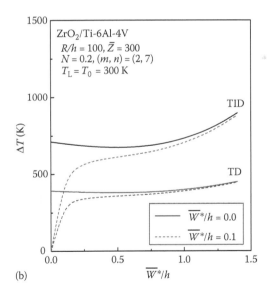

(b)

FIGURE 5.18
Effect of material properties on the thermal postbuckling behavior of FGM cylindrical shells under heat conduction: (a) (Si₃N₄/SUS304) shells; (b) (ZrO₂/Ti-6Al-4V) shells.

Figure 5.20 compares the thermal postbuckling load–deflection curves for the same two FGM cylindrical shells under heat conduction and uniform temperature field. It can be seen that the buckling temperature and post-buckling thermal loads are much lower under uniform temperature field for the Si₃N₄/SUS304 shell. In contrast, for the ZrO₂/Ti-6Al-4V shell, the

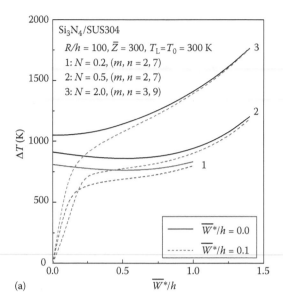

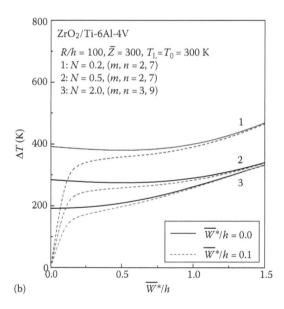

FIGURE 5.19

Effect of volume fraction index N on the thermal postbuckling behavior of FGM cylindrical shells under heat conduction: (a) $(Si_3N_4/SUS304)$ shells; (b) $(ZrO_2/Ti-6Al-4V)$ shells.

buckling temperature is slightly higher, but the postbuckling thermal loads are lower under uniform temperature field.

It is noted that in all these figures $\overline{W}^*/h = 0.05$ denotes the dimensionless maximum initial geometric imperfection of the shell. Unlike solutions of

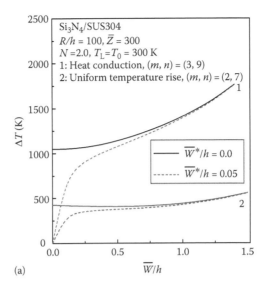

(a)

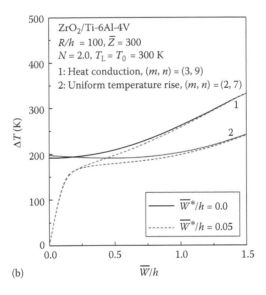

(b)

FIGURE 5.20

Comparisons of thermal postbuckling behavior of FGM cylindrical shells under heat conduction and uniform temperature field: (a) (Si_3N_4/SUS304) shells; (b) (ZrO_2/Ti-6Al-4V) shells.

Mirzavand et al. (2005) and Mirzavand and Eslami (2006), no buckling temperature could be found for imperfect FGM cylindrical shells. From Figures 5.18 through 5.20, it can be seen that the thermal postbuckling equilibrium path is stable or weakly unstable and the shell structure is virtually imperfection-insensitive for both T-ID and T-D cases.

References

Amazigo J.C. (1974), Asymptotic analysis of the buckling of externally pressurized cylinders with random imperfections, *Quarterly of Applied Mathematics*, **31**, 429–442.

Ambrose H.S., Walter, R., and Goldie, B. (1937), Torsion tests of tubes, NACA TR-601.

Batdorf S.B. (1947), A simplified method of elastic-stability analysis for thin cylindrical shells, NACA TR-874.

Budiansky B. (1969), Post-buckling behavior of cylinders in torsion, in *Theory of Thin Shells* (ed. F.I. Niordson), Springer, Berlin, pp. 212–233.

Budiansky B. and Amazigo J.C. (1968), Initial post-buckling behavior of cylindrical shells under external pressure, *Journal of Mathematics and Physics*, **47**, 223–235.

Djerroud M., Chahrour I., and Reynouard J.M. (1991), Buckling of thin cylindrical shells subjected to external pressure, in *Buckling of Shell Structures, on Land, in the Sea and in the Air* (ed. J.F. Jullien), Elsevier Applied Science, London, pp. 190–200.

Donnell L.H. (1933), Stability of thin-walled tubes under torsion, NACA Report No. 479.

Donnell L.H. and Wan C.C. (1950), Effect of imperfections on buckling of thin cylinders and columns under axial compression, *Journal of Applied Mechanics ASME*, **17**, 73–83.

Dym C.L. and Hoff N.J. (1968), Perturbation solutions for the buckling problems of axially compressed thin cylindrical shells of infinite or finite length, *Journal of Applied Mechanics ASME*, **35**, 754–762.

Ekstrom R.E. (1963), Buckling of cylindrical shells under combined torsion and hydrostatic pressure, *Experimental Mechanics*, **3**, 192–197.

Hautala K.T. (1998), Buckling of axially compressed cylindrical shells made of austenitic stainless steels at ambient and elevated temperatures, Dr. Ing. thesis, University of Essen, Germany.

Hutchinson J.W. and Amazigo J.C. (1967), Imperfection-sensitivity of eccentrically stiffened cylindrical shells, *AIAA Journal*, **5**, 392–401.

Kadoli R. and Ganesan N. (2006), Buckling and free vibration analysis of functionally graded cylindrical shells subjected to a temperature-specified boundary condition, *Journal of Sound and Vibration*, **289**, 450–480.

von Kármán T. and Tsien H.S. (1941), The buckling of thin cylindrical shells under axial compression, *Journal of the Aerospace Sciences*, **8**, 303–312.

Kasagi A. and Sridharan S. (1993), Buckling and postbuckling analysis of thick composite cylindrical shells under hydrostatic pressure, *Composites Engineering*, **3**, 467–487.

Koiter W.T. (1945), On the stability of elastic equilibrium (in Dutch), PhD thesis, Delft, H. J. Paris, Amsterdam; also NASA TTF-10, 833, 1967.

Koiter W.T. (1963), Elastic stability and postbuckling behavior, in *Nonlinear Problems* (ed. R.E. Langer), University of Wisconsin Press, Madison, pp. 257–275.

Loo T.-T. (1954), Effects of large deflections and imperfections on the elastic buckling of cylinders under torsion and axial compression, *Proceedings of the 2nd U.S. National Congress on Applied Mechanics*, University of Michigan, Ann Arbor, MI, pp. 345–357.

Loy C.T., Lam K.Y., and Reddy J.N. (1999), Vibration of functionally graded cylindrical shells, *International Journal of Mechanical Sciences*, **41**, 309–324.

Mirzavand B. and Eslami M.R. (2006), Thermal buckling of imperfect functionally graded cylindrical shells on the Wan–Donnell model, *Journal of Thermal Stresses*, **29**, 37–55.

Mirzavand B. and Eslami M.R. (2007), Thermal buckling of simply supported piezoelectric FGM cylindrical shells, *Journal of Thermal Stresses*, **30**, 1117–1135.

Mirzavand B., Eslami M.R., and Shahsiah R. (2005), Effect of imperfections on thermal buckling of functionally graded cylindrical shells, *AIAA Journal*, **43**, 2073–2076.

Nash W.A. (1957), Buckling of initially imperfect cylindrical shells subjected to torsion, *Journal of Applied Mechanics ASME*, **24**, 125–130.

Nash W.A. (1959), An experimental analysis of the buckling of thin initially imperfect cylindrical shells subject to torsion, *Proceedings of the Society for Experimental Stress Analysis*, **16**(2), pp. 55–68.

Ng T.Y., Lam K.Y., Liew K.M., and Reddy J.N. (2001), Dynamic stability analysis of functionally graded cylindrical shells under periodic axial loading, *International Journal of Solids and Structures*, **38**, 1295–1309.

Pradhan S.C., Loy C.T., Lam K.Y., and Reddy J.N. (2000), Vibration characteristics of functionally graded cylindrical shells under various boundary conditions, *Applied Acoustics*, **61**, 111–129.

Reddy J.N. and Liu C.F. (1985), A higher-order shear deformation theory of laminated elastic shells, *International Journal of Engineering Science*, **23**, 319–330.

Seleim S.S. and Roorda J. (1987), Theoretical and experimental results on the post-buckling of ring-stiffened cylinders, *Mechanics of Structures and Machines*, **15**, 69–87.

Shahsiah R. and Eslami M.R. (2003a), Thermal buckling of functionally graded cylindrical shell, *Journal of Thermal Stresses*, **26**, 277–294.

Shahsiah R. and Eslami M.R. (2003b), Functionally graded cylindrical shell thermal instability based on improved Donnell equations, *AIAA Journal*, **41**, 1819–1826.

Shen H.-S. (2002), Postbuckling analysis of axially-loaded functionally graded cylindrical shells in thermal environments, *Composites Science and Technology*, **62**, 977–987.

Shen H.-S. (2003), Postbuckling analysis of pressure-loaded functionally graded cylindrical shells in thermal environments, *Engineering Structures*, **25**, 487–497.

Shen H.-S. (2004), Thermal postbuckling behavior of functionally graded cylindrical shells with temperature-dependent properties, *International Journal of Solids and Structures*, **41**, 1961–1974.

Shen H.-S. (2005), Postbuckling of axially-loaded FGM hybrid cylindrical shells in thermal environments, *Composites Science and Technology*, **65**, 1675–1690.

Shen H.-S. (2007), Thermal postbuckling of shear deformable FGM cylindrical shells with temperature-dependent properties, *Mechanics of Advanced Materials and Structures*, **14**, 439–452.

Shen H.-S. (2008a), Boundary layer theory for the buckling and postbuckling of an anisotropic laminated cylindrical shell, Part I: Prediction under axial compression, *Composite Structures*, **82**, 346–361.

Shen H.-S. (2008b), Boundary layer theory for the buckling and postbuckling of an anisotropic laminated cylindrical shell, Part II: Prediction under external pressure, *Composite Structures*, **82**, 362–370.

Shen H.-S. (2008c), Boundary layer theory for the buckling and postbuckling of an anisotropic laminated cylindrical shell, Part III: Prediction under torsion, *Composite Structures*, **82**, 371–381.

Shen H.-S. and Chen T.-Y. (1988), A boundary layer theory for the buckling of thin cylindrical shells under external pressure, *Applied Mathematics and Mechanics*, **9**, 557–571.

Shen H.-S. and Chen T.-Y. (1990), A boundary layer theory for the buckling of thin cylindrical shells under axial compression, in *Advances in Applied Mathematics and Mechanics in China* (eds. W.Z. Chien and Z.Z. Fu), Vol. 2, pp. 155–172, International Academic Publishers, Beijing, China.

Shen H.-S. and Li Q.S. (2002), Thermomechanical postbuckling of shear deformable laminated cylindrical shells with local geometric imperfections, *International Journal of Solids and Structures*, **39**, 4525–4542.

Shen H.-S. and Noda N. (2005), Postbuckling of FGM cylindrical shells under combined axial and radial mechanical loads in thermal environments, *International Journal of Solids and Structures*, **42**, 4641–4662.

Shen H.-S. and Noda N. (2007), Postbuckling of pressure-loaded FGM hybrid cylindrical shells in thermal environments, *Composite Structures*, **77**, 546–560.

Sheng G.G. and Wang X. (2008), Thermal vibration, buckling and dynamic stability of functionally graded cylindrical shells embedded in an elastic medium, *Journal of Reinforced Plastics and Composites*, **27**, 117–134.

Sofiyev A.H. (2007), Vibration and stability of composite cylindrical shells containing a FG layer subjected to various loads, *Structural Engineering and Mechanics*, **27**, 365–391.

Sofiyev A.H. and Schnak, E. (2004), The stability of functionally graded cylindrical shells under linearly increasing dynamic torsional loading, *Engineering Structures*, **26**, 1321–1331.

Stein M. (1962), The effect on the buckling of perfect cylinders of prebuckling deformations and stresses induced by edge support, NASA TN D-1510.

Stein M. (1964), The influence of prebuckling deformations and stresses in the buckling of perfect cylinders, NASA TRR-190.

Suer H.S. and Harris L.A. (1959), The stability of thin-walled cylinders under combined torsion and external lateral or hydrostatic pressure, *Journal of Applied Mechanics ASME*, **26**, 138–140.

Woo J., Meguid S.A., Stranart J.C., and Liew K.M. (2005), Thermomechanical postbuckling analysis of moderately thick functionally graded plates and shallow shells, *International Journal of Mechanical Sciences*, **47**, 1147–1171.

Wu L., Jiang Z., and Liu J. (2005), Thermoelastic stability of functionally graded cylindrical shells, *Composite Structures*, **70**, 60–68.

Yamaki N. (1984), *Elastic Stability of Circular Cylindrical Shells*. Elsevier Science Publishers, B.V., North Holland, Amsterdam.

Yamaki N. and Kodama S. (1980), Perturbation analysis for the postbuckling and imperfection sensitivity of circular cylindrical shells under torsion, in *Theory of Shells* (eds. W.T. Koiter and G.K. Mikhailov), North Holland Publishing Company, Amsterdam, pp. 635–667.

Yamaki N. and Matsuda K. (1975), Postbuckling behavior of circular cylindrical shells under torsion, *Ingenieur-Archiv*, **45**, 79–89.

Zhang C.-L. and Shen H.-S. (2006), Buckling and postbuckling analysis of single-walled carbon nanotubes in thermal environments via molecular dynamics simulation, *Carbon*, **44**, 2608–2616.

Appendix A

In Equations 2.14 and 2.16

$$(\gamma_{110}, \gamma_{112}, \gamma_{114}) = (4/3h^2)[F_{11}^*, (F_{12}^* + F_{21}^* + 4F_{66}^*)/2, F_{22}^*]/D_{11}^*,$$

$$(\gamma_{120}, \gamma_{122}) = [D_{11}^* - 4F_{11}^*/3h^2, (D_{12}^* + 2D_{66}^*) - 4(F_{12}^* + 2F_{66}^*)/3h^2]/D_{11}^*$$

$$(\gamma_{131}, \gamma_{133}) = \left[(D_{12}^* + 2D_{66}^*) - 4(F_{21}^* + 2F_{66}^*)/3h^2, D_{22}^* - 4F_{22}^*/3h^2\right]/D_{11}^*$$

$$(\gamma_{140}, \gamma_{142}, \gamma_{144}) = [B_{21}^*, (B_{11}^* + B_{22}^* - 2B_{66}^*), B_{12}^*]/[D_{11}^* D_{22}^* A_{11}^* A_{22}^*]^{1/4}$$

$$(\gamma_{212}, \gamma_{214}) = (A_{12}^* + A_{66}^*/2, A_{11}^*)/A_{22}^*$$

$$(\gamma_{220}, \gamma_{222}) = \left[B_{21}^* - 4E_{21}^*/3h^2, (B_{11}^* - B_{66}^*) - 4(E_{11}^* - E_{66}^*)/3h^2\right]/[D_{11}^* D_{22}^* A_{11}^* A_{22}^*]^{1/4}$$

$$(\gamma_{231}, \gamma_{233}) = \left[(B_{22}^* - B_{66}^*) - 4(E_{22}^* - E_{66}^*)/3h^2, B_{12}^* - 4E_{12}^*/3h^2\right]/[D_{11}^* D_{22}^* A_{11}^* A_{22}^*]^{1/4}$$

$$(\gamma_{240}, \gamma_{242}, \gamma_{244}) = (4/3h^2)[[E_{21}^*, (E_{11}^* + E_{22}^* - 2E_{66}^*), E_{12}^*]/[D_{11}^* D_{22}^* A_{11}^* A_{22}^*]^{1/4}$$

$$(\gamma_{31}, \gamma_{41}) = (a^2/\pi^2)\left[A_{55} - 8D_{55}/h^2 + 16F_{55}/h^4, A_{44} - 8D_{44}/h^2 + 16F_{44}/h^4\right]/D_{11}^*,$$

$$(\gamma_{310}, \gamma_{312}) = (4/3h^2)\left[F_{11}^* - 4H_{11}^*/3h^2, (F_{21}^* + 2F_{66}^*) - 4(H_{12}^* + 2H_{66}^*)/3h^2\right]/D_{11}^*$$

$$(\gamma_{320}, \gamma_{322}) = \left(D_{11}^* - 8F_{11}^*/3h^2 + 16H_{11}^*/9h^4, D_{66}^* - 8F_{66}^*/3h^2 + 16H_{66}^*/9h^4\right)/D_{11}^*$$

$$\gamma_{331} = \left[(D_{12}^* + D_{66}^*) - 4(F_{12}^* + F_{21}^* + 2F_{66}^*)/3h^2 + 16(H_{12}^* + H_{66}^*)/9h^4\right]/D_{11}^*$$

$$(\gamma_{411}, \gamma_{413}) = (4/3h^2)\left[(F_{12}^* + 2F_{66}^*) - 4(H_{12}^* + 2H_{66}^*)/3h^2, F_{22}^* - 4H_{22}^*/3h^2\right]/D_{11}^*$$

$$(\gamma_{430}, \gamma_{432}) = \left(D_{66}^* - 8F_{66}^*/3h^2 + 16H_{66}^*/9h^4, D_{22}^* - 8F_{22}^*/3h^2 + 16H_{22}^*/9h^4\right)/D_{11}^*$$

$$(\gamma_{511}, \gamma_{522}) = \left(B_{11}^* - 4E_{11}^*/3h^2, B_{22}^* - 4E_{22}^*/3h^2\right)/[D_{11}^* D_{22}^* A_{11}^* A_{22}^*]^{1/4}$$

$$(\gamma_{611}, \gamma_{622}) = (4/3h^2)(E_{11}^*, E_{22}^*)/[D_{11}^* D_{22}^* A_{11}^* A_{22}^*]^{1/4}$$

$$(\gamma_{711}, \gamma_{722}) = (B_{11}^*, B_{22}^*)/[D_{11}^* D_{22}^* A_{11}^* A_{22}^*]^{1/4}$$

$$(\gamma_{812}, \gamma_{821}) = \left(D_{12}^* - 4F_{12}^*/3h^2, D_{12}^* - 4F_{21}^*/3h^2\right)/D_{11}^*$$

$$(\gamma_{912}, \gamma_{921}) = (4/3h^2)(F_{12}^*, F_{21}^*)/D_{11}^* \tag{A.1}$$

Appendix B

In Equation 2.9

$$A_x^T = \frac{h}{1-\nu}\left\{(\alpha_m - \alpha_c)(E_m - E_c)\frac{1}{2N+1} + [\alpha_c(E_m - E_c) + (\alpha_m - \alpha_c)E_c]\frac{1}{N+1} + \alpha_c E_c\right\}$$

$$D_x^T = \frac{h^2}{1-\nu}\left\{(\alpha_m - \alpha_c)(E_m - E_c)\frac{N}{2(N+1)(2N+1)}\right.$$

$$\left. + [\alpha_c(E_m - E_c) + (\alpha_m - \alpha_c)E_c]\frac{N}{2(N+1)(N+2)}\right\}$$

$$F_x^T = \frac{h^4}{1-\nu}\left\{(\alpha_m - \alpha_c)(E_m - E_c)\left[\frac{1}{8(2N+1)} - \frac{3}{8(N+1)(2N+1)}\right.\right.$$

$$\left. + \frac{3}{2(N+1)(2N+1)(2N+3)} - \frac{3}{2(N+1)(N+2)(2N+1)(2N+3)}\right]$$

$$+ [\alpha_c(E_m - E_c) + (\alpha_m - \alpha_c)E_c]\left[\frac{1}{8(N+1)} - \frac{3}{4(N+1)(N+2)}\right.$$

$$\left.\left. + \frac{3}{(N+1)(N+2)(N+3)} - \frac{6}{(N+1)(N+2)(N+3)(N+4)}\right]\right\} \tag{B.1}$$

Appendix C

In Equation 2.19

$$\begin{bmatrix} M_x^{(1)} & S_x^{(1)} \\ M_y^{(1)} & S_y^{(1)} \end{bmatrix} = \frac{16}{\pi^2} \frac{\Delta T}{(\overline{W}/h)} \begin{bmatrix} \gamma_{T3} & (\gamma_{T3} - \gamma_{T6}) \\ \gamma_{T4} & (\gamma_{T4} - \gamma_{T7}) \end{bmatrix} A_{11}^{(1)} \qquad (C.1)$$

and in Equations 2.32 through 2.34

$$A_W^{(0)} = -\pi^4 \Delta T \left[(\gamma_{T3} m^2 + \gamma_{T4} n^2 \beta^2) - \Theta_{11}^* \right]$$

$$A_W^{(1)} = \pi^4 C_{11} [S_{11} - C_{W1}]$$

$$A_W^{(2)} = -\pi^4 \Theta_{W22} \times \frac{h}{[D_{11}^* D_{22}^* A_{11}^* A_{22}^*]^{1/4}}$$

$$A_W^{(3)} = \pi^4 C_{11} \left\{ (\alpha_{313} + \alpha_{331})[S_{11} - C_{W1}] + \frac{1}{16} \gamma_{14} \gamma_{24} \Theta_2 \right\} \times \frac{h^2}{[D_{11}^* D_{22}^* A_{11}^* A_{22}^*]^{1/2}}$$

$$A_{MX}^{(0)} = -92416 \Delta T (\gamma_{T3} - \Theta_{X11}^*)/11025 + C_{X0}$$

$$A_{MX}^{(1)} = \pi^2 \Theta_{X11}$$

$$A_{MX}^{(2)} = -\pi^2 \Theta_{X22} \times \frac{h}{[D_{11}^* D_{22}^* A_{11}^* A_{22}^*]^{1/4}}$$

$$A_{MX}^{(3)} = \pi^2 [\Theta_{X11}(\alpha_{313} + \alpha_{331}) - \Theta_{X33}] \times \frac{h^2}{[D_{11}^* D_{22}^* A_{11}^* A_{22}^*]^{1/2}}$$

$$A_{MY}^{(0)} = -92416 \Delta T (\gamma_{T4} - \Theta_{Y11}^*)/11025 + C_{Y0}$$

$$A_{MY}^{(1)} = \pi^2 \Theta_{Y11}$$

$$A_{MY}^{(2)} = -\pi^2 \Theta_{Y22} \times \frac{h}{[D_{11}^* D_{22}^* A_{11}^* A_{22}^*]^{1/4}}$$

$$A_{MY}^{(3)} = \pi^2 [\Theta_{Y11}(\alpha_{313} + \alpha_{331}) - \Theta_{Y33}] \times \frac{h^2}{[D_{11}^* D_{22}^* A_{11}^* A_{22}^*]^{1/2}} \qquad (C.2)$$

in which

$$\Theta_2 = \frac{m^4}{\gamma_7} + \frac{n^4 \beta^4}{\gamma_6} + C_{22}, \quad S_{11} = g_{08} + \gamma_{14} \gamma_{24} m^2 n^2 \beta^2 \frac{g_{05} g_{07}}{g_{06}},$$

$$S_{13} = g_{138} + \gamma_{14} \gamma_{24} 9 m^2 n^2 \beta^2 \frac{g_{135} g_{137}}{g_{136}}, \quad S_{31} = g_{318} + \gamma_{14} \gamma_{24} 9 m^2 n^2 \beta^2 \frac{g_{315} g_{317}}{g_{316}}$$

$$\Theta_{W22} = \frac{1}{3C_{11}}\gamma_{14}\gamma_{24}m^2n^2\beta^2\left(\frac{\gamma_8}{\gamma_6} + \frac{\gamma_9}{\gamma_7} + 4\frac{g_{05}}{g_{06}}\right), \quad \Theta_{11}^* = g_{08}^* + \gamma_{14}\gamma_{24}\frac{g_{05}^*g_{07}}{g_{06}},$$

$$\gamma_6 = 1 + \gamma_{14}\gamma_{24}\gamma_{230}^2\frac{4m^2}{\gamma_{41} + \gamma_{322}4m^2}, \quad \gamma_7 = \gamma_{24}^2 + \gamma_{14}\gamma_{24}\gamma_{223}^2\frac{4n^2\beta^2}{\gamma_{31} + \gamma_{322}4n^2\beta^2},$$

$$\gamma_8 = \gamma_{140} - \gamma_{120}\gamma_{220}\frac{4m^2}{\gamma_{31} + \gamma_{320}4m^2}, \quad \gamma_9 = \gamma_{144} - \gamma_{133}\gamma_{233}\frac{4n^2\beta^2}{\gamma_{41} + \gamma_{432}4n^2\beta^2},$$

$$\gamma_{18} = \gamma_{722} - \gamma_{812}\gamma_{220}\frac{4m^2}{\gamma_{31} + \gamma_{320}4m^2}, \quad \gamma_{19} = \gamma_{711} - \gamma_{821}\gamma_{233}\frac{4n^2\beta^2}{\gamma_{41} + \gamma_{432}4n^2\beta^2},$$

$$J_{13} = S_{13} - C_{13}, \quad J_{31} = S_{31} - C_{31},$$

$$\alpha_{313} = \frac{1}{16}\gamma_{14}\gamma_{24}\frac{m^2}{\gamma_7 J_{13}}, \quad \alpha_{331} = \frac{1}{16}\gamma_{14}\gamma_{24}\frac{n^2\beta^2}{\gamma_6 J_{31}},$$

$$\Theta_{X11} = (\gamma_{110}m^2 + \gamma_{921}n^2\beta^2) + \gamma_{14}\gamma_{24}(\gamma_{140}m^2 + \gamma_{711}n^2\beta^2)\frac{g_{05}}{g_{06}}$$

$$+ \gamma_{120}m^2\left(\frac{g_{04}}{g_{00}} - \gamma_{14}\gamma_{24}\frac{g_{02}g_{05}}{g_{00}g_{06}}\right) + \gamma_{821}n^2\beta^2\left(\frac{g_{03}}{g_{00}} - \gamma_{14}\gamma_{24}\frac{g_{01}g_{05}}{g_{00}g_{06}}\right)$$

$$\Theta_{X22} = \frac{1}{8}\gamma_{14}\gamma_{24}\left(\frac{m^2\gamma_{19}}{\gamma_7} + \frac{n^2\beta^2\gamma_8}{\gamma_6} + C_{X2}\right), \quad \Theta_{X33} = \frac{1}{16}\gamma_{14}\gamma_{24}(\Theta_{X13} + \Theta_{X31})$$

$$\Theta_{X13} = \gamma_{14}\gamma_{24}\frac{m^4}{\gamma_7 J_{13}}\left[(\gamma_{140}m^2 + \gamma_{711}9n^2\beta^2) - \frac{m^2\gamma_{120}g_{132} + 9n^2\beta^2\gamma_{821}g_{131}}{g_{130}}\right]\frac{g_{135}}{g_{136}}$$

$$+ \frac{m^4}{\gamma_7 J_{13}}\left[(\gamma_{110}m^2 + \gamma_{921}9n^2\beta^2) + \frac{m^2\gamma_{120}g_{134} + 9n^2\beta^2\gamma_{821}g_{133}}{g_{130}}\right]$$

$$\Theta_{X31} = \gamma_{14}\gamma_{24}\frac{n^4\beta^4}{\gamma_6 J_{31}}\left[(\gamma_{140}9m^2 + \gamma_{711}n^2\beta^2) - \frac{9m^2\gamma_{120}g_{312} + n^2\beta^2\gamma_{821}g_{311}}{g_{310}}\right]\frac{g_{315}}{g_{316}}$$

$$+ \frac{n^4\beta^4}{\gamma_6 J_{31}}\left[(\gamma_{110}9m^2 + \gamma_{921}n^2\beta^2) + \frac{9m^2\gamma_{120}g_{314} + n^2\beta^2\gamma_{821}g_{313}}{g_{310}}\right]$$

$$\Theta_{Y11} = (\gamma_{912}m^2 + \gamma_{114}n^2\beta^2) + \gamma_{14}\gamma_{24}(\gamma_{722}m^2 + \gamma_{144}n^2\beta^2)\frac{g_{05}}{g_{06}}$$

$$+ \gamma_{812}m^2\left(\frac{g_{04}}{g_{00}} - \gamma_{14}\gamma_{24}\frac{g_{02}g_{05}}{g_{00}g_{06}}\right) + \gamma_{133}n^2\beta^2\left(\frac{g_{03}}{g_{00}} - \gamma_{14}\gamma_{24}\frac{g_{01}g_{05}}{g_{00}g_{06}}\right)$$

$$\Theta_{Y22} = \frac{1}{8}\gamma_{14}\gamma_{24}\left(\frac{m^2\gamma_9}{\gamma_7} + \frac{n^2\beta^2\gamma_{18}}{\gamma_6} + C_{Y2}\right), \quad \Theta_{Y33} = \frac{1}{16}\gamma_{14}\gamma_{24}(\Theta_{Y13} + \Theta_{Y31})$$

$$\Theta_{Y13} = \gamma_{14}\gamma_{24}\frac{m^4}{\gamma_7 J_{13}}\left[(\gamma_{722}m^2 + \gamma_{144}9n^2\beta^2) - \frac{m^2\gamma_{812}g_{132} + 9n^2\beta^2\gamma_{133}g_{131}}{g_{130}}\right]\frac{g_{135}}{g_{136}}$$

$$+ \frac{m^4}{\gamma_7 J_{13}}\left[(\gamma_{912}m^2 + \gamma_{114}9n^2\beta^2) + \frac{m^2\gamma_{812}g_{134} + 9n^2\beta^2\gamma_{133}g_{133}}{g_{130}}\right]$$

$$\Theta_{Y31} = \gamma_{14}\gamma_{24}\frac{n^4\beta^4}{\gamma_6 J_{31}}\left[(\gamma_{722}9m^2 + \gamma_{144}n^2\beta^2) - \frac{9m^2\gamma_{812}g_{312} + n^2\beta^2\gamma_{133}g_{311}}{g_{310}}\right]\frac{g_{315}}{g_{316}}$$

$$+ \frac{n^4\beta^4}{\gamma_6 J_{31}}\left[(\gamma_{912}9m^2 + \gamma_{114}n^2\beta^2) + \frac{9m^2\gamma_{812}g_{314} + n^2\beta^2\gamma_{133}g_{313}}{g_{310}}\right]$$

$$\Theta_{X11}^* = \gamma_{120}m^2\frac{g_{04}^*}{g_{00}} + \gamma_{821}n^2\beta^2\frac{g_{03}^*}{g_{00}}, \quad \Theta_{Y11}^* = \gamma_{812}m^2\frac{g_{04}^*}{g_{00}} + \gamma_{133}n^2\beta^2\frac{g_{03}^*}{g_{00}}$$

$$g_{00} = (\gamma_{31} + \gamma_{320}m^2 + \gamma_{322}n^2\beta^2)(\gamma_{41} + \gamma_{430}m^2 + \gamma_{432}n^2\beta^2) - \gamma_{331}^2 m^2 n^2\beta^2$$

$$g_{01} = (\gamma_{31} + \gamma_{320}m^2 + \gamma_{322}n^2\beta^2)(\gamma_{231}m^2 + \gamma_{233}n^2\beta^2) - \gamma_{331}n^2\beta^2(\gamma_{220}m^2 + \gamma_{222}n^2\beta^2)$$

$$g_{02} = (\gamma_{41} + \gamma_{430}m^2 + \gamma_{432}n^2\beta^2)(\gamma_{220}m^2 + \gamma_{222}n^2\beta^2) - \gamma_{331}m^2(\gamma_{231}m^2 + \gamma_{233}n^2\beta^2)$$

$$g_{03} = (\gamma_{31} + \gamma_{320}m^2 + \gamma_{322}n^2\beta^2)(\gamma_{41} - \gamma_{411}m^2 - \gamma_{413}n^2\beta^2)$$
$$- \gamma_{331}m^2(\gamma_{31} - \gamma_{310}m^2 - \gamma_{312}n^2\beta^2)$$

$$g_{03}^* = (\gamma_{31} + \gamma_{320}m^2 + \gamma_{322}n^2\beta^2)(\gamma_{T4} - \gamma_{T7}) - \gamma_{331}m^2(\gamma_{T3} - \gamma_{T6})$$

$$g_{04} = (\gamma_{41} + \gamma_{430}m^2 + \gamma_{432}n^2\beta^2)(\gamma_{31} - \gamma_{310}m^2 - \gamma_{312}n^2\beta^2)$$
$$- \gamma_{331}n^2\beta^2(\gamma_{41} - \gamma_{411}m^2 - \gamma_{413}n^2\beta^2)$$

$$g_{04}^* = (\gamma_{41} + \gamma_{430}m^2 + \gamma_{432}n^2\beta^2)(\gamma_{T3} - \gamma_{T6}) - \gamma_{331}n^2\beta^2(\gamma_{T4} - \gamma_{T7})$$

$$g_{05} = (\gamma_{240}m^4 + \gamma_{242}m^2 n^2\beta + \gamma_{244}n^4\beta^4)$$
$$+ \frac{m^2(\gamma_{220}m^2 + \gamma_{222}n^2\beta^2)g_{04} + n^2\beta^2(\gamma_{231}m^2 + \gamma_{233}n^2\beta^2)g_{03}}{g_{00}},$$

$$g_{05}^* = \frac{m^2(\gamma_{220}m^2 + \gamma_{222}n^2\beta^2)g_{04}^* + n^2\beta^2(\gamma_{231}m^2 + \gamma_{233}n^2\beta^2)g_{03}^*}{g_{00}},$$

$$g_{06} = (m^4 + 2\gamma_{212}m^2 n^2\beta^2 + \gamma_{214}n^4\beta^4)$$
$$+ \gamma_{14}\gamma_{24}\frac{m^2(\gamma_{220}m^2 + \gamma_{222}n^2\beta^2)g_{02} + n^2\beta^2(\gamma_{231}m^2 + \gamma_{233}n^2\beta^2)g_{01}}{g_{00}}$$

$$g_{07} = (\gamma_{140}m^4 + \gamma_{142}m^2 n^2\beta + \gamma_{144}n^4\beta^4)$$
$$- \frac{m^2(\gamma_{120}m^2 + \gamma_{122}n^2\beta^2)g_{02} + n^2\beta^2(\gamma_{131}m^2 + \gamma_{133}n^2\beta^2)g_{01}}{g_{00}}$$

$$g_{08} = (\gamma_{110}m^4 + 2\gamma_{112}m^2 n^2\beta^2 + \gamma_{114}n^4\beta^4)$$
$$+ \frac{m^2(\gamma_{120}m^2 + \gamma_{122}n^2\beta^2)g_{04} + n^2\beta^2(\gamma_{131}m^2 + \gamma_{133}n^2\beta^2)g_{03}}{g_{00}}$$

$$g_{08}^* = \frac{m^2(\gamma_{120}m^2 + \gamma_{122}n^2\beta^2)g_{04}^* + n^2\beta^2(\gamma_{131}m^2 + \gamma_{133}n^2\beta^2)g_{03}^*}{g_{00}}$$

$$g_{130} = (\gamma_{31} + \gamma_{320}m^2 + \gamma_{322}9n^2\beta^2)(\gamma_{41} + \gamma_{430}m^2 + \gamma_{432}9n^2\beta^2) - \gamma_{331}^2 9m^2 n^2\beta^2$$

$$g_{131} = (\gamma_{31} + \gamma_{320}m^2 + \gamma_{322}9n^2\beta^2)(\gamma_{231}m^2 + \gamma_{233}9n^2\beta^2) - \gamma_{331}9n^2\beta^2(\gamma_{220}m^2 + \gamma_{222}n^2\beta^2)$$

$$g_{132} = (\gamma_{41} + \gamma_{430}m^2 + \gamma_{432}9n^2\beta^2)(\gamma_{220}m^2 + \gamma_{222}9n^2\beta^2) - \gamma_{331}m^2(\gamma_{231}m^2 + \gamma_{233}9n^2\beta^2)$$

$$g_{133} = \left(\gamma_{31} + \gamma_{320}m^2 + \gamma_{322}9n^2\beta^2\right)\left(\gamma_{41} - \gamma_{411}m^2 - \gamma_{413}9n^2\beta^2\right)$$
$$- \gamma_{331}m^2\left(\gamma_{31} - \gamma_{310}m^2 - \gamma_{312}9n^2\beta^2\right)$$

$$g_{134} = \left(\gamma_{41} + \gamma_{430}m^2 + \gamma_{432}9n^2\beta^2\right)\left(\gamma_{31} - \gamma_{310}m^2 - \gamma_{312}9n^2\beta^2\right)$$
$$- \gamma_{331}9n^2\beta^2\left(\gamma_{41} - \gamma_{411}m^2 - \gamma_{413}9n^2\beta^2\right)$$

$$g_{135} = \left(\gamma_{240}m^4 + 9\gamma_{242}m^2n^2\beta + 81\gamma_{244}n^4\beta^4\right)$$
$$+ \frac{m^2\left(\gamma_{220}m^2 + \gamma_{222}9n^2\beta^2\right)g_{134} + 9n^2\beta^2\left(\gamma_{231}m^2 + \gamma_{233}9n^2\beta^2\right)g_{133}}{g_{130}},$$

$$g_{136} = \left(m^4 + 18\gamma_{212}m^2n^2\beta^2 + 81\gamma_{214}n^4\beta^4\right)$$
$$+ \gamma_{14}\gamma_{24}\frac{m^2\left(\gamma_{220}m^2 + \gamma_{222}9n^2\beta^2\right)g_{132} + 9n^2\beta^2\left(\gamma_{231}m^2 + \gamma_{233}9n^2\beta^2\right)g_{131}}{g_{130}}$$

$$g_{137} = \left(\gamma_{140}m^4 + 9\gamma_{142}m^2n^2\beta + 81\gamma_{144}n^4\beta^4\right)$$
$$- \frac{m^2\left(\gamma_{120}m^2 + \gamma_{122}9n^2\beta^2\right)g_{132} + 9n^2\beta^2\left(\gamma_{131}m^2 + \gamma_{133}9n^2\beta^2\right)g_{131}}{g_{130}}$$

$$g_{138} = \left(\gamma_{110}m^4 + 18\gamma_{112}m^2n^2\beta^2 + \gamma_{114}81n^4\beta^4\right)$$
$$+ \frac{m^2\left(\gamma_{120}m^2 + \gamma_{122}9n^2\beta^2\right)g_{134} + 9n^2\beta^2\left(\gamma_{131}m^2 + \gamma_{133}9n^2\beta^2\right)g_{133}}{g_{130}}$$

$$g_{310} = \left(\gamma_{31} + \gamma_{320}9m^2 + \gamma_{322}n^2\beta^2\right)\left(\gamma_{41} + \gamma_{430}9m^2 + \gamma_{432}n^2\beta^2\right) - \gamma_{331}^2 9m^2n^2\beta^2$$

$$g_{311} = \left(\gamma_{31} + \gamma_{320}9m^2 + \gamma_{322}n^2\beta^2\right)\left(\gamma_{231}9m^2 + \gamma_{233}n^2\beta^2\right) - \gamma_{331}n^2\beta^2\left(\gamma_{220}9m^2 + \gamma_{222}n^2\beta^2\right)$$

$$g_{312} = \left(\gamma_{41} + \gamma_{430}9m^2 + \gamma_{432}n^2\beta^2\right)\left(\gamma_{220}9m^2 + \gamma_{222}n^2\beta^2\right) - \gamma_{331}9m^2\left(\gamma_{231}9m^2 + \gamma_{233}n^2\beta^2\right)$$

$$g_{313} = \left(\gamma_{31} + \gamma_{320}9m^2 + \gamma_{322}n^2\beta^2\right)\left(\gamma_{41} - \gamma_{411}9m^2 - \gamma_{413}n^2\beta^2\right)$$
$$- \gamma_{331}9m^2\left(\gamma_{31} - \gamma_{310}9m^2 - \gamma_{312}n^2\beta^2\right)$$

$$g_{314} = \left(\gamma_{41} + \gamma_{430}9m^2 + \gamma_{432}n^2\beta^2\right)\left(\gamma_{31} - \gamma_{310}9m^2 - \gamma_{312}n^2\beta^2\right)$$
$$- \gamma_{331}n^2\beta^2\left(\gamma_{41} - \gamma_{411}9m^2 - \gamma_{413}n^2\beta^2\right)$$

$$g_{315} = \left(81\gamma_{240}m^4 + 9\gamma_{242}m^2n^2\beta + \gamma_{244}n^4\beta^4\right)$$
$$+ \frac{9m^2\left(\gamma_{220}m^2 + \gamma_{222}n^2\beta^2\right)g_{314} + n^2\beta^2\left(\gamma_{231}9m^2 + \gamma_{233}n^2\beta^2\right)g_{313}}{g_{310}},$$

$$g_{316} = \left(81m^4 + 18\gamma_{212}m^2n^2\beta^2 + \gamma_{214}n^4\beta^4\right)$$
$$+ \gamma_{14}\gamma_{24}\frac{9m^2\left(\gamma_{220}9m^2 + \gamma_{222}n^2\beta^2\right)g_{312} + n^2\beta^2\left(\gamma_{231}9m^2 + \gamma_{233}n^2\beta^2\right)g_{311}}{g_{310}},$$

$$g_{317} = \left(81\gamma_{140}m^4 + 9\gamma_{142}m^2n^2\beta + \gamma_{144}n^4\beta^4\right)$$
$$- \frac{9m^2\left(\gamma_{120}9m^2 + \gamma_{122}n^2\beta^2\right)g_{312} + n^2\beta^2\left(\gamma_{131}9m^2 + \gamma_{133}n^2\beta^2\right)g_{311}}{g_{310}},$$

$$g_{318} = \left(81\gamma_{110}m^4 + 18\gamma_{112}m^2n^2\beta^2 + \gamma_{114}n^4\beta^4\right)$$
$$+ \frac{9m^2\left(\gamma_{120}9m^2 + \gamma_{122}n^2\beta^2\right)g_{314} + n^2\beta^2\left(\gamma_{131}9m^2 + \gamma_{133}n^2\beta^2\right)g_{313}}{g_{310}} \tag{C.3}$$

and

$$C_{11} = \frac{\pi^2}{16} mn \quad \text{(for uniform load)} \tag{C.4a}$$

$$C_{11} = 1 \quad \text{(for sinusoidal load)} \tag{C.4b}$$

in the above equations, for the case of movable edges

$$C_{W1} = C_{X0} = C_{Y0} = C_{X2} = C_{Y2} = C_{22} = C_{13} = C_{31} = 0 \tag{C.5a}$$

and for the case of immovable edges

$$C_{W1} = \gamma_{14}(\gamma_{T1}m^2 + \gamma_{T2}n^2\beta^2)\Delta T,$$

$$C_{X0} = \pi^2 \gamma_{14}(\gamma_{711}\gamma_{T1} + \gamma_{140}\gamma_{T2})\Delta T \times \frac{[D_{11}^* D_{22}^* A_{11}^* A_{22}^*]^{1/4}}{h}$$

$$C_{Y0} = \pi^2 \gamma_{14}(\gamma_{144}\gamma_{T1} + \gamma_{722}\gamma_{T2})\Delta T \times \frac{[D_{11}^* D_{22}^* A_{11}^* A_{22}^*]^{1/4}}{h}$$

$$C_{22} = 2\frac{(m^4 + \gamma_{24}^2 n^4\beta^4) + 2\gamma_5 m^2 n^2\beta^2}{\gamma_{24}^2 - \gamma_5^2}$$

$$C_{13} = \gamma_{14}(\gamma_{T1}m^2 + 9\gamma_{T2}n^2\beta^2)\Delta T, \quad C_{31} = \gamma_{14}(9\gamma_{T1}m^2 + \gamma_{T2}n^2\beta^2)\Delta T$$

$$C_{X2} = \gamma_{711}\frac{m^2 + \gamma_5 n^2\beta^2}{\gamma_{24}^2 - \gamma_5^2} + \gamma_{140}\frac{\gamma_5 m^2 + \gamma_{24}^2 n^2\beta^2}{\gamma_{24}^2 - \gamma_5^2},$$

$$C_{Y2} = \gamma_{144}\frac{m^2 + \gamma_5 n^2\beta^2}{\gamma_{24}^2 - \gamma_5^2} + \gamma_{722}\frac{\gamma_5 m^2 + \gamma_{24}^2 n^2\beta^2}{\gamma_{24}^2 - \gamma_5^2} \tag{C.5b}$$

Appendix D

In Equation 2.44, when $\Delta T \neq 0$

$$
\begin{aligned}
A_x^{\mathrm{T}} = {}& \frac{h}{1-\nu_{\mathrm{f}}}\frac{T_{\mathrm{U}}-T_0}{\Delta T}\left\{(\alpha_{\mathrm{m}}-\alpha_{\mathrm{c}})(E_{\mathrm{m}}-E_{\mathrm{c}})\frac{1}{2N+1}\right. \\
& \left. + [\alpha_{\mathrm{c}}(E_{\mathrm{m}}-E_{\mathrm{c}})+E_{\mathrm{c}}(\alpha_{\mathrm{m}}-\alpha_{\mathrm{c}})]\frac{1}{N+1}+\alpha_{\mathrm{c}}E_{\mathrm{c}}\right\} \\
& + \frac{h}{1-\nu_{\mathrm{f}}}\frac{T_{\mathrm{L}}-T_{\mathrm{U}}}{C\Delta T}\left\{(\alpha_{\mathrm{m}}-\alpha_{\mathrm{c}})(E_{\mathrm{m}}-E_{\mathrm{c}})\left[\frac{1}{2N+2}-\frac{\kappa_{\mathrm{mc}}}{(N+1)(3N+2)\kappa_{\mathrm{c}}}\right.\right. \\
& + \frac{\kappa_{\mathrm{mc}}^2}{(2N+1)(4N+2)\kappa_{\mathrm{c}}^2}-\frac{\kappa_{\mathrm{mc}}^3}{(3N+1)(5N+2)\kappa_{\mathrm{c}}^3}+\frac{\kappa_{\mathrm{mc}}^4}{(4N+1)(6N+2)\kappa_{\mathrm{c}}^4} \\
& \left.-\frac{\kappa_{\mathrm{mc}}^5}{(5N+1)(7N+2)\kappa_{\mathrm{c}}^5}\right]+[\alpha_{\mathrm{c}}(E_{\mathrm{m}}-E_{\mathrm{c}})+E_{\mathrm{c}}(\alpha_{\mathrm{m}}-\alpha_{\mathrm{c}})]\left[\frac{1}{N+2}\right. \\
& -\frac{\kappa_{\mathrm{mc}}}{(N+1)(2N+2)\kappa_{\mathrm{c}}}+\frac{\kappa_{\mathrm{mc}}^2}{(2N+1)(3N+2)\kappa_{\mathrm{c}}^2}-\frac{\kappa_{\mathrm{mc}}^3}{(3N+1)(4N+2)\kappa_{\mathrm{c}}^3} \\
& \left.+\frac{\kappa_{\mathrm{mc}}^4}{(4N+1)(5N+2)\kappa_{\mathrm{c}}^4}-\frac{\kappa_{\mathrm{mc}}^5}{(5N+1)(6N+2)\kappa_{\mathrm{c}}^5}\right] \\
& +\alpha_{\mathrm{c}}E_{\mathrm{c}}\left[\frac{1}{2}-\frac{\kappa_{\mathrm{mc}}}{(N+1)(N+2)\kappa_{\mathrm{c}}}+\frac{\kappa_{\mathrm{mc}}^2}{(2N+1)(2N+2)\kappa_{\mathrm{c}}^2}\right. \\
& \left.\left.-\frac{\kappa_{\mathrm{mc}}^3}{(3N+1)(3N+2)\kappa_{\mathrm{c}}^3}+\frac{\kappa_{\mathrm{mc}}^4}{(4N+1)(4N+2)\kappa_{\mathrm{c}}^4}-\frac{\kappa_{\mathrm{mc}}^5}{(5N+1)(5N+2)\kappa_{\mathrm{c}}^5}\right]\right\}
\end{aligned}
$$

$$\text{(D.1)}$$

$$
D_x^{\mathrm{T}} = \frac{h^2}{1 - \nu_{\mathrm{f}}} \frac{T_{\mathrm{U}} - T_0}{\Delta T} \left\{ (\alpha_{\mathrm{m}} - \alpha_{\mathrm{c}})(E_{\mathrm{m}} - E_{\mathrm{c}}) \left[\frac{1}{2(2N + 1)} - \frac{1}{(2N + 1)(2N + 2)} \right] \right.
$$

$$
\left. + [\alpha_{\mathrm{c}}(E_{\mathrm{m}} - E_{\mathrm{c}}) + E_{\mathrm{c}}(\alpha_{\mathrm{m}} - \alpha_{\mathrm{c}})] \left[\frac{1}{2(N + 1)} - \frac{1}{(N + 1)(N + 2)} \right] \right\}
$$

$$
+ \frac{h^2}{1 - \nu_{\mathrm{f}}} \frac{T_{\mathrm{L}} - T_{\mathrm{U}}}{C\Delta T} \left\{ (\alpha_{\mathrm{m}} - \alpha_{\mathrm{c}})(E_{\mathrm{m}} - E_{\mathrm{c}}) \left(\left[\frac{1}{2(2N + 2)} - \frac{1}{(2N + 2)(2N + 3)} \right] \right. \right.
$$

$$
- \frac{\kappa_{\mathrm{mc}}}{(N + 1)\kappa_{\mathrm{c}}} \left[\frac{1}{2(3N + 2)} - \frac{1}{(3N + 2)(3N + 3)} \right]
$$

$$
+ \frac{\kappa_{\mathrm{mc}}^2}{(2N + 1)\kappa_{\mathrm{c}}^2} \left[\frac{1}{2(4N + 2)} - \frac{1}{(4N + 2)(4N + 3)} \right]
$$

$$
- \frac{\kappa_{\mathrm{mc}}^3}{(3N + 1)\kappa_{\mathrm{c}}^3} \left[\frac{1}{2(5N + 2)} - \frac{1}{(5N + 2)(5N + 3)} \right]
$$

$$
+ \frac{\kappa_{\mathrm{mc}}^4}{(4N + 1)\kappa_{\mathrm{c}}^4} \left[\frac{1}{2(6N + 2)} - \frac{1}{(6N + 2)(6N + 3)} \right]
$$

$$
\left. - \frac{\kappa_{\mathrm{mc}}^5}{(5N + 1)\kappa_{\mathrm{c}}^5} \left[\frac{1}{2(7N + 2)} - \frac{1}{(7N + 2)(7N + 3)} \right] \right)
$$

$$
+ [\alpha_{\mathrm{c}}(E_{\mathrm{m}} - E_{\mathrm{c}}) + E_{\mathrm{c}}(\alpha_{\mathrm{m}} - \alpha_{\mathrm{c}})] \left(\left[\frac{1}{2(N + 2)} - \frac{1}{(N + 2)(N + 3)} \right] \right.
$$

$$
- \frac{\kappa_{\mathrm{mc}}}{(N + 1)\kappa_{\mathrm{c}}} \left[\frac{1}{2(2N + 2)} - \frac{1}{(2N + 2)(2N + 3)} \right]
$$

$$
+ \frac{\kappa_{\mathrm{mc}}^2}{(2N + 1)\kappa_{\mathrm{c}}^2} \left[\frac{1}{2(3N + 2)} - \frac{1}{(3N + 2)(3N + 3)} \right]
$$

$$
- \frac{\kappa_{\mathrm{mc}}^3}{(3N + 1)\kappa_{\mathrm{c}}^3} \left[\frac{1}{2(4N + 2)} - \frac{1}{(4N + 2)(4N + 3)} \right]
$$

$$
+ \frac{\kappa_{\mathrm{mc}}^4}{(4N + 1)\kappa_{\mathrm{c}}^4} \left[\frac{1}{2(5N + 2)} - \frac{1}{(5N + 2)(5N + 3)} \right]
$$

$$
\left. - \frac{\kappa_{\mathrm{mc}}^5}{(5N + 1)\kappa_{\mathrm{c}}^5} \left[\frac{1}{2(6N + 2)} - \frac{1}{(6N + 2)(6N + 3)} \right] \right)
$$

$$
+ \alpha_{\mathrm{c}} E_{\mathrm{c}} \left(\frac{1}{12} - \frac{\kappa_{\mathrm{mc}}}{(N + 1)\kappa_{\mathrm{c}}} \left[\frac{1}{2(N + 2)} - \frac{1}{(N + 2)(N + 3)} \right] \right.
$$

$$
+ \frac{\kappa_{\mathrm{mc}}^2}{(2N + 1)\kappa_{\mathrm{c}}^2} \left[\frac{1}{2(2N + 2)} - \frac{1}{(2N + 2)(2N + 3)} \right]
$$

$$
- \frac{\kappa_{\mathrm{mc}}^3}{(3N + 1)\kappa_{\mathrm{c}}^3} \left[\frac{1}{2(3N + 2)} - \frac{1}{(3N + 2)(3N + 3)} \right]
$$

$$
+ \frac{\kappa_{\mathrm{mc}}^4}{(4N + 1)\kappa_{\mathrm{c}}^4} \left[\frac{1}{2(4N + 2)} - \frac{1}{(4N + 2)(4N + 3)} \right]
$$

$$
\left. - \frac{\kappa_{\mathrm{mc}}^5}{(5N + 1)\kappa_{\mathrm{c}}^5} \left[\frac{1}{2(5N + 2)} - \frac{1}{(5N + 2)(5N + 3)} \right] \right)
$$

$$
\tag{D.2}
$$

$$F_x^{\mathrm{T}} = \frac{h^4}{1-\nu_{\mathrm{f}}} \frac{T_{\mathrm{U}} - T_0}{\Delta T} \left\{ (\alpha_{\mathrm{m}} - \alpha_{\mathrm{c}})(E_{\mathrm{m}} - E_{\mathrm{c}}) \left[\frac{1}{8(2N+1)} - \frac{3}{4(2N+1)(2N+2)} \right. \right.$$

$$+ \frac{3}{(2N+1)(2N+2)(2N+3)} - \frac{6}{(2N+1)(2N+2)(2N+3)(2N+4)} \Bigg]$$

$$+ \left[\alpha_{\mathrm{c}}(E_{\mathrm{m}} - E_{\mathrm{c}}) + E_{\mathrm{c}}(\alpha_{\mathrm{m}} - \alpha_{\mathrm{c}}) \right] \left[\frac{1}{8(N+1)} - \frac{3}{4(N+1)(N+2)} \right.$$

$$+ \left. \left. \frac{3}{(N+1)(N+2)(N+3)} - \frac{6}{(N+1)(N+2)(N+3)(N+4)} \right] \right\}$$

$$+ \frac{h^4}{1-\nu_{\mathrm{f}}} \frac{T_{\mathrm{L}} - T_{\mathrm{U}}}{C\Delta T} \left\{ (\alpha_{\mathrm{m}} - \alpha_{\mathrm{c}})(E_{\mathrm{m}} - E_{\mathrm{c}}) \left(\left[\frac{1}{8(2N+2)} \right. \right. \right.$$

$$- \frac{3}{4(2N+2)(2N+3)} + \frac{3}{(2N+2)(2N+3)(2N+4)}$$

$$- \frac{6}{(2N+2)(2N+3)(2N+4)(2N+5)} \Bigg] - \frac{\kappa_{\mathrm{mc}}}{(N+1)\kappa_{\mathrm{c}}} \left[\frac{1}{8(3N+2)} \right.$$

$$- \frac{3}{4(3N+2)(3N+3)} + \frac{3}{(3N+2)(3N+3)(3N+4)}$$

$$- \frac{6}{(3N+2)(3N+3)(3N+4)(3N+5)} \Bigg] + \frac{\kappa_{\mathrm{mc}}^2}{(2N+1)\kappa_{\mathrm{c}}^2}$$

$$\times \left[\frac{1}{8(4N+2)} - \frac{3}{4(4N+2)(4N+3)} + \frac{3}{(4N+2)(4N+3)(4N+4)} \right.$$

$$- \frac{6}{(4N+2)(4N+3)(4N+4)(4N+5)} \Bigg] - \frac{\kappa_{\mathrm{mc}}^3}{(3N+1)\kappa_{\mathrm{c}}^3} \left[\frac{1}{8(5N+2)} \right.$$

$$- \frac{3}{4(5N+2)(5N+3)} + \frac{3}{(5N+2)(5N+3)(5N+4)}$$

$$- \frac{6}{(5N+2)(5N+3)(5N+4)(5N+5)} \Bigg] + \frac{\kappa_{\mathrm{mc}}^4}{(4N+1)\kappa_{\mathrm{c}}^4} \left[\frac{1}{8(6N+2)} \right.$$

$$- \frac{3}{4(6N+2)(6N+3)} + \frac{3}{(6N+2)(6N+3)(6N+4)}$$

$$- \frac{6}{(6N+2)(6N+3)(6N+4)(6N+5)} \Bigg] - \frac{\kappa_{\mathrm{mc}}^5}{(5N+1)\kappa_{\mathrm{c}}^5} \left[\frac{1}{8(7N+2)} \right.$$

$$- \frac{3}{4(7N+2)(7N+3)} + \frac{3}{(7N+2)(7N+3)(7N+4)}$$

$$- \frac{6}{(7N+2)(7N+3)(7N+4)(7N+5)} \Bigg] \right) + \left[\alpha_{\mathrm{c}}(E_{\mathrm{m}} - E_{\mathrm{c}}) + E_{\mathrm{c}}(\alpha_{\mathrm{m}} - \alpha_{\mathrm{c}}) \right]$$

$$\times \left(\left[\frac{1}{8(N+2)} - \frac{3}{4(N+2)(N+3)} + \frac{3}{(N+2)(N+3)(N+4)} - \frac{6}{(N+2)(N+3)(N+4)(N+5)} \right] \right.$$

$$- \frac{\kappa_{\mathrm{mc}}}{(N+1)\kappa_{\mathrm{c}}} \left[\frac{1}{8(2N+2)} - \frac{3}{4(2N+2)(2N+3)} + \frac{3}{(2N+2)(2N+3)(2N+4)} \right.$$

$$\left. - \frac{6}{(2N+2)(2N+3)(2N+4)(2N+5)} \right] + \frac{\kappa_{\mathrm{mc}}^2}{(2N+1)\kappa_{\mathrm{c}}^2} \left[\frac{1}{8(3N+2)} - \frac{3}{4(3N+2)(3N+3)} \right.$$

$$\left. + \frac{3}{(3N+2)(3N+3)(3N+4)} - \frac{6}{(3N+2)(3N+3)(3N+4)(3N+5)} \right]$$

$$- \frac{\kappa_{\mathrm{mc}}^3}{(3N+1)\kappa_{\mathrm{c}}^3} \left[\frac{1}{8(4N+2)} - \frac{3}{4(4N+2)(4N+3)} + \frac{3}{(4N+2)(4N+3)(4N+4)} \right.$$

$$\left. - \frac{6}{(4N+2)(4N+3)(4N+4)(4N+5)} \right] + \frac{\kappa_{\mathrm{mc}}^4}{(4N+1)\kappa_{\mathrm{c}}^4} \left[\frac{1}{8(5N+2)} - \frac{3}{4(5N+2)(5N+3)} \right.$$

$$\left. + \frac{3}{(5N+2)(5N+3)(5N+4)} - \frac{6}{(5N+2)(5N+3)(5N+4)(5N+5)} \right]$$

$$- \frac{\kappa_{\mathrm{mc}}^5}{(5N+1)\kappa_{\mathrm{c}}^5} \left[\frac{1}{8(6N+2)} - \frac{3}{4(6N+2)(6N+3)} + \frac{3}{(6N+2)(6N+3)(6N+4)} \right.$$

$$\left. \left. - \frac{6}{(6N+2)(6N+3)(6N+4)(6N+5)} \right] \right) + \alpha_{\mathrm{c}} E_{\mathrm{c}} \left(\frac{1}{80} - \frac{\kappa_{\mathrm{mc}}}{(N+1)\kappa_{\mathrm{c}}} \left[\frac{1}{8(N+2)} \right. \right.$$

$$\left. - \frac{3}{4(N+2)(N+3)} + \frac{3}{(N+2)(N+3)(N+4)} - \frac{6}{(N+2)(N+3)(N+4)(N+5)} \right]$$

$$+ \frac{\kappa_{\mathrm{mc}}^2}{(2N+1)\kappa_{\mathrm{c}}^2} \left[\frac{1}{8(2N+2)} - \frac{3}{4(2N+2)(2N+3)} + \frac{3}{(2N+2)(2N+3)(2N+4)} \right.$$

$$\left. - \frac{6}{(2N+2)(2N+3)(2N+4)(2N+5)} \right] - \frac{\kappa_{\mathrm{mc}}^3}{(3N+1)\kappa_{\mathrm{c}}^3} \left[\frac{1}{8(3N+2)} - \frac{3}{4(3N+2)(3N+3)} \right.$$

$$\left. + \frac{3}{(3N+2)(3N+3)(3N+4)} - \frac{6}{(3N+2)(3N+3)(3N+4)(3N+5)} \right]$$

$$+ \frac{\kappa_{\mathrm{mc}}^4}{(4N+1)\kappa_{\mathrm{c}}^4} \left[\frac{1}{8(4N+2)} - \frac{3}{4(4N+2)(4N+3)} + \frac{3}{(4N+2)(4N+3)(4N+4)} \right.$$

$$\left. - \frac{6}{(4N+2)(4N+3)(4N+4)(4N+5)} \right] - \frac{\kappa_{\mathrm{mc}}^5}{(5N+1)\kappa_{\mathrm{c}}^5} \left[\frac{1}{8(5N+2)} \right.$$

$$- \frac{3}{4(5N+2)(5N+3)} + \frac{3}{(5N+2)(5N+3)(5N+4)}$$

$$\left. \left. \left. - \frac{6}{(5N+2)(5N+3)(5N+4)(5N+5)} \right] \right) \right\} \tag{D.3}$$

Appendix E

In Equation 2.45

$$\begin{bmatrix} M_x^0 & S_x^0 \\ M_y^0 & S_y^0 \end{bmatrix} = \frac{16}{\pi^2} \begin{bmatrix} \gamma_{T3} & (\gamma_{T3} - \gamma_{T6}) \\ \gamma_{T4} & (\gamma_{T4} - \gamma_{T7}) \end{bmatrix} \Delta T \times \frac{h}{[D_{11}^* D_{22}^* A_{11}^* A_{22}^*]^{1/4}} \tag{E.1}$$

and in Equations 2.46 through 2.49

$$A_W^{(1)} = \frac{[D_{11}^* D_{22}^* A_{11}^* A_{22}^*]^{1/4}}{h}$$

$$A_W^{(3)} = \frac{1}{16} \gamma_{14} \gamma_{24} \left(\frac{m^4}{J_{13} \gamma_7} + \frac{n^4 \beta^4}{J_{31} \gamma_6} \right) \times \frac{[D_{11}^* D_{22}^* A_{11}^* A_{22}^*]^{1/4}}{h}$$

$$A_{MX}^{(0)} = C_{X0} - 16\Delta T \left(\frac{5{,}776}{11{,}025} \gamma_{T3} - \frac{1}{mn} \frac{(\gamma_{T3} - \gamma_{T6})m^2 g_{104} + (\gamma_{T4} - \gamma_{T7})n^2 \beta^2 g_{103}}{g_{00}} \right)$$

$$A_{MX}^{(1)} = \pi^2 \Theta_{X11} \times \frac{[D_{11}^* D_{22}^* A_{11}^* A_{22}^*]^{1/4}}{h}$$

$$A_{MX}^{(2)} = -\pi^2 \Theta_{X22} \times \frac{[D_{11}^* D_{22}^* A_{11}^* A_{22}^*]^{1/4}}{h}$$

$$A_{MX}^{(3)} = -\pi^2 \Theta_{X33} \times \frac{[D_{11}^* D_{22}^* A_{11}^* A_{22}^*]^{1/4}}{h}$$

$$A_{MY}^{(0)} = \pi^2 \gamma_{14}(\gamma_{144} \gamma_{T1} + \gamma_{722} \gamma_{T2}) \Delta T \times \frac{[D_{11}^* D_{22}^* A_{11}^* A_{22}^*]^{1/4}}{h}$$
$$- 16\Delta T \left(\frac{5{,}776}{11{,}025} \gamma_{T4} - \frac{1}{mn} \frac{(\gamma_{T3} - \gamma_{T6})m^2 g_{204} + (\gamma_{T4} - \gamma_{T7})n^2 \beta^2 g_{203}}{g_{00}} \right)$$

$$A_{MY}^{(1)} = \pi^2 \Theta_{Y11} \times \frac{[D_{11}^* D_{22}^* A_{11}^* A_{22}^*]^{1/4}}{h}$$

$$A_{MY}^{(2)} = -\pi^2 \Theta_{Y22} \times \frac{[D_{11}^* D_{22}^* A_{11}^* A_{22}^*]^{1/4}}{h}$$

$$A_{MY}^{(3)} = -\pi^2 \Theta_{Y33} \times \frac{[D_{11}^* D_{22}^* A_{11}^* A_{22}^*]^{1/4}}{h}$$

$$\lambda = \frac{16}{\pi^2 mn G_{08}} \Delta T \left(\left(\gamma_{T3} m^2 + \gamma_{T4} n^2 \beta^2 \right) - \frac{(\gamma_{T3} - \gamma_{T6}) m^2 g_{102} + (\gamma_{T4} - \gamma_{T7}) n^2 \beta^2 g_{101}}{g_{00}} \right)$$

$$\times \frac{h}{[D_{11}^* D_{22}^* A_{11}^* A_{22}^*]^{1/4}}$$

$$\Theta_2 = \frac{4}{3\pi^2 G_{08}} \gamma_{14} \gamma_{24} m^2 n^2 \beta^2 \left(\frac{\gamma_8}{\gamma_6} + \frac{\gamma_9}{\gamma_7} + 4 \frac{g_{05}}{g_{06}} \right)$$

$$\Theta_3 = 2\Theta_2^2 - \frac{1}{16 G_{08}} \gamma_{14} \gamma_{24} \left(\frac{m^4}{\gamma_7} + \frac{n^4 \beta^4}{\gamma_6} + C_{33} \right) \tag{E.2}$$

Appendix F

In Equations 3.33 and 3.34

$$\left(\lambda_x^{(0)}, \lambda_x^{(2)}, \lambda_x^{(4)}\right) = \frac{1}{4\beta^2 \gamma_{14} C_{11}} (S_0, S_2, S_4), \quad \delta_x^{(0)} = C_{00}\lambda_x - \delta_x^P,$$

$$\delta_x^{(2)} = \frac{1}{32\beta^2} C_{11}(1 + 2\mu),$$

$$\delta_x^{(4)} = \frac{1}{256\beta^2} \gamma_{14}\gamma_{24}C_{11}^2 \left(\frac{m^4}{J_{13}\gamma_{24}^2} + \frac{n^4\beta^4}{J_{31}}\right)(1 + \mu)^2(1 + 2\mu)^2 \quad \text{(F.1)}$$

in which (with g_{08}, g_{138}, g_{318}, etc. are defined as in Appendix C)

$$S_0 = \frac{\Theta_{11}}{(1 + \mu)} - S_0^P, \quad S_2 = \frac{1}{16}\gamma_{14}\gamma_{24}\Theta_2(1 + 2\mu),$$

$$S_4 = \frac{1}{256}\gamma_{14}^2\gamma_{24}^2 C_{11}(C_{24} - C_{44}),$$

$$\Theta_{11} = g_{08}, \quad \Theta_{13} = g_{138}, \quad \Theta_{31} = g_{318}$$

$$\Theta_2 = \left(\frac{m^4}{\gamma_{24}^2} + n^4\beta^4 + C_{22}\right), \quad C_{24} = 2(1 + \mu)^2(1 + 2\mu)^2\Theta_2\left(\frac{m^4}{J_{13}\gamma_{24}^2} + \frac{n^4\beta^4}{J_{31}}\right),$$

$$C_{44} = (1 + \mu)(1 + 2\mu)[2(1 + \mu)^2 + (1 + 2\mu)]\left(\frac{m^8}{J_{13}\gamma_{24}^4} + \frac{n^8\beta^8}{J_{31}}\right)$$

$$J_{13} = \Theta_{13}C_{11}(1 + \mu) - \Theta_{11}C_{13} + J^P, \quad J_{31} = \Theta_{31}C_{11}(1 + \mu) - \Theta_{11}C_{31} - J^P \quad \text{(F.2)}$$

in the above equations, for the case of four edges movable

$$C_{00} = \gamma_{24}, \quad C_{11} = C_{13} = m^2, \quad C_{31} = 9m^2, \quad C_{22} = 0, \quad S_0^P = J^P = 0,$$

$$\delta_x^P = \frac{1}{4\beta^2\gamma_{24}}\left[(\gamma_{24}^2\gamma_{T1} - \gamma_5\gamma_{T2})\Delta T + (\gamma_{24}^2\gamma_{P1} - \gamma_5\gamma_{P2})\Delta V\right] \quad \text{(F.3)}$$

and for the case of unloaded edges immovable

$$C_{00} = \frac{1}{\gamma_{24}}(\gamma_{24}^2 - \gamma_5^2)$$

$$C_{11} = m^2 + \gamma_5 n^2 \beta^2, \quad C_{13} = m^2 + 9\gamma_5 n^2 \beta^2, \quad C_{31} = 9m^2 + \gamma_5 n^2 \beta^2, \quad C_{22} = 2n^4 \beta^4,$$

$$S_0^P = \gamma_{14} n^2 \beta^2 [(\gamma_{T2} - \gamma_5 \gamma_{T1})\Delta T + (\gamma_{P2} - \gamma_5 \gamma_{P1})\Delta V],$$

$$\delta_x^P = \frac{C_{00}}{4\beta^2}(\gamma_{T1}\Delta T + \gamma_{P1}\Delta V),$$

$$J^P = 8\gamma_{14} m^2 n^2 \beta^2 (1 + \mu)[(\gamma_{T2} - \gamma_5 \gamma_{T1})\Delta T + (\gamma_{P2} - \gamma_5 \gamma_{P1})\Delta V]. \tag{F.4}$$

Appendix G

In Equation 3.53

$$(\lambda_T^{(0)}, \lambda_T^{(2)}, \lambda_T^{(4)}) = \frac{1}{\gamma_{14}C_{11}}(S_0, S_2, S_4) \tag{G.1}$$

in which [with g_{08}, g_{138}, g_{318}, etc. are defined as in Appendix C]

$$S_0 = \frac{\Theta_{11}}{(1+\mu)} - S_0^P, \quad S_2 = \frac{1}{16}\frac{\gamma_{14}}{\gamma_{24}}\Theta_2(1+2\mu), \quad S_4 = \frac{1}{256}\frac{\gamma_{14}^2}{\gamma_{24}^2}C_{11}(C_{24} - C_{44})$$

$$S_0^P = \gamma_{14}n^2\beta^2(\gamma_{P2} - \gamma_5\gamma_{P1})\Delta V, \quad \Theta_{11} = g_{08}, \quad \Theta_{13} = g_{138}, \quad \Theta_{31} = g_{318}$$

$$\Theta_2 = \frac{(3\gamma_{24}^2 - \gamma_5^2)\left(m^4 + \gamma_{24}^2 n^4\beta^4\right) + 4\gamma_5\gamma_{24}^2 m^2 n^2\beta^2}{\gamma_{24}^2 - \gamma_5^2}$$

$$C_{24} = 2(1+\mu)^2(1+2\mu)^2\Theta_2\left(\frac{m^4}{J_{13}} + \frac{\gamma_{24}^2 n^4\beta^4}{J_{31}}\right)$$

$$C_{44} = (1+\mu)(1+2\mu)[2(1+\mu)^2 + (1+2\mu)]\left(\frac{m^8}{J_{13}} + \frac{\gamma_{24}^4 n^8\beta^8}{J_{31}}\right)$$

$$J_{13} = \Theta_{13}C_{11}(1+\mu) - \Theta_{11}C_{13} - \gamma_{14}(1+\mu)[C_{11}S_{13} - C_{13}S_{11}]\Delta V$$

$$J_{31} = \Theta_{31}C_{11}(1+\mu) - \Theta_{11}C_{31} - \gamma_{14}(1+\mu)[C_{11}S_{31} - C_{31}S_{11}]\Delta V$$

$$S_{11} = \left(\gamma_{P1}m^2 + \gamma_{P2}n^2\beta^2\right), \quad S_{13} = \left(\gamma_{P1}m^2 + 9\gamma_{P2}n^2\beta^2\right),$$

$$S_{31} = \left(9\gamma_{P1}m^2 + \gamma_{P2}n^2\beta^2\right) \tag{G.2}$$

in the above equations, for the case of uniform temperature rise

$$C_{11} = \left(\gamma_{T1}m^2 + \gamma_{T2}n^2\beta^2\right), \quad C_{13} = \left(\gamma_{T1}m^2 + 9\gamma_{T2}n^2\beta^2\right),$$

$$C_{31} = \left(9\gamma_{T1}m^2 + \gamma_{T2}n^2\beta^2\right)$$

$$C_7 = \gamma_{T1}, \quad C_8 = \gamma_{T2}, \quad C_9 = C_{10} = 1.0 \tag{G.3}$$

and for the case of in-plane parabolic temperature variation

$$C_{11} = \left(\gamma_{T1}m^2 + \gamma_{T2}n^2\beta^2\right)\left(\frac{T_1}{T_2} + \frac{4}{9}\right) + \frac{m^2}{\beta^2}\frac{\gamma_{T2} - \gamma_5\gamma_{T1}}{\gamma_{24}^2}\left(\frac{4}{3\pi^2 n^2} + \frac{4}{\pi^4 n^4}\right)$$

$$+ n^2\beta^2\left(\gamma_{24}^2\gamma_{T1} - \gamma_5\gamma_{T2}\right)\left(\frac{4}{3\pi^2 m^2} + \frac{4}{\pi^4 m^4}\right)$$

$$C_{13} = \left(\gamma_{T1}m^2 + 9\gamma_{T2}n^2\beta^2\right)\left(\frac{T_1}{T_2} + \frac{4}{9}\right) + \frac{m^2}{\beta^2}\frac{\gamma_{T2} - \gamma_5\gamma_{T1}}{\gamma_{24}^2}\left(\frac{4}{27\pi^2 n^2} + \frac{4}{81\pi^4 n^4}\right)$$

$$+ 9n^2\beta^2\left(\gamma_{24}^2\gamma_{T1} - \gamma_5\gamma_{T2}\right)\left(\frac{4}{3\pi^2 m^2} + \frac{4}{\pi^4 m^4}\right)$$

$$C_{31} = \left(9\gamma_{T1}m^2 + \gamma_{T2}n^2\beta^2\right)\left(\frac{T_1}{T_2} + \frac{4}{9}\right) + \frac{9m^2}{\beta^2}\frac{\gamma_{T2} - \gamma_5\gamma_{T1}}{\gamma_{24}^2}\left(\frac{4}{3\pi^2 n^2} + \frac{4}{\pi^4 n^4}\right)$$

$$+ n^2\beta^2\left(\gamma_{24}^2\gamma_{T1} - \gamma_5\gamma_{T2}\right)\left(\frac{4}{27\pi^2 m^2} + \frac{4}{81\pi^4 m^4}\right)$$

$$C_7 = \gamma_{T1}\left(\frac{T_0}{T_1} + \frac{4}{9}\right) + \frac{4}{5}\frac{\gamma_{T2} - \gamma_5\gamma_{T1}}{\beta^2\gamma_{24}^2},$$

$$C_8 = \gamma_{T2}\left(\frac{T_0}{T_1} + \frac{4}{9}\right) + \frac{4}{5}\beta^2\left(\gamma_{24}^2\gamma_{T1} - \gamma_5\gamma_{T2}\right)$$

$$C_9 = 1 - \frac{C_5}{24\pi}\left(\frac{3}{5}\pi^4 - \frac{\pi^2}{n^2} - \frac{3}{n^4}\right), \quad C_{10} = 1 - \frac{C_6}{24\pi}\left(\frac{3}{5}\pi^4 - \frac{\pi^2}{m^2} - \frac{3}{m^4}\right) \quad \text{(G.4)}$$

Appendix H

In Equation 3.76

$$(\lambda_p^{(0)}, \lambda_p^{(2)}, \lambda_p^{(4)}) = \frac{1}{4\beta^2 \gamma_{14} C_{11}} (S_0, S_2, S_4), \quad \delta_x^{(0)} = \gamma_{24}\lambda_p - \delta_x^{T},$$

$$\delta_x^{(2)} = \frac{1}{32\beta^2} C_{11}(1 + 2\mu),$$

$$\delta_x^{(4)} = \frac{1}{256\beta^2} \gamma_{14}\gamma_{24}C_{11}^2 \left(\frac{m^4}{J_{13}\gamma_{24}^2} + \frac{n^4\beta^4}{J_{31}}\right)(1 + \mu)^2(1 + 2\mu)^2 \quad \text{(H.1)}$$

and in Equation 3.77

$$(\lambda_T^{(0)}, \lambda_T^{(2)}, \lambda_T^{(4)}) = \frac{1}{\gamma_{14}C_{11}} (S_0, S_2, S_4) \quad \text{(H.2)}$$

in which [with g_{08}, g_{138}, g_{318}, etc. are defined as in Appendix C]

$$S_0 = \frac{\Theta_{11}}{(1 + \mu)}, \quad S_2 = \frac{1}{16}\frac{\gamma_{14}}{\gamma_{24}}\Theta_2(1 + 2\mu), \quad S_4 = \frac{1}{256}\frac{\gamma_{14}^2}{\gamma_{24}^2}C_{11}(C_{24} - C_{44})$$

$$\Theta_{11} = g_{08}, \quad \Theta_{13} = g_{138}, \quad \Theta_{31} = g_{318}$$

$$C_{24} = 2(1 + \mu)^2(1 + 2\mu)^2\Theta_2\left(\frac{m^4}{J_{13}} + \frac{\gamma_{24}^2 n^4\beta^4}{J_{31}}\right)$$

$$C_{44} = (1 + \mu)(1 + 2\mu)[2(1 + \mu)^2 + (1 + 2\mu)]\left(\frac{m^8}{J_{13}} + \frac{\gamma_{24}^4 n^8\beta^8}{J_{31}}\right)$$

$$J_{13} = \Theta_{13}C_{11}(1 + \mu) - \Theta_{11}C_{13}, \quad J_{31} = \Theta_{31}C_{11}(1 + \mu) - \Theta_{11}C_{31} \quad \text{(H.3)}$$

in the above equations, for the case of compressive postbuckling

$$\Theta_2 = (m^4 + \gamma_{24}^2 n^4\beta^4), \quad \delta_x^T = \frac{1}{4\beta^2\gamma_{24}}\left[(\gamma_{24}^2\gamma_{T1} - \gamma_5\gamma_{T2})\Delta T\right],$$

$$C_{11} = C_{13} = m^2, \quad C_{31} = 9m^2 \quad \text{(H.4)}$$

and for the case of thermal postbuckling due to heat conduction

$$\Theta_2 = \frac{(3\gamma_{24}^2 - \gamma_5^2)(m^4 + \gamma_{24}^2 n^4 \beta^4) + 4\gamma_5 \gamma_{24}^2 m^2 n^2 \beta^2}{\gamma_{24}^2 - \gamma_5^2}$$

$$C_{11} = (\gamma_{T1} m^2 + \gamma_{T2} n^2 \beta^2), \quad C_{13} = (\gamma_{T1} m^2 + 9\gamma_{T2} n^2 \beta^2),$$

$$C_{31} = (9\gamma_{T1} m^2 + \gamma_{T2} n^2 \beta^2) \tag{H.5}$$

Appendix I

In Equation 4.19

$$w_{kl} = \sqrt[3]{-\frac{q_{kl}}{2} + \sqrt{\frac{(q_{kl})^2}{4} + \frac{(p_{kl})^3}{27}}} + \sqrt[3]{-\frac{q_{kl}}{2} - \sqrt{\frac{(q_{kl})^2}{4} + \frac{(p_{kl})^3}{27}}}$$

$$f_{kl} = c_{31}^{(k,l)} w_{kl}^2 + c_{32}^{(k,l)} w_{kl} + c_{33}^{(k,l)}$$

$$(\psi_x)_{kl} = c_{11}^{(k,l)} w_{kl} + c_{12}^{(k,l)} f_{kl} + c_{13}^{(k,l)}, \quad (\psi_y)_{kl} = c_{21}^{(k,l)} w_{kl} + c_{22}^{(k,l)} f_{kl} + c_{23}^{(k,l)} \qquad \text{(I.1)}$$

where

$$\left(c_{11}^{(k,l)}, c_{12}^{(k,l)}, c_{13}^{(k,l)} \right) = \frac{1}{b_{32}^{(k,l)} b_{43}^{(k,l)} - b_{42}^{(k,l)} b_{33}^{(k,l)}}$$
$$\times \left(b_{41}^{(k,l)} b_{33}^{(k,l)} - b_{31}^{(k,l)} b_{43}^{(k,l)}, \, b_{44}^{(k,l)} b_{33}^{(k,l)} - b_{34}^{(k,l)} b_{43}^{(k,l)}, \, y_3 b_{43}^{(k,l)} - y_4 b_{33}^{(k,l)} \right)$$

$$\left(c_{21}^{(k,l)}, c_{22}^{(k,l)}, c_{23}^{(k,l)} \right) = \frac{1}{b_{33}^{(k,l)} b_{42}^{(k,l)} - b_{43}^{(k,l)} b_{32}^{(k,l)}}$$
$$\times \left(b_{41}^{(k,l)} b_{32}^{(k,l)} - b_{31}^{(k,l)} b_{42}^{(k,l)}, \, b_{44}^{(k,l)} b_{32}^{(k,l)} - b_{34}^{(k,l)} b_{42}^{(k,l)}, \, y_3 b_{42}^{(k,l)} - y_4 b_{32}^{(k,l)} \right)$$

$$\left(c_{31}^{(k,l)}, c_{32}^{(k,l)}, c_{33}^{(k,l)} \right) = -\frac{1}{b_{24}^{(k,l)} + b_{22}^{(k,l)} c_{12}^{(k,l)} + b_{23}^{(k,l)} c_{22}^{(k,l)}}$$
$$\times \left(\frac{16\gamma_{24} kl\beta^2}{3\pi^2}, \, b_{21}^{(k,l)} + b_{22}^{(k,l)} c_{11}^{(k,l)} + b_{23}^{(k,l)} c_{21}^{(k,l)}, \, b_{22}^{(k,l)} c_{13}^{(k,l)} + b_{23}^{(k,l)} c_{23}^{(k,l)} \right)$$

$$d_1^{(k,l)} = -\left(b_{12}^{(k,l)} c_{12}^{(k,l)} c_{31}^{(k,l)} + b_{13}^{(k,l)} c_{22}^{(k,l)} c_{31}^{(k,l)} + b_{14}^{(k,l)} c_{31}^{(k,l)} - s^{(k,l)} c_{32}^{(k,l)} \right) \Big/ \left(s^{(k,l)} c_{31}^{(k,l)} \right)$$

$$d_2^{(k,l)} = \left(b_{11}^{(k,l)} + b_{12}^{(k,l)} c_{11}^{(k,l)} + b_{12}^{(k,l)} c_{12}^{(k,l)} c_{32}^{(k,l)} + b_{13}^{(k,l)} c_{21}^{(k,l)} + b_{13}^{(k,l)} c_{22}^{(k,l)} c_{32}^{(k,l)} \right.$$
$$\left. + b_{14}^{(k,l)} c_{32}^{(k,l)} - s^{(k,l)} c_{33}^{(k,l)} \right) \Big/ \left(s^{(k,l)} c_{31}^{(k,l)} \right)$$

$$d_3^{(k,l)} = \left(b_{12}^{(k,l)} c_{13}^{(k,l)} + b_{13}^{(k,l)} c_{23}^{(k,l)} + b_{14}^{(k,l)} c_{33}^{(k,l)} - y_1 \right) \Big/ \left(s^{(k,l)} c_{31}^{(k,l)} \right)$$

$$p_{kl} = -d_2^{(k,l)} + \left(d_1^{(k,l)} \right)^2 \Big/ 3, \quad q_{kl} = -d_3^{(k,l)} - \frac{2}{27} \left(d_1^{(k,l)} \right)^2 + \frac{1}{3} d_1^{(k,l)} d_2^{(k,l)}$$

$$s^{(k,l)} = \frac{32\gamma_{14} kl\beta^2}{3\pi^2}$$

$$b_{11}^{(i,j)} = \gamma_{110}(im)^4 + 2\gamma_{112}(im)^2(jn)^2\beta^2 + \gamma_{114}(jn)^4\beta^4 - \gamma_{14}\beta^2(p_y m^2 + p_x n^2)$$

$$b_{12}^{(i,j)} = -\left[\gamma_{120}(im)^3 + \gamma_{122}im(jn)^2\beta^2\right]$$

$$b_{13}^{(i,j)} = -\left[\gamma_{131}(im)^2 jn\beta + \gamma_{133}(jn)^3\beta^3\right]b_{14}^{(i,j)}$$

$$= \gamma_{14}\left[\gamma_{140}(im)^4 + \gamma_{142}(im)^2(jn)^2\beta^2 + \gamma_{144}(jn)^4\beta^4\right]$$

$$b_{21}^{(i,j)} = -\gamma_{24}\left[\gamma_{240}(im)^4 + \gamma_{242}(im)^2(jn)^2\beta^2 + \gamma_{244}(jn)^4\beta^4\right]$$

$$b_{22}^{(i,j)} = \gamma_{24}\left[\gamma_{220}(im)^3 + \gamma_{222}(im)(jn)^2\beta^2\right]$$

$$b_{23}^{(i,j)} = \gamma_{24}\left[\gamma_{231}(im)^2(jn)\beta + \gamma_{233}(jn)^3\beta^3\right]$$

$$b_{24}^{(i,j)} = m^4 + 2\gamma_{212}(im)^2(jn)^2\beta^2 + \gamma_{214}(jn)^4\beta^4$$

$$b_{31}^{(i,j)} = \gamma_{31}(im) - \gamma_{310}(im)^3 - \gamma_{312}(im)(jn)^2\beta^2$$

$$b_{32}^{(i,j)} = \gamma_{31} + r_{320}(im)^2 + \gamma_{322}(jn)^2\beta^2$$

$$b_{33}^{(i,j)} = \gamma_{331}(im)(jn)\beta$$

$$b_{34}^{(i,j)} = -\gamma_{24}\left[\gamma_{220}(im)^3 + \gamma_{222}(im)(jn)^2\beta^2\right]$$

$$b_{41}^{(i,j)} = \gamma_{41}(jn)\beta - \gamma_{411}(im)^2(jn)\beta - \gamma_{413}(jn)^3\beta^3$$

$$b_{42}^{(i,j)} = \gamma_{331}(im)(jn)\beta$$

$$b_{43}^{(i,j)} = \gamma_{41} + r_{430}(im)^2 + \gamma_{432}(jn)^2\beta^2$$

$$b_{44}^{(i,j)} = -\gamma_{14}\left[\gamma_{231}(im)^2(jn)\beta + \gamma_{233}(jn)^3\beta^3\right]$$

$$y_1^{(k,l)} = M_x^{(0)}k/l + \beta^2 M_y^{(0)}l/k, \quad y_3^{(k,l)} = -S_x^{(0)}\Big/l, \quad y_4^{(k,l)} = \beta S_y^{(0)}\Big/k \tag{I.2}$$

and in Equation 4.20

$$\begin{bmatrix} M_x^{(0)} & S_x^{(0)} \\ M_y^{(0)} & S_y^{(0)} \end{bmatrix} = \frac{16hT_1}{\pi^2[D_{11}^* D_{22}^* A_{11}^* A_{22}^*]^{1/4}} \begin{bmatrix} \gamma_{T3} & (\gamma_{T3} - \gamma_{T6}) \\ \gamma_{T4} & (\gamma_{T4} - \gamma_{T7}) \end{bmatrix} \tag{I.3}$$

Appendix J

In Equations 4.31 through 4.35 (with $i, j = 1, 3$)

$$B_{00}^{(0)} = -\frac{1}{(\gamma_5^2 - \gamma_{24}^2)\beta^2} \left\{ \left[(\gamma_{24}^2 \gamma_{T1} - \gamma_5 \gamma_{T2}) + \gamma_5(\gamma_{T2} - \gamma_5 \gamma_{T1}) \right] T_1 \right.$$

$$- \frac{4}{\pi^2} \sum_{k=1,3,\ldots} \sum_{l=1,3,\ldots} \frac{1}{kl} \left[(\gamma_5^2 - \gamma_{24}^2) n^2 \beta^2 f_{kl} - \gamma_{24}(\gamma_{511} + \gamma_5 \gamma_{220}) m \psi_{kl} \right.$$

$$\left. \left. - \gamma_{24}(\gamma_{233} + \gamma_5 \gamma_{522}) n\beta \psi_{kl} + \gamma_{24}(\gamma_{611} m^2 + \gamma_{244} n^2 \beta^2) w_{kl} + \gamma_5 \gamma_{24}(\gamma_{240} m^2 + \gamma_{622} n^2 \beta^2) w_{kl} \right] \right\}$$

$$b_{00}^{(0)} = -\frac{1}{(\gamma_5^2 - \gamma_{24}^2)} \left\{ \left[\gamma_5(\gamma_{24}^2 \gamma_{T1} - \gamma_5 \gamma_{T2}) + \gamma_{24}^2(\gamma_{T2} - \gamma_5 \gamma_{T1}) \right] T_1 \right.$$

$$- \frac{4}{\pi^2} \sum_{k=1,3,\ldots} \sum_{l=1,3,\ldots} \frac{1}{kl} \left[(\gamma_5^2 - \gamma_{24}^2) m^2 f_{kl} - \gamma_{24}(\gamma_5 \gamma_{511} + \gamma_{24}^2 \gamma_{220}) m \psi_{kl} \right.$$

$$\left. \left. - \gamma_{24}(\gamma_5 \gamma_{233} + \gamma_{24}^2 \gamma_{522}) n\beta \psi_{kl} + \gamma_5 \gamma_{24}(\gamma_{611} m^2 + \gamma_{244} n^2 \beta^2) w_{kl} + \gamma_{24}^3(\gamma_{240} m^2 + \gamma_{622} n^2 \beta^2) w_{kl} \right] \right\}$$

$$g_{11}^{(i,j)} = \frac{k_{23}^{(i,j)} k_{31}^{(i,j)} - k_{33}^{(i,j)} k_{21}^{(i,j)}}{k_{22}^{(i,j)} k_{33}^{(i,j)} - k_{32}^{(i,j)} k_{23}^{(i,j)}}, \quad g_{21}^{(i,j)} = \frac{k_{22}^{(i,j)} k_{31}^{(i,j)} - k_{32}^{(i,j)} k_{21}^{(i,j)}}{k_{23}^{(i,j)} k_{32}^{(i,j)} - k_{33}^{(i,j)} k_{22}^{(i,j)}}$$

$$g_{31}^{(i,j)} = a_1^{(i,j)} + b_1^{(i,j)} g_{11}^{(i,j)} + c_1^{(i,j)} g_{21}^{(i,j)}$$

$$g_{402} = \frac{\gamma_{24} m^2 n^2 \beta^2}{2 \left(16 \gamma_{214} n^4 \beta^4 + \left((64 \gamma_{14} \gamma_{24} \gamma_{223}^2 n^6 \beta^6) / (\gamma_{41} + 4\gamma_{432} n^2 \beta^2) \right) \right)},$$

$$g_{420} = \frac{\gamma_{24} m^2 n^2 \beta^2}{2 \left(16 m^4 + \left((64 \gamma_{14} \gamma_{24} \gamma_{220}^2 m^6) / (\gamma_{31} + 4\gamma_{320} m^2) \right) \right)}$$

$$g_{12} = -\frac{8 \gamma_{220} \gamma_{14} m^3}{\gamma_{31} + 4\gamma_{320} m^2} \cdot g_{420}, \quad g_{22} = -\frac{8 \gamma_{233} \gamma_{14} n^3 \beta^3}{\gamma_{41} + 4\gamma_{432} n^2 \beta^2} \cdot g_{402}$$

$$g_{441} = 8 g_{12} \gamma_{120} m^3 + 16 g_{420} \gamma_{14} \gamma_{140} m^4 + g_{31}^{(1,1)} \gamma_{14} \beta^2 m^2 n^2$$

$$g_{442} = 8 g_{22} \gamma_{133} n^3 \beta^3 + 16 g_{420} \gamma_{14} \gamma_{144} n^4 \beta^4 + g_{31}^{(1,1)} \gamma_{14} \beta^2 m^2 n^2$$

$$B_{00}^{(2)} = \frac{\gamma_{24}(m^2 + \gamma_5 n^2 \beta^2)}{8(\gamma_5^2 - \gamma_{24}^2)\beta^2}, \quad b_{00}^{(2)} = \frac{\gamma_{24}(\gamma_5 m^2 + \gamma_{24}^2 n^2 \beta^2)}{8(\gamma_5^2 - \gamma_{24}^2)}$$

$$g_{311} = \frac{\gamma_{14}\beta^2 \left(m^2 B_{00}^{(2)} + n^2 b_{00}^{(2)} \right) - 2m^2 n^2 \gamma_{14}\beta^2 (g_{402} + g_{420})}{k_{11}^{(1,1)} + k_{12}^{(1,1)} g_{11}^{(1,1)} + k_{13}^{(1,1)} g_{21}^{(1,1)} - \gamma_{14}\beta^2 \left(m^2 B_{00}^{(0)} + n^2 b_{00}^{(0)} + s^{(1,1)} \right)}$$

$$g_{331} = \frac{2\gamma_{14} m^2 n^2 \beta^2}{k_{11}^{(3,1)} + k_{12}^{(3,1)} g_{11}^{(3,1)} + k_{13}^{(3,1)} g_{21}^{(3,1)} - \gamma_{14}\beta^2 \left(9m^2 B_{00}^{(0)} + n^2 b_{00}^{(0)} + s^{(3,1)} \right)} g_{420}$$

$$g_{313} = \frac{2\gamma_{14} m^2 n^2 \beta^2}{k_{11}^{(1,3)} + k_{12}^{(1,3)} g_{11}^{(1,3)} + k_{13}^{(1,3)} g_{21}^{(1,3)} - \gamma_{14}\beta^2 \left(m^2 B_{00}^{(0)} + 9n^2 b_{00}^{(0)} + s^{(1,3)} \right)} g_{402}$$

$$g_1 = \frac{-b_2 k_{13}^{(1,1)}\left(k_{12}^{(1,1)}k_{33}^{(1,1)} - k_{32}^{(1,1)}k_{13}^{(1,1)}\right) + b_3 k_{13}^{(1,1)}\left(k_{12}^{(1,1)}k_{23}^{(1,1)} - k_{22}^{(1,1)}k_{13}^{(1,1)}\right)}{\left(k_{11}^{(1,1)}k_{23}^{(1,1)} - k_{21}^{(1,1)}k_{13}^{(1,1)}\right)\left(k_{12}^{(1,1)}k_{33}^{(1,1)} - k_{32}^{(1,1)}k_{13}^{(1,1)}\right) - \left(k_{11}^{(1,1)}k_{33}^{(1,1)} - k_{31}^{(1,1)}k_{13}^{(1,1)}\right)\left(k_{12}^{(1,1)}k_{23}^{(1,1)} - k_{22}^{(1,1)}k_{13}^{(1,1)}\right)}$$

$$g_2 = \frac{-b_2 k_{13}^{(1,1)}\left(k_{11}^{(1,1)}k_{33}^{(1,1)} - k_{31}^{(1,1)}k_{13}^{(1,1)}\right) + b_3 k_{13}^{(1,1)}\left(k_{11}^{(1,1)}k_{23}^{(1,1)} - k_{21}^{(1,1)}k_{13}^{(1,1)}\right)}{\left(k_{12}^{(1,1)}k_{23}^{(1,1)} - k_{22}^{(1,1)}k_{13}^{(1,1)}\right)\left(k_{11}^{(1,1)}k_{33}^{(1,1)} - k_{31}^{(1,1)}k_{13}^{(1,1)}\right) - \left(k_{12}^{(1,1)}k_{33}^{(1,1)} - k_{32}^{(1,1)}k_{13}^{(1,1)}\right)\left(k_{11}^{(1,1)}k_{23}^{(1,1)} - k_{21}^{(1,1)}k_{13}^{(1,1)}\right)}$$

$$g_3 = -\frac{k_{11}^{(1,1)}g_1 + k_{12}^{(1,1)}g_2}{k_{13}^{(1,1)}}, \quad g_4 = a_1^{(1,1)}g_1 + b_1^{(1,1)}g_2 + c_1^{(1,1)}g_3$$

$$g_{41} = k_{11}^{(1,1)} + k_{12}^{(1,1)}g_{11}^{(1,1)} + k_{13}^{(1,1)}g_{21}^{(1,1)} - \gamma_{14}\beta^2\left(m^2 B_{00}^{(0)} + n^2 b_{00}^{(0)} + s^{(1,1)}\right)$$

$$g_{42} = -\gamma_{14}\beta^2\left[m^2 B_{00}^{(2)} + n^2 b_{00}^{(2)} - 2m^2 n^2 (g_{402} + g_{420})\right]$$

$$g_{43} = -\left(\gamma_{170} - \gamma_{171}m^2 - \gamma_{171}n^2\beta^2 - \gamma_{80}mg_{11}^{(1,1)} - \gamma_{80}n\beta g_{21}^{(1,1)}\right)$$

$$g_{44} = -\frac{2}{\pi^2 mn}\left[g_{441}(1-\cos m\pi)\left(\frac{2}{3} + \frac{1}{3}\cos 3m\pi - \cos m\pi\right)\right.$$
$$\left. + g_{442}(1-\cos n\pi)\left(\frac{2}{3} + \frac{1}{3}\cos 3n\pi - \cos n\pi\right)\right] - \gamma_{14}\beta^2\left(B_{00}^{(2)}m^2 + b_{00}^{(2)}n^2\right)w_{mn} \qquad (J.1)$$

where

$$s^{(m,n)} = \frac{2}{\pi^2}\sum_{k=1,3,\dots}\sum_{l=1,3,\dots}\left[2(k^2 n^2 + l^2 m^2)\left(\frac{1}{k} - \frac{1}{2k+4m} - \frac{1}{2k-4im}\right)\left(\frac{1}{l} - \frac{1}{2l+4n} - \frac{1}{2l-4n}\right)\right.$$
$$\left. - klmn\left(\frac{1}{2m+k} + \frac{1}{2m-k}\right)\left(\frac{1}{2n+l} + \frac{1}{2n-l}\right)\left(g_{13}^{(k,l)}w_{kl} + f_{kl}\right)\right]$$
$$b_2 = \gamma_{90}m + \gamma_{10}g_{11}^{(1,1)}, \quad b_3 = \gamma_{90}n\beta + \gamma_{10}g_{21}^{(1,1)} \qquad (J.2)$$

In the above equations

$$k_{11}^{(i,j)} = \gamma_{110}(im)^4 + 2\gamma_{112}(im)^2(jn)^2\beta^2 + \gamma_{114}(jn)^4\beta^4 + \gamma_{14}\left[\gamma_{140}(im)^4 + \gamma_{142}(im)^2(jn)^2\beta^2\right.$$
$$\left. + \gamma_{144}(jn)^4\beta^4\right]a_1^{(i,j)} - \alpha_1\gamma_{14}\beta^2\left(B_{00}^{(0)}m^2 + b_{00}^{(0)}n^2\right)$$

$$k_{12}^{(i,j)} = -\left[\gamma_{120}(im)^3 + \gamma_{122}im(jn)^2\beta^2\right] + \gamma_{14}\left[\gamma_{140}(im)^4 + \gamma_{142}(im)^2(jn)^2\beta^2 + \gamma_{144}(jn)^4\beta^4\right]b_1^{(i,j)}$$

$$k_{13}^{(i,j)} = -\left[\gamma_{131}(im)^2 jn\beta + \gamma_{133}(jn)^3\beta^3\right] + \gamma_{14}\left[\gamma_{140}(im)^4 + \gamma_{142}(im)^2(jn)^2\beta^2 + \gamma_{144}(jn)^4\beta^4\right]c_1^{(i,j)}$$

$$k_{21}^{(i,j)} = \gamma_{31}im - \gamma_{310}(im)^3 - \gamma_{312}im(jn)^2\beta^2 - \gamma_{14}\left[\gamma_{220}(im)^3 + \gamma_{222}(im)(jn)^2\beta^2\right]a_1^{(i,j)}$$

$$k_{22}^{(i,j)} = \gamma_{31} + r_{320}(im)^2 + \gamma_{322}(jn)^2\beta^2 - \gamma_{14}\left[\gamma_{220}(im)^3 + \gamma_{222}(im)(jn)^2\beta^2\right]b_1^{(i,j)}$$

$$k_{23}^{(i,j)} = \gamma_{331}(im)(jn)\beta - \gamma_{14}\left[\gamma_{220}(im)^3 + \gamma_{222}(im)(jn)^2\beta^2\right]c_1^{(i,j)}$$

$$k_{31}^{(i,j)} = \gamma_{41}jn\beta - \gamma_{411}(im)^2 jn\beta - \gamma_{413}(jn)^3\beta^3 - \gamma_{14}\left[\gamma_{231}(im)^2(jn)\beta + \gamma_{233}(jn)^3\beta^3\right]a_1^{(i,j)}$$

$$k_{32}^{(i,j)} = \gamma_{331}(im)(jn)\beta - \gamma_{14}\left[\gamma_{231}(im)^2(jn)\beta + \gamma_{233}(jn)^3\beta^3\right]b_1^{(i,j)}$$

$$k_{33}^{(i,j)} = \gamma_{41} + r_{430}(im)^2 + \gamma_{432}(jn)^2\beta^2 - \gamma_{14}\left[\gamma_{231}(im)^2(jn)\beta + \gamma_{233}(jn)^3\beta^3\right]c_1^{(i,j)} \qquad (J.3)$$

in which, when w_{kl}, ψ_{xkl}, ψ_{ykl} and f_{kl} are considered, $\alpha_1 = 1$, otherwise $\alpha_1 = 0$, and

$$a_1^{(i,j)} = -\frac{\gamma_{24}}{m^4 + 2\gamma_{212}(im)^2(jn)^2\beta^2 + \gamma_{214}(jn)^4\beta^4}$$

$$\times \left\{ \gamma_{240}(im)^4 + \gamma_{242}(im)^2(jn)^2\beta^2 + \gamma_{244}(jn)^4\beta^4 + \frac{2\beta^2}{\pi^2}\sum_{k=1,3\ldots}\sum_{l=1,3,\ldots} \right.$$

$$\times \left[2(k^2n^2 + l^2m^2)\left(\frac{1}{k} - \frac{1}{2k+4m} - \frac{1}{2k-4m}\right)\left(\frac{1}{l} - \frac{1}{2l+4n} - \frac{1}{2l-4n}\right) \right.$$

$$\left. \left. - mnkl\left(\frac{1}{2m+k} + \frac{1}{2m-k}\right)\left(\frac{1}{2n+l} - \frac{1}{2n-l}\right)\right]w_{kl} \right\}$$

$$b_1^{(i,j)} = -\frac{\gamma_{24}\left[\gamma_{220}(im)^3 + \gamma_{222}(im)(jn)^2\beta^2\right]}{m^4 + 2\gamma_{212}(im)^2(jn)^2\beta^2 + \gamma_{214}(jn)^4\beta^4}$$

$$c_1^{(i,j)} = -\frac{\gamma_{24}\left[\gamma_{231}(im)^2(jn)\beta + \gamma_{233}(jn)^3\beta^3\right]}{m^4 + 2\gamma_{212}(im)^2(jn)^2\beta^2 + \gamma_{214}(jn)^4\beta^4} \tag{J.4}$$

Appendix K

In Equations 4.73 through 4.76 (with other symbols can be found in Appendix J)

$$g_1 = \frac{b_3\left(k_{12}^{(1,1)}k_{23}^{(1,1)} - k_{22}^{(1,1)}k_{13}^{(1,1)}\right) - b_2\left(k_{12}^{(1,1)}k_{33}^{(1,1)} - k_{13}^{(1,1)}k_{32}^{(1,1)}\right)}{g_{xxk3}}$$

$$g_2 = \frac{b_2\left(k_{11}^{(1,1)}k_{33}^{(1,1)} - k_{13}^{(1,1)}k_{31}^{(1,1)}\right) - b_3\left(k_{11}^{(1,1)}k_{23}^{(1,1)} - k_{13}^{(1,1)}k_{21}^{(1,1)}\right)}{g_{xxk3}}$$

$$g_{41} = k_{11}^{(1,1)} + k_{12}^{(1,1)}g_{11}^{(1,1)} + k_{13}^{(1,1)}g_{21}^{(1,1)} - \gamma_{14}\beta^2\left(m^2 g_{xxk1} + n^2 g_{xxk2} + s^{(1,1)}\right)$$

$$g_{42} = 2\gamma_{14}\beta^2 m^2 n^2 (g_{402} + g_{420})$$

$$g_{44} = -\frac{2}{\pi^2 mn}\left[g_{441}(1 - \cos n\pi)\left(\frac{2}{3} + \frac{1}{3}\cos 3m\pi - \cos m\pi\right)\right.$$

$$\left. + g_{442}(1 - \cos m\pi)\left(\frac{2}{3} + \frac{1}{3}\cos 3n\pi - \cos n\pi\right)\right] \tag{K.1}$$

where

$$\begin{aligned}
s^{(i,j)} = \frac{2}{\pi^2}\Bigg\{ &\left[2(k^2 n^2 + l^2 m^2)\left(\frac{1}{k} - \frac{1}{2k + 4m} - \frac{1}{2k - 4m}\right)\left(\frac{1}{l} - \frac{1}{2l + 4n} - \frac{1}{2l - 4n}\right)\right. \\
&\left. - mnkl\left(\frac{1}{2m + k} + \frac{1}{2m - k}\right)\left(\frac{1}{2n + l} + \frac{1}{2n - l}\right)\right]\left(\varepsilon A_{11}^{(1)}\right) \\
&+ \left[2(k^2 n^2 + 9l^2 m^2)\left(\frac{1}{k} - \frac{1}{2k + 4m} - \frac{1}{2k - 4m}\right)\left(\frac{1}{3l} - \frac{1}{6l + 4n} - \frac{1}{6l - 4n}\right)\right. \\
&\left. - 3mnkl\left(\frac{1}{2m + k} + \frac{1}{2m - k}\right)\left(\frac{1}{2n + 3l} + \frac{1}{2n - 3l}\right)\right]\left(\varepsilon^3 A_{13}^{(3)}\right) \\
&+ \left[2(9k^2 n^2 + l^2 m^2)\left(\frac{1}{3k} - \frac{1}{6k + 4m} - \frac{1}{6k - 4m}\right)\left(\frac{1}{l} - \frac{1}{2l + 4n} - \frac{1}{2l - 4n}\right)\right. \\
&\left. - 3mnkl\left(\frac{1}{2m + 3k} + \frac{1}{2m - 3k}\right)\left(\frac{1}{2n + l} + \frac{1}{2n - l}\right)\right]\left(\varepsilon^3 A_{31}^{(3)}\right)\Bigg\}g_{31}^{(i,j)}
\end{aligned}$$

$$g_{xxk1} = B_{00}^{(0)} + \varepsilon^2 B_{00}^{(2)} + \varepsilon^4 B_{00}^{(4)}, \quad g_{xxk2} = b_{00}^{(0)} + \varepsilon^2 b_{00}^{(2)} + \varepsilon^4 b_{00}^{(4)}$$

$$g_{xxk3} = k_{11}^{(1,1)} k_{22}^{(1,1)} k_{33}^{(1,1)} + k_{12}^{(1,1)} k_{23}^{(1,1)} k_{31}^{(1,1)} + k_{21}^{(1,1)} k_{32}^{(1,1)} k_{13}^{(1,1)} - k_{23}^{(1,1)} k_{32}^{(1,1)} k_{11}^{(1,1)}$$
$$- k_{22}^{(1,1)} k_{31}^{(1,1)} k_{13}^{(1,1)} - k_{21}^{(1,1)} k_{12}^{(1,1)} k_{33}^{(1,1)}$$

$$g_{xxk4} = -\frac{\gamma_{24}}{m^4 + 2\gamma_{212}(im)^2(jn)^2\beta^2 + \gamma_{214}(jn)^4\beta^4}$$

$$k_{11}^{(i,j)} = \gamma_{110}(im)^4 + 2\gamma_{112}(im)^2(jn)^2\beta^2 + \gamma_{114}(jn)^4\beta^4$$
$$+ \gamma_{14}[\gamma_{140}(im)^4 + \gamma_{142}(im)^2(jn)^2\beta^2 + \gamma_{144}(jn)^4\beta^4]a_1^{(i,j)}$$

$$a_1^{(i,j)} = g_{xxk4} \times \left[\gamma_{240}(im)^4 + \gamma_{242}(im)^2(jn)^2\beta^2 + \gamma_{244}(jn)^4\beta^4\right]$$

$$+ g_{xxk4} \times \frac{2\beta^2}{\pi^2}\left\{\left[2(k^2n^2 + l^2m^2)\left(\frac{1}{k} - \frac{1}{2k+4m} - \frac{1}{2k-4m}\right)\right.\right.$$

$$\times \left(\frac{1}{l} - \frac{1}{2l+4n} - \frac{1}{2l-4n}\right) - mnkl\left(\frac{1}{2m+k} + \frac{1}{2m-k}\right)\left(\frac{1}{2n+l} + \frac{1}{2n-l}\right)\bigg](\varepsilon A_{11}^{(1)})\right\}$$

$$+ \left[2(k^2n^2 + 9l^2m^2) \times \left(\frac{1}{k} - \frac{1}{2k+4m} - \frac{1}{2k-4m}\right)\left(\frac{1}{3l} - \frac{1}{6l+4n} - \frac{1}{6l-4n}\right)\right.$$

$$- 3mnkl\left(\frac{1}{2m+k} + \frac{1}{2m-k}\right)\left(\frac{1}{2n+3l} + \frac{1}{2n-3l}\right)\bigg](\varepsilon^3 A_{13}^{(3)})$$

$$+ \left\{\left[2(9k^2n^2 + l^2m^2) \times \left(\frac{1}{3k} - \frac{1}{6k+4m} - \frac{1}{6k-4m}\right)\left(\frac{1}{l} - \frac{1}{2l+4n} - \frac{1}{2l-4n}\right)\right.\right.$$

$$- 3mnkl\left(\frac{1}{2m+3k} + \frac{1}{2m-3k}\right)\left(\frac{1}{2n+l} + \frac{1}{2n-l}\right)\bigg](\varepsilon^3 A_{31}^{(3)})\right\} \tag{K.2}$$

and $A_{11}^{(1)}\varepsilon$ may be solved from

$$\lambda_i = \lambda_i^{(0)} + \lambda_i^{(2)}\left(A_{11}^{(1)}\varepsilon\right)^2 + \lambda_i^{(4)}\left(A_{11}^{(1)}\varepsilon\right)^4 + \cdots \quad (i = p, T)$$

in which (with other symbols can be found in Appendices F and G)

$$\left(\lambda_P^{(0)}, \lambda_P^{(2)}, \lambda_P^{(4)}\right) = \frac{1}{4\beta^2\gamma_{14}C_{11}}(S_0, S_2, S_4), \quad \left(\lambda_T^{(0)}, \lambda_T^{(2)}, \lambda_T^{(4)}\right) = \frac{1}{\gamma_{14}C_{11}}(S_0, S_2, S_4)$$

$$S_0 = \frac{\Theta_{11}}{(1 + \mu)}, \quad S_2 = \frac{1}{16}\frac{\gamma_{14}}{\gamma_{24}}\Theta_2(1 + 2\mu), \quad S_4 = -\frac{1}{256}\frac{\gamma_{14}^2}{\gamma_{24}^2}C_{11}C_{44}$$

$$\varepsilon^3 A_{13}^{(3)} = \frac{m^4}{16J_{13}}C_{11}\left(A_{11}^{(1)}\varepsilon\right)^3, \quad \varepsilon^3 A_{31}^{(3)} = \frac{n^4\beta^4}{16J_{31}}C_{11}\left(A_{11}^{(1)}\varepsilon\right)^3 \tag{K.3}$$

Appendix L

In Equations 5.51 through 5.54

$$\Theta_1 = \frac{1}{C_3}\left[\gamma_{14}\gamma_{24}\frac{m^4(1+\mu)}{16n^2\beta^2 g_{09}g_{06}}\varepsilon^{-1} - \gamma_{24}\gamma_{14}\frac{m^2 g_{11}}{32n^2\beta^2 g_{09}} + \frac{2\gamma_5}{\gamma_{24}}\lambda_{\mathrm{p}}^{(2)}\right]$$

$$\Theta_2 = \frac{1}{\gamma_{24}}\left[(\gamma_{T2} - \gamma_5\gamma_{T1})\Delta T\right] + \frac{2\gamma_5}{\gamma_{24}}\lambda_{\mathrm{p}}^{(0)}$$

$$\lambda_{\mathrm{p}}^{(0)} = \frac{1}{2}\left\{\frac{\gamma_{24}m^2}{(1+\mu)g_{06}}\varepsilon^{-1} + \gamma_{24}\frac{g_{05}+(1+\mu)g_{07}}{(1+\mu)^2 g_{06}}\right.$$
$$+ \frac{1}{\gamma_{14}(1+\mu)m^2}\left[g_{08} + \gamma_{14}\gamma_{24}\frac{g_{05}}{g_{06}}\frac{(1+\mu)g_{07}-\mu(2+\mu)g_{05}}{(1+\mu)^2}\right]\varepsilon$$
$$- \frac{\mu}{(1+\mu)^2}\frac{g_{05}}{\gamma_{14}m^4}\left[1 + \frac{g_{05}}{(1+\mu)m^2}\varepsilon\right]$$
$$\left.\times\left[g_{08} + \gamma_{14}\gamma_{24}\frac{g_{05}}{g_{06}}\frac{(1+\mu)g_{07}+g_{05}}{(1+\mu)^2}(2+\mu)\right]\varepsilon^2\right\}$$

$$\lambda_{\mathrm{p}}^{(2)} = \frac{1}{8}\left\{\gamma_{14}\gamma_{24}^2\frac{m^6(2+\mu)}{2g_{09}g_{06}^2}\varepsilon^{-1} + \gamma_{14}\gamma_{24}^2\frac{m^4}{2g_{09}g_{06}}\right.$$
$$\times\left[\frac{g_{05}}{g_{06}} + \frac{g_{07}}{g_{06}}(1+\mu) + g_{12}(1+\mu) - \frac{1}{1+\mu}g_{11}\right]$$
$$- \frac{1}{4}\gamma_{24}m^2 g_{13}(1+2\mu)\varepsilon + \gamma_{14}\gamma_{24}^2\frac{m^2 g_{11}}{2g_{09}}\left[\frac{g_{05}}{g_{06}}\frac{1}{1+\mu} - \frac{g_{07}}{g_{06}} - g_{12}\right]\varepsilon$$
$$+ \gamma_{14}\gamma_{24}^2\frac{m^2 g_{05}}{2g_{09}g_{06}}\left[\frac{2(1+\mu)^2-(1+2\mu)}{2(1+\mu)^2}g_{14} + \frac{\mu}{1+\mu}\frac{g_{05}}{g_{06}}\right]$$
$$\left.\times(2+\mu)\varepsilon + \gamma_{24}\frac{m^2 n^4\beta^4}{g_{06}}\frac{S_2}{S_1}\varepsilon\right\}$$

$$\lambda_{\mathrm{p}}^{(4)} = \frac{1}{128}\gamma_{14}^2\gamma_{24}^3\frac{m^{10}(1+\mu)}{g_{09}^2 g_{06}^3}\frac{S_3}{S_{13}}\varepsilon^{-1}$$

$$\delta_x^{(0)} = \frac{1}{\gamma_{24}}\left[\gamma_{24}^2 - \frac{2}{\pi}\frac{\gamma_5^2}{\gamma_{24}}\left(\vartheta b_{01}^{(2)} - \phi b_{10}^{(2)}\right)\varepsilon^{1/2}\right]\lambda_{\mathrm{p}} + \left[\frac{\gamma_5^2}{2\pi\gamma_{24}^2}\frac{b_{11}}{\vartheta}\varepsilon^{1/2}\right]\lambda_{\mathrm{p}}^2$$

$$\delta_x^{(T)} = \frac{1}{2\gamma_{24}}\left[(\gamma_{24}^2\gamma_{T1} - \gamma_5\gamma_{T2})\Delta T\right]$$

$$\delta_x^{(2)} = \frac{1}{16}\left[m^2(1+2\mu)\varepsilon - 2g_{05}\varepsilon^2 + \frac{g_{05}^2}{m^2}\varepsilon^3\right]$$

$$\delta_p^{(4)} = \frac{1}{128}\left\{\frac{b_{11}}{32\pi\vartheta}\gamma_{14}^2\gamma_{24}^2\frac{m^8(1+\mu)^2}{n^4\beta^4 g_{09}^2 g_{06}^2}\varepsilon^{-3/2} + m^2 n^4\beta^4(1+\mu)^2\left(\frac{S_4}{S_1}\right)^2\varepsilon^3\right\} \quad \text{(L.1)}$$

in the above equations

$$S_1 = g_{06}(1+\mu) - 4m^2 C_2 g_{10}, \quad S_{13} = g_{136}C_9 - g_{06}(1+\mu)$$

$$S_2 = g_{06}\left(5 + 11\mu + 4\mu^2\right) + 8m^4(1+\mu)(2+\mu)g_{10}$$

$$S_3 = g_{136}\left(6 + 6\mu + \mu^2\right) + g_{06}\left(6 - \mu^2\right)(1+\mu)$$

$$S_4 = g_{06}(1+2\mu) + 8m^4(1+\mu)g_{10}$$

$$C_3 = 1 - \frac{g_{05}}{m^2}\varepsilon, \quad a_{01}^{(1)} = 1, \quad a_{10}^{(1)} = \frac{\vartheta}{\phi}g_{17}, \quad b_{01}^{(2)} = \gamma_{24}g_{19}, \quad b_{10}^{(2)} = \gamma_{24}\frac{\vartheta}{\phi}g_{20}$$

$$b_{11} = \frac{1}{b}\left[\left(a_{10}^{(1)}\right)^2\phi^2 b + a_{10}^{(1)}2\vartheta\phi c + \left(2\vartheta^4 - \vartheta^2\phi^2 + \phi^4\right)\right] \quad \text{(L.2)}$$

and

$$g_{00} = \left(\gamma_{31} + \gamma_{320}m^2 + \gamma_{322}n^2\beta^2\right)\left(\gamma_{41} + \gamma_{430}m^2 + \gamma_{432}n^2\beta^2\right) - \gamma_{331}^2 m^2 n^2\beta^2$$

$$g_{01} = \left(\gamma_{41} + \gamma_{430}m^2 + \gamma_{432}n^2\beta^2\right)\left(\gamma_{220}m^2 + \gamma_{222}n^2\beta^2\right) - \gamma_{331}n^2\beta^2$$
$$\times \left(\gamma_{231}m^2 + \gamma_{233}n^2\beta^2\right)$$

$$g_{02} = \left(\gamma_{31} + \gamma_{320}m^2 + \gamma_{322}n^2\beta^2\right)\left(\gamma_{231}m^2 + \gamma_{233}n^2\beta^2\right) - \gamma_{331}m^2$$
$$\times \left(\gamma_{220}m^2 + \gamma_{222}n^2\beta^2\right)$$

$$g_{03} = \left(\gamma_{31} + \gamma_{320}m^2 + \gamma_{322}n^2\beta^2\right)\left(\gamma_{41} - \gamma_{411}m^2 - \gamma_{413}n^2\beta^2\right)$$
$$- \gamma_{331}m^2\left(\gamma_{31} - \gamma_{310}m^2 - \gamma_{312}n^2\beta^2\right)$$

$$g_{04} = \left(\gamma_{41} + \gamma_{430}m^2 + \gamma_{432}n^2\beta^2\right)\left(\gamma_{31} - \gamma_{310}m^2 - \gamma_{312}n^2\beta^2\right)$$
$$- \gamma_{331}n^2\beta^2\left(\gamma_{41} - \gamma_{411}m^2 - \gamma_{413}n^2\beta^2\right)$$

$$g_{05} = \left(\gamma_{240}m^4 + 2\gamma_{242}m^2 n^2\beta^2 + \gamma_{244}n^4\beta^4\right)$$
$$+ \frac{m^2\left(\gamma_{220}m^2 + \gamma_{222}n^2\beta^2\right)g_{04} + n^2\beta^2\left(\gamma_{231}m^2 + \gamma_{233}n^2\beta^2\right)g_{03}}{g_{00}}$$

$$g_{06} = \left(m^4 + 2\gamma_{212}m^2n^2\beta^2 + \gamma_{214}n^4\beta^4\right)$$
$$+ \gamma_{14}\gamma_{24}\frac{m^2\left(\gamma_{220}m^2 + \gamma_{222}n^2\beta^2\right)g_{01} + n^2\beta^2\left(\gamma_{231}m^2 + \gamma_{233}n^2\beta^2\right)g_{02}}{g_{00}}$$

$$g_{07} = \left(\gamma_{140}m^4 + 2\gamma_{142}m^2n^2\beta^2 + \gamma_{144}n^4\beta^4\right)$$
$$- \frac{m^2\left(\gamma_{120}m^2 + \gamma_{122}n^2\beta^2\right)g_{01} + n^2\beta^2\left(\gamma_{131}m^2 + \gamma_{133}n^2\beta^2\right)g_{02}}{g_{00}}$$

$$g_{08} = \left(\gamma_{110}m^4 + 2\gamma_{112}m^2n^2\beta^2 + \gamma_{114}n^4\beta^4\right)$$
$$+ \frac{m^2\left(\gamma_{120}m^2 + \gamma_{122}n^2\beta^2\right)g_{04} + n^2\beta^2\left(\gamma_{131}m^2 + \gamma_{133}n^2\beta^2\right)g_{03}}{g_{00}}$$

$$g_{10} = 1 + \gamma_{14}\gamma_{24}\gamma_{220}^2\frac{4m^2}{\gamma_{31} + \gamma_{320}4m^2}$$

$$g_{12} = \frac{\gamma_{244}\left(\gamma_{41} + \gamma_{432}4n^2\beta^2\right) + \gamma_{233}\left(\gamma_{41} - \gamma_{413}4n^2\beta^2\right)}{\gamma_{214}\left(\gamma_{41} + \gamma_{432}4n^2\beta^2\right) + \gamma_{14}\gamma_{24}\gamma_{233}^2 4n^2\beta^2}$$

$$g_{12}^* = \frac{\gamma_{214}\left(\gamma_{41} - \gamma_{413}4n^2\beta^2\right) - \gamma_{14}\gamma_{24}\gamma_{233}\gamma_{244}4n^2\beta^2}{\gamma_{214}\left(\gamma_{41} + \gamma_{432}4n^2\beta^2\right) + \gamma_{14}\gamma_{24}\gamma_{233}^2 4n^2\beta^2}$$

$$g_{09} = \gamma_{114} + \gamma_{133}g_{12}^* + \gamma_{14}\gamma_{24}\gamma_{144}g_{12}$$

$$g_{13} = \frac{\gamma_{41} + \gamma_{432}4n^2\beta^2}{\gamma_{214}\left(\gamma_{41} + \gamma_{432}4n^2\beta^2\right) + \gamma_{14}\gamma_{24}\gamma_{233}^2 4n^2\beta^2}$$

$$g_{14} = -\frac{\gamma_{144}\left(\gamma_{41} + \gamma_{432}4n^2\beta^2\right) - \gamma_{133}\gamma_{233}4n^2\beta^2}{\gamma_{214}\left(\gamma_{41} + \gamma_{432}4n^2\beta^2\right) + \gamma_{14}\gamma_{24}\gamma_{233}^2 4n^2\beta^2}$$

$$g_{11} = g_{14}(1 + 2\mu) + 2\frac{g_{05}}{g_{06}}$$

$$g_{130} = \left(\gamma_{31} + \gamma_{320}m^2 + \gamma_{322}9n^2\beta^2\right)\left(\gamma_{41} + \gamma_{430}m^2 + \gamma_{432}9n^2\beta^2\right) - \gamma_{331}^2 9m^2n^2\beta^2$$

$$g_{131} = \left(\gamma_{41} + \gamma_{430}m^2 + \gamma_{432}9n^2\beta^2\right)\left(\gamma_{220}m^2 + \gamma_{222}9n^2\beta^2\right)$$
$$- \gamma_{331}9n^2\beta^2\left(\gamma_{231}m^2 + \gamma_{233}9n^2\beta\right)^2$$

$$g_{132} = \left(\gamma_{31} + \gamma_{320}m^2 + \gamma_{322}9n^2\beta^2\right)\left(\gamma_{231}m^2 + \gamma_{233}9n^2\beta^2\right) - \gamma_{331}m^2$$
$$\times \left(\gamma_{220}m^2 + \gamma_{222}9n^2\beta^2\right)$$

$$g_{136} = \left(m^4 + 18\gamma_{212}m^2n^2\beta^2 + \gamma_{214}81n^4\beta^4\right)$$
$$+ \gamma_{14}\gamma_{24}\frac{m^2\left(\gamma_{220}m^2 + \gamma_{222}9n^2\beta^2\right)g_{131} + 9n^2\beta^2\left(\gamma_{231}m^2 + \gamma_{233}9n^2\beta^2\right)g_{132}}{g_{130}}$$

$$g_{15} = \gamma_{220}(\gamma_{310} + \gamma_{120}) - \gamma_{320}(\gamma_{140} + \gamma_{240})$$

$$g_{16} = \left(\gamma_{320} + \gamma_{14}\gamma_{24}\gamma_{220}^2\right)(\gamma_{320}\gamma_{110} - \gamma_{310}\gamma_{120})$$
$$+ \gamma_{14}\gamma_{24}(\gamma_{320}\gamma_{140} - \gamma_{120}\gamma_{220})(\gamma_{320}\gamma_{240} - \gamma_{310}\gamma_{220})$$

$$g_{17} = \frac{(\gamma_{310} + \gamma_{14}\gamma_{24}\gamma_{220}\gamma_{240})b - \gamma_{14}\gamma_{24}\gamma_{220}}{(\gamma_{310} + \gamma_{14}\gamma_{24}\gamma_{220}\gamma_{240})b + \gamma_{14}\gamma_{24}\gamma_{220}}$$

$$g_{19} = \frac{\gamma_{320}}{\gamma_{320} + \gamma_{14}\gamma_{24}\gamma_{220}^2} \frac{\left(2\vartheta^2 g_{17} - c\right)}{b^2} + \frac{\gamma_{310}\gamma_{220} - \gamma_{320}\gamma_{240}}{\gamma_{320} + \gamma_{14}\gamma_{24}\gamma_{220}^2}$$

$$g_{20} = -\frac{\gamma_{320}}{\gamma_{320} + \gamma_{14}\gamma_{24}\gamma_{220}^2} \frac{\left(2\phi^2 g_{17} + c\right)}{b^2} + \frac{2}{b}\frac{\gamma_{320}g_{17}}{\gamma_{320} + \gamma_{14}\gamma_{24}\gamma_{220}^2}$$

$$-\frac{2\gamma_{310}\gamma_{320} - (\gamma_{310}\gamma_{220}g_{17} - \gamma_{320}\gamma_{240})[(\gamma_{310} + \gamma_{14}\gamma_{24}\gamma_{220}\gamma_{240})b + \gamma_{14}\gamma_{24}\gamma_{220}]}{(\gamma_{320} + \gamma_{14}\gamma_{24}\gamma_{220}^2)[(\gamma_{310} + \gamma_{14}\gamma_{24}\gamma_{220}\gamma_{240})b + \gamma_{14}\gamma_{24}\gamma_{220}]}$$

$$(L.3)$$

Appendix M

In Equation 5.87 through 5.90

$$\Theta_3 = \frac{1}{C_3}\left[C_2 + \frac{1}{\gamma_{24}}\left(1 - \frac{1}{2}a\gamma_5\right)\lambda_q^{(2)}\right]$$

$$\Theta_4 = \frac{1}{\gamma_{24}}[(\gamma_{T2} - \gamma_5\gamma_{T1})\Delta T]\varepsilon - \frac{1}{\gamma_{24}}\left(1 - \frac{1}{2}a\gamma_5\right)\lambda_q^{(0)}$$

$$\lambda_q^{(0)} = \left\{\frac{\gamma_{24}m^4}{C_1(1+\mu)g_{06}} + \frac{\gamma_{24}m^2}{C_1(1+\mu)^2}\frac{g_{05} + (1+\mu)g_{07}}{g_{06}}\varepsilon\right.$$

$$+ \varepsilon^2 \frac{1}{\gamma_{14}C_1(1+\mu)}\left[g_{08} + \gamma_{14}\gamma_{24}\frac{g_{05}}{g_{06}}\frac{(1+\mu)g_{07} - \mu g_{05}}{(1+\mu)^2}\right]$$

$$\left.\times\left[1 - \frac{\mu g_{05}}{(1+\mu)m^2}\varepsilon\left(1 - \frac{\mu g_{05}}{(1+\mu)m^2}\varepsilon\right)\right]\right\}$$

$$\lambda_q^{(2)} = \frac{1}{4}\frac{m^4 n^2\beta^2}{g_{06}}\left\{4\gamma_{24}(1+\mu) + \frac{1}{4}\frac{\gamma_{24}g_{06}g_{13}}{n^2\beta^2 C_1}(1+2\mu)\right.$$

$$- \frac{\gamma_{24}n^2\beta^2 g_{06}}{C_1(1+\mu)g_{06} - 2am^6 g_{10}}\left[2(1+\mu)^2 + \frac{1}{2}\frac{am^2}{C_1}(1+2\mu)\right.$$

$$\left.\left.+ 2\frac{(1+2\mu)g_{06} + 8m^4 g_{10}(1+\mu)}{g_{06}}\right]\right\}$$

$$\delta_x^{(0)} = \frac{1}{\gamma_{24}}\left[\left(\frac{1}{2}a\gamma_{24}^2 - \gamma_5\right) + \frac{2}{\pi}\frac{\gamma_5}{\gamma_{24}}\left(1 - \frac{1}{2}a\gamma_5\right)\left(\vartheta b_{01}^{(5/2)} - \phi b_{10}^{(5/2)}\right)\varepsilon^{1/2}\right]\lambda_q$$

$$+ \left[\frac{1}{\pi(3)^{3/4}\gamma_{24}^2}\frac{b_{11}}{\vartheta}\left(1 - \frac{1}{2}a\gamma_5\right)^2\varepsilon\right]\lambda_q^2$$

$$\delta_x^{(T)} = \frac{(3)^{3/4}}{4\gamma_{24}}[(\gamma_{24}^2\gamma_{T1} - \gamma_5\gamma_{T2})\Delta T]\varepsilon^{-1/2}$$

$$\delta_x^{(2)} = \frac{1}{32}(3)^{3/4}\left[m^2(1+2\mu)\varepsilon^{-3/2} - 2g_{05}\varepsilon^{-1/2} + \frac{g_{05}^2}{m^2}\varepsilon^{1/2}\right]$$

$$\text{(M.1)}$$

in the above equations

$$C_1 = n^2\beta^2 + \frac{1}{2}am^2, \quad C_3 = 1 - \frac{g_{05}}{m^2}\varepsilon$$

$$C_2 = \frac{1}{8}n^2\beta^2\left[(1+2\mu) + C_1(1+\mu)\frac{g_{06}(1+2\mu) + 8m^4(1+\mu)g_{10}}{C_1(1+\mu)g_{06} - 2am^6 g_{10}}\right] - \frac{1}{4}C_1(1+\mu)^2$$

$$a_{01}^{(3/2)} = 1, \quad a_{10}^{(3/2)} = \frac{\vartheta}{\phi} g_{17}, \quad b_{01}^{(5/2)} = \gamma_{24} g_{19}, \quad b_{10}^{(5/2)} = \gamma_{24} \frac{\vartheta}{\phi} g_{20}$$

$$b_{11} = \frac{1}{b} \left[\left(a_{10}^{(3/2)} \right)^2 \phi^2 b + a_{10}^{(3/2)} 2\vartheta \phi c + \left(2\vartheta^4 - \vartheta^2 \phi^2 + \phi^4 \right) \right] \qquad \text{(M.2)}$$

Appendix N

In Equations 5.122 through 5.126

$$\Theta_5 = \frac{1}{\gamma_{24}}[(\gamma_{T2} - \gamma_5\gamma_{T1})\Delta T] + \left(\frac{2\gamma_5}{\gamma_{24}}\lambda_P^{(0)}\right)\varepsilon^{1/4}$$

$$\Theta_6 = \frac{1}{C_3}\left\{\frac{\gamma_{14}\gamma_{24}m^4(1+\mu)}{16n^2\,\beta^2 g_{09}g_{210}}\varepsilon^{-1} - \frac{\gamma_{14}\gamma_{24}m^2 g_{11}}{32n^2\beta^2 g_{09}}\right.$$

$$\left. + \frac{1}{8}(n^2\beta^2 + k^2\beta^2)\frac{g_{210}^2 + g_{220}^2}{g_{210}^2}(1+2\mu)\varepsilon + \left(\frac{2\gamma_5}{\gamma_{24}}\lambda_P^{(2)}\right)\varepsilon^{1/4}\right\} \qquad (N.1)$$

$$\lambda_s^{(0)} = -\frac{m}{2n\beta}\left\{\frac{\gamma_{24}m^2 g_{220}}{(g_{210}^2 - g_{220}^2)(1+\mu)}\varepsilon^{-5/4} + \frac{\gamma_{24}}{(1+\mu)^2(g_{210}^2 - g_{220}^2)}\right.$$

$$\times [(1+\mu)(g_{210}g_{32} + g_{220}g_{31}) + (g_{220}g_{310} + g_{210}g_{320})]\varepsilon^{-1/4}$$

$$+ \frac{1}{m^2(1+\mu)}\left[\frac{g_{120}}{\gamma_{14}} + \frac{\gamma_{24}}{(1+\mu)}\frac{g_{31}(g_{220}g_{310} + g_{210}g_{320}) + g_{32}(g_{220}g_{320} + g_{210}g_{310})}{g_{210}^2 - g_{220}^2}\right.$$

$$\left. - \frac{\gamma_{24}\mu}{(1+\mu)^2}\frac{2g_{210}g_{310}g_{320} + g_{220}(g_{310}^2 + g_{320}^2)}{g_{210}^2 - g_{220}^2}\right]\varepsilon^{3/4}$$

$$- \frac{\mu}{m^4(1+\mu)^2}\left[\frac{g_{110}g_{320} + g_{120}g_{310}}{\gamma_{14}} + \frac{\gamma_{24}g_{31}}{(1+\mu)}\frac{2g_{210}g_{310}g_{320} + g_{220}(g_{310}^2 + g_{320}^2)}{g_{210}^2 - g_{220}^2}\right.$$

$$+ \frac{\gamma_{24}g_{32}}{(1+\mu)}\frac{g_{210}(g_{310}^2 + g_{320}^2) + 2g_{220}g_{310}g_{320}}{g_{210}^2 - g_{220}^2}$$

$$\left. - \frac{\gamma_{24}\mu}{(1+\mu)^2}\frac{g_{220}g_{310}(g_{310}^2 + 3g_{320}^2) + g_{210}g_{320}(3g_{310}^2 + g_{320}^2)}{g_{210}^2 - g_{220}^2}\right]\varepsilon^{7/4}$$

$$+ \frac{\mu^2}{m^6(1+\mu)^3}\left[\frac{g_{120}(g_{310}^2 + g_{320}^2) + 2g_{110}g_{310}g_{320}}{\gamma_{14}}\right.$$

$$+ \frac{\gamma_{24}g_{31}}{(1+\mu)}\frac{g_{220}g_{310}(g_{310}^2 + 3g_{320}^2) + g_{210}g_{320}(3g_{310}^2 + g_{320}^2)}{g_{210}^2 - g_{220}^2}$$

$$+ \frac{\gamma_{24}g_{32}}{(1+\mu)}\frac{g_{220}g_{320}(3g_{310}^2 + g_{320}^2) + g_{210}g_{310}(g_{310}^2 + 3g_{320}^2)}{g_{210}^2 - g_{220}^2}$$

$$\left. - \frac{\gamma_{24}\mu}{(1+\mu)^2}\frac{4g_{210}g_{310}g_{320}(g_{310}^2 + g_{320}^2) + g_{220}(g_{310}^4 + 6g_{310}^2 g_{320}^2 + g_{320}^4)}{g_{210}^2 - g_{220}^2}\right]\varepsilon^{11/4}\right\}$$

$$\lambda_s^{(2)} = -\frac{m}{2n\beta}\left\{\frac{\gamma_{14}\gamma_{24}^2 m^6 g_{210}g_{220}}{2g_{09}\left(g_{210}^2 - g_{220}^2\right)^2}\varepsilon^{-5/4} - \frac{\gamma_{14}\gamma_{24}^2 m^4}{8g_{09}g_{210}\left(g_{210}^2 - g_{220}^2\right)^2}\right.$$

$$\times\left[(3g_{210}^2 + g_{220}^2)g_{220}(g_{310} - g_{31}) + (g_{210}^2 + 3g_{220}^2)g_{210}(g_{320} - g_{32})\right]\varepsilon^{-1/4}$$

$$+\frac{\gamma_{14}\gamma_{24}^2 m^4 g_{320}\left(g_{210}^2 - 3g_{220}^2\right)}{2g_{09}\left(g_{210}^2 - g_{220}^2\right)^2(1+\mu)}\varepsilon^{-1/4}$$

$$-\frac{\gamma_{14}\gamma_{24}^2 m^4 g_{220}}{4g_{09}\left(g_{210}^2 - g_{220}^2\right)(1+\mu)}\left[g_{14}(1+2\mu) - g_{302}(1+\mu)^2\right]\varepsilon^{-1/4}$$

$$-\frac{\mu\gamma_{14}\gamma_{24}m^2}{8g_{09}g_{210}\left(g_{210}^2 - g_{220}^2\right)}\left[\frac{g_{120}\left(g_{210}^2 + g_{220}^2\right) + 2g_{110}g_{210}g_{220}}{\gamma_{14}}\right.$$

$$+\frac{\gamma_{24}}{1+\mu}\frac{g_{220}\left(g_{310}^2 + g_{320}^2\right)\left(3g_{210}^2 + g_{220}^2\right) + 2g_{210}g_{310}g_{320}\left(g_{210}^2 + 3g_{220}^2\right)}{g_{210}^2 - g_{220}^2}$$

$$+\frac{\gamma_{24}}{1+\mu}\frac{g_{210}\left(g_{210}^2 + 3g_{220}^2\right)(g_{32}g_{310} + g_{31}g_{320}) + g_{220}(3g_{210}^2 + g_{220}^2)(g_{31}g_{310} + g_{32}g_{320})}{g_{210}^2 - g_{220}^2}\right]\varepsilon^{3/4}$$

$$-\frac{\gamma_{24}m^2}{8}\frac{g_{220}}{g_{210}}g_{13}(1+2\mu)\varepsilon^{3/4} + \frac{2\gamma_{24}m^2 n^4\beta^4 g_{220}}{\left(g_{210}^2 - g_{220}^2\right)(1+\mu)}$$

$$\times\frac{\left(g_{210}^2 - g_{220}^2\right)\left[2(1+\mu)^2 + 3(1+2\mu)\right] + g_{210}g_{200}(1+\mu)}{4\left(g_{210}^2 - g_{220}^2\right)(1+\mu) - g_{210}g_{200}}\varepsilon^{3/4}$$

$$-\frac{\gamma_{14}\gamma_{24}^2 m^2 g_{220}}{8g_{09}\left(g_{210}^2 - g_{220}^2\right)}\left[g_{210}g_{302}g_{11} - 4g_{310}g_{14}\right]\varepsilon^{3/4}$$

$$+\frac{\gamma_{14}\gamma_{24}^2 m^2 g_{320}g_{302}\left(g_{210}^2 + g_{220}^2\right)}{4g_{09}g_{210}\left(g_{210}^2 - g_{220}^2\right)}\varepsilon^{3/4} - \frac{\gamma_{14}\gamma_{24}^2 m^2 g_{14}(1+2\mu)}{8g_{09}g_{210}\left(g_{210}^2 - g_{220}^2\right)(1+\mu)^2}$$

$$\times\left[(g_{210}^2 + g_{220}^2)g_{320} + 2g_{210}g_{220}g_{310}\right]\varepsilon^{3/4}$$

$$-\frac{\gamma_{14}\gamma_{24}^2 m^2 g_{320}}{2g_{09}g_{210}\left(g_{210}^2 - g_{220}^2\right)(1+\mu)}\left[\frac{(3g_{210}^2 + g_{220}^2)g_{220}g_{320} + (g_{210}^2 + 3g_{220}^2)g_{210}g_{310}}{g_{210}^2 - g_{220}^2}\right]\varepsilon^{3/4}$$

$$-\frac{\gamma_{14}\gamma_{24}^2 m^2\left(3g_{210}^2 + g_{220}^2\right)g_{220}g_{11}(g_{31} - g_{310})}{16g_{09}\left(g_{210}^2 - g_{220}^2\right)^2(1+\mu)}\varepsilon^{3/4}$$

$$\left.-\frac{\gamma_{14}\gamma_{24}^2 m^2\left(g_{210}^2 + 3g_{220}^2\right)g_{210}g_{11}(g_{32} - g_{320})}{16g_{09}\left(g_{210}^2 - g_{220}^2\right)^2(1+\mu)^2}\varepsilon^{3/4}\right\}$$

$$\lambda_s^{(4)} = -\frac{m}{2n\beta} \frac{\gamma_{14}^2 \gamma_{24}^3 m^{10}(1+\mu)}{64 g_{09}^2 \left(g_{210}^2 - g_{220}^2\right)} \left\{ 4\frac{g_{220}\left(g_{210}^2 + 3g_{220}^2\right)}{\left(g_{210}^2 - g_{220}^2\right)^2} - \frac{g_{220}}{g_{210}^2} \right.$$

$$+ \frac{(1+\mu)}{\left(g_{210}^2 - g_{220}^2\right)\left(g_{23}^2 - g_{24}^2\right)g_{210}^2} \left[g_{24}\left(g_{210}^2 - g_{220}^2\right)^2 + 2g_{210}g_{220}g_{23}\left(3g_{210}^2 + g_{220}^2\right) \right.$$

$$\left. - 2g_{210}^2 g_{24}\left(g_{210}^2 + 3g_{220}^2\right)\right]$$

$$+ \left(\frac{(1+\mu)}{g_{210}^2}\frac{g_{220}g_{23}\left(3g_{210}^2 - g_{220}^2\right) - g_{210}g_{24}\left(g_{210}^2 + g_{220}^2\right)}{g_{23}^2 - g_{24}^2}\right.$$

$$\left. + 2\frac{g_{220}}{g_{210}}\frac{g_{210}g_{23} - g_{220}g_{24}}{g_{23}^2 - g_{24}^2} + 6\frac{g_{220}}{g_{210}}\right)R_1$$

$$- \left(\frac{(1+\mu)}{g_{210}^2}\frac{g_{220}g_{24}\left(3g_{210}^2 - g_{220}^2\right) - g_{210}g_{23}\left(g_{210}^2 + g_{220}^2\right)}{g_{23}^2 - g_{24}^2}\right.$$

$$\left.\left. - 2\frac{g_{220}}{g_{210}}\frac{g_{220}g_{23} - g_{210}g_{24}}{g_{23}^2 - g_{24}^2} + \frac{g_{210}^2 + g_{220}^2}{g_{210}^2}\right)R_2 \right\}\varepsilon^{-5/4} \tag{N.2}$$

in which

$$\lambda_p^{(0)} = \frac{k\beta}{m}\lambda_s^{(0)} + \lambda_{xp}^{(0)}, \quad \lambda_p^{(2)} = \frac{k\beta}{m}\lambda_s^{(2)} + \lambda_{xp}^{(2)}, \quad \lambda_p^{(4)} = \frac{k\beta}{m}\lambda_s^{(4)} + \lambda_{xp}^{(4)}, \tag{N.3}$$

and k can be determined through equation

$$\lambda_{xp}^{(0)} - \lambda_{xp}^{(2)}\left(A_{11}^{(2)}\varepsilon\right)^2 + \lambda_{xp}^{(4)}\left(A_{11}^{(2)}\varepsilon\right)^4 = 0 \tag{N.4}$$

where

$$
\begin{aligned}
\lambda_{\mathrm{xp}}^{(0)} = \frac{1}{2}\Bigg\{ & \frac{\gamma_{24}m^2 g_{210}}{(g_{210}^2 - g_{220}^2)(1+\mu)}\varepsilon^{-5/4} + \frac{\gamma_{24}}{(1+\mu)^2(g_{210}^2 - g_{220}^2)}[(1+\mu)(g_{210}g_{31} + g_{220}g_{32}) \\
& + (g_{220}g_{320} + g_{210}g_{310})]\varepsilon^{-1/4} \\
& + \frac{1}{m^2(1+\mu)}\Bigg[\frac{g_{110}}{\gamma_{14}} + \frac{\gamma_{24}}{(1+\mu)}\frac{g_{31}(g_{220}g_{320} + g_{210}g_{310}) + g_{32}(g_{220}g_{310} + g_{210}g_{320})}{g_{210}^2 - g_{220}^2} \\
& - \frac{\gamma_{24}\mu}{(1+\mu)^2}\frac{2g_{220}g_{310}g_{320} + g_{210}(g_{310}^2 + g_{320}^2)}{g_{210}^2 - g_{220}^2}\Bigg]\varepsilon^{3/4} \\
& - \frac{\mu}{m^4(1+\mu)^2}\Bigg[\frac{g_{110}g_{310} + g_{120}g_{320}}{\gamma_{14}} + \frac{\gamma_{24}g_{31}}{(1+\mu)}\frac{2g_{220}g_{310}g_{320} + g_{210}(g_{310}^2 + g_{320}^2)}{g_{210}^2 - g_{220}^2} \\
& + \frac{\gamma_{24}g_{32}}{(1+\mu)}\frac{g_{220}(g_{310}^2 + g_{320}^2) + 2g_{210}g_{310}g_{320}}{g_{210}^2 - g_{220}^2} \\
& - \frac{\gamma_{24}\mu}{(1+\mu)^2}\frac{g_{210}g_{310}(g_{310}^2 + 3g_{320}^2) + g_{220}g_{320}(3g_{310}^2 + g_{320}^2)}{g_{210}^2 - g_{220}^2}\Bigg]\varepsilon^{7/4} \\
& + \frac{\mu^2}{m^6(1+\mu)^3}\Bigg[\frac{g_{110}(g_{310}^2 + g_{320}^2) + 2g_{120}g_{310}g_{320}}{\gamma_{14}} \\
& + \frac{\gamma_{24}g_{31}}{(1+\mu)}\frac{g_{220}g_{320}(3g_{310}^2 + g_{320}^2) + g_{210}g_{310}(g_{310}^2 + 3g_{320}^2)}{g_{210}^2 - g_{220}^2} \\
& + \frac{\gamma_{24}g_{32}}{(1+\mu)}\frac{g_{220}g_{310}(g_{310}^2 + 3g_{320}^2) + g_{210}g_{320}(3g_{310}^2 + g_{320}^2)}{g_{210}^2 - g_{220}^2} \\
& - \frac{\gamma_{24}\mu}{(1+\mu)^2}\frac{4g_{220}g_{310}g_{320}(g_{310}^2 + g_{320}^2) + g_{210}(g_{310}^4 + 6g_{310}^2 g_{320}^2 + g_{320}^4)}{g_{210}^2 - g_{220}^2}\Bigg]\varepsilon^{11/4}\Bigg\}
\end{aligned}
$$

$$
\lambda_{xp}^{(2)} = \frac{1}{2}\left\{ \frac{\gamma_{14}\gamma_{24}^2 m^6 \left(g_{210}^2 + g_{220}^2\right)}{4g_{09}\left(g_{210}^2 - g_{220}^2\right)^2} \varepsilon^{-5/4} \right.
$$

$$
- \frac{\gamma_{14}\gamma_{24}^2 m^4}{8g_{09}g_{210}\left(g_{210}^2 - g_{220}^2\right)^2}\left[\left(3g_{210}^2 + g_{220}^2\right)g_{220}\left(g_{320} - g_{32}\right)\right.
$$

$$
\left. + \left(g_{210}^2 + 3g_{220}^2\right)g_{210}\left(g_{310} - g_{31}\right)\right]\varepsilon^{-1/4} + \frac{\gamma_{14}\gamma_{24}^2 m^4 g_{220}g_{320}\left(3g_{210}^2 + g_{220}^2\right)}{2g_{09}g_{210}\left(g_{210}^2 - g_{220}^2\right)^2(1+\mu)}\varepsilon^{-1/4}
$$

$$
- \frac{\gamma_{14}\gamma_{24}^2 m^4\left(g_{210}^2 + g_{220}^2\right)}{8g_{09}g_{210}\left(g_{210}^2 - g_{220}^2\right)(1+\mu)}\left[g_{14}(1+2\mu) - g_{302}(1+\mu)^2\right]\varepsilon^{-1/4}
$$

$$
- \frac{\mu\gamma_{14}\gamma_{24}m^2}{8g_{09}g_{210}\left(g_{210}^2 - g_{220}^2\right)}\left[\frac{g_{110}\left(g_{210}^2 + g_{220}^2\right) + 2g_{120}g_{210}g_{220}}{\gamma_{14}}\right.
$$

$$
+ \frac{\gamma_{24}}{1+\mu}\frac{g_{210}\left(g_{310}^2 + g_{320}^2\right)\left(g_{210}^2 + 3g_{220}^2\right) + 2g_{220}g_{310}g_{320}\left(3g_{210}^2 + g_{220}^2\right)}{g_{210}^2 - g_{220}^2}
$$

$$
\left.+ \frac{\gamma_{24}}{1+\mu}\frac{g_{210}\left(g_{210}^2 + 3g_{220}^2\right)\left(g_{31} + g_{310} + g_{32}g_{320}\right) + g_{220}\left(3g_{210}^2 + g_{220}^2\right)\left(g_{32}g_{310} + g_{31}g_{320}\right)}{g_{210}^2 - g_{220}^2}\right]\varepsilon^{3/4}
$$

$$
- \frac{\gamma_{24}m^2}{16}\frac{g_{210}^2 + g_{220}^2}{g_{210}^2}g_{13}(1+2\mu)\varepsilon^{3/4} + \frac{\gamma_{24}m^2 n^4\beta^4\left(g_{210}^2 + g_{220}^2\right)}{g_{210}\left(g_{210}^2 - g_{220}^2\right)(1+\mu)}
$$

$$
\times \frac{\left(g_{210}^2 - g_{220}^2\right)\left[2(1+\mu)^2 + 3(1+2\mu)\right] + g_{210}g_{200}(1+\mu)}{4\left(g_{210}^2 - g_{220}^2\right)(1+\mu) - g_{210}g_{200}}\varepsilon^{3/4}
$$

$$
- \frac{\gamma_{14}\gamma_{24}^2 m^2\left(g_{210}^2 + g_{220}^2\right)}{16g_{09}g_{210}\left(g_{210}^2 - g_{220}^2\right)}\left[g_{210}g_{302}g_{11} - 4g_{310}g_{14}\right]\varepsilon^{3/4}
$$

$$
+ \frac{\gamma_{14}\gamma_{24}^2 m^2 g_{320}g_{302}g_{220}}{2g_{09}\left(g_{210}^2 - g_{220}^2\right)}\varepsilon^{3/4} - \frac{\gamma_{14}\gamma_{24}^2 m^2 g_{14}(1+2\mu)}{8g_{09}g_{210}\left(g_{210}^2 - g_{220}^2\right)(1+\mu)^2}
$$

$$
\times \left[\left(g_{210}^2 + g_{220}^2\right)g_{310} + 2g_{210}g_{220}g_{320}\right]\varepsilon^{3/4}
$$

$$
- \frac{\gamma_{14}\gamma_{24}^2 m^2 g_{320}}{2g_{09}g_{210}\left(g_{210}^2 - g_{220}^2\right)(1+\mu)}\left[\frac{\left(3g_{210}^2 + g_{220}^2\right)g_{220}g_{310} + \left(g_{210}^2 + 3g_{220}^2\right)g_{210}g_{320}}{g_{210}^2 - g_{220}^2}\right]\varepsilon^{3/4}
$$

$$
- \frac{\gamma_{14}\gamma_{24}^2 m^2\left(3g_{210}^2 + g_{220}^2\right)g_{220}g_{11}\left(g_{32} - g_{320}\right)}{16g_{09}\left(g_{210}^2 - g_{220}^2\right)^2(1+\mu)}\varepsilon^{3/4}
$$

$$
\left.- \frac{\gamma_{14}\gamma_{24}^2 m^2\left(g_{210}^2 + 3g_{220}^2\right)g_{210}g_{11}\left(g_{31} - g_{310}\right)}{16g_{09}\left(g_{210}^2 - g_{220}^2\right)^2(1+\mu)}\varepsilon^{3/4}\right\}
$$

$$\lambda_{\text{xp}}^{(4)} = \frac{1}{2} \frac{\gamma_{14}^2 \gamma_{24}^3 m^{10}(1+\mu)}{64 g_{09}^2 g_{210} \left(g_{210}^2 - g_{220}^2\right)} \left\{ 4 \frac{g_{220}^2 \left(3g_{210}^2 + g_{220}^2\right)}{\left(g_{210}^2 - g_{220}^2\right)^2} - 1 \right.$$

$$+ \frac{(1+\mu)}{\left(g_{210}^2 - g_{220}^2\right)\left(g_{23}^2 - g_{24}^2\right)g_{210}}$$

$$\times \left[g_{23}\left(g_{210}^2 + g_{220}^2\right)\left(3g_{210}^2 + g_{220}^2\right) - 2g_{24}g_{210}g_{220}\left(2g_{210}^2 + g_{220}^2\right)\right]$$

$$+ \left((1+\mu)\frac{g_{210}^2 + g_{220}^2}{g_{210}} \frac{g_{210}g_{23} - g_{220}g_{24}}{g_{23}^2 - g_{24}^2} \right.$$

$$\left. + 2g_{210}\frac{g_{210}g_{23} - g_{220}g_{24}}{g_{23}^2 - g_{24}^2} + 3\frac{g_{210}^2 + g_{220}^2}{g_{210}} \right) R_1$$

$$+ \left((1+\mu)\frac{g_{210}^2 + g_{220}^2}{g_{210}} \frac{g_{220}g_{23} - g_{210}g_{24}}{g_{23}^2 - g_{24}^2} \right.$$

$$\left. + 2g_{210}\frac{g_{220}g_{23} - g_{210}g_{24}}{g_{23}^2 - g_{24}^2} - 2g_{220} \right) R_2 \right\} \varepsilon^{-5/4} \qquad (\text{N.5})$$

$$\delta_x^{(0)} = \frac{1}{\gamma_{24}} \left[2\gamma_{24}^2 - \frac{4}{\pi}\left(\vartheta b_{01}^{(9/4)} - \phi b_{10}^{(9/4)}\right)\frac{\gamma_5^2}{\gamma_{24}}\varepsilon^{1/2} \right]\lambda_p,$$

$$\delta_x^{(\text{T})} = \frac{1}{2\gamma_{24}}\left[\left(\gamma_{24}^2 \gamma_{T1} - \gamma_5 \gamma_{T2}\right)\Delta T\right]\varepsilon^{-1/4}$$

$$\delta_x^{(2)} = \frac{1}{8}\left[\frac{b_{11}}{2\pi\vartheta}\frac{g_{210}^2 + g_{220}^2}{g_{210}^2}\varepsilon^{1/4} + m^2(1+2\mu)\frac{g_{210}^2 + g_{220}^2}{g_{210}^2}\varepsilon^{3/4} \right.$$

$$- 2\frac{g_{310}\left(g_{210}^2 + g_{220}^2\right) - 2g_{320}g_{210}g_{220}}{g_{210}^2}\varepsilon^{7/4}$$

$$+ \frac{3}{m^2}\frac{\left(g_{210}^2 + g_{220}^2\right)\left(g_{310}^2 + g_{320}^2\right) - 4g_{210}g_{220}g_{310}g_{320}}{g_{210}^2}\varepsilon^{11/4}$$

$$\left. - \frac{2}{m^4}\frac{g_{310}\left(g_{210}^2 + g_{220}^2\right)\left(g_{310}^2 + 3g_{320}^2\right) - 2g_{320}g_{210}g_{220}\left(3g_{310}^2 + g_{320}^2\right)}{g_{210}^2}\varepsilon^{15/4} \right]$$

$$\delta_x^{(4)} = \frac{1}{32}\left\{ \frac{b_{11}}{64\pi\vartheta}\frac{\gamma_{14}^2 \gamma_{24}^2 m^8(1+\mu)^2}{n^4\beta^4 g_{09}^2 g_{210}^2}\varepsilon^{-7/4} \right.$$

$$+ \frac{\gamma_{14}\gamma_{24}}{g_{09}g_{210}}\frac{g_{21}^2 - g_{22}^2}{g_{210}^2}(1+\mu)^2 m^4(m^2 - g_{310}\varepsilon)\varepsilon^{3/4}$$

$$+ 2m^2 n^4\beta^4(1+\mu)^2\left(\frac{g_{210}^2 - g_{220}^2}{g_{210}^2}\right)^2$$

$$\left. \times \left[\frac{2\left(g_{210}^2 - g_{220}^2\right)(1+2\mu) + g_{210}g_{200}(1+\mu)}{4\left(g_{210}^2 - g_{220}^2\right)(1+\mu) - g_{210}g_{200}}\right]^2\varepsilon^{11/4} \right\} \qquad (\text{N.6})$$

$$\gamma^{(0)} = \frac{\gamma_{266}}{\gamma_{24}} \lambda_s$$

$$\gamma^{(2)} = -\frac{1}{4}\left[\left(\frac{k\beta}{m}\frac{g_{210}^2 + g_{220}^2}{g_{210}^2} + \frac{n\beta}{m}\frac{2g_{220}}{g_{210}}\right)m^2(1+2\mu)\varepsilon^{3/4}\right.$$

$$-\left(\frac{k\beta}{m}\frac{g_{310}(g_{210}^2 + g_{220}^2) - 2g_{320}g_{210}g_{220}}{g_{210}^2}\right.$$

$$\left.+\frac{n\beta}{m}\frac{g_{320}(g_{210}^2 + g_{220}^2) - 2g_{310}g_{210}g_{220}}{g_{210}^2}\right)\varepsilon^{7/4}$$

$$+2\left(\frac{k\beta}{m}\frac{(g_{210}^2 + g_{220}^2)(g_{310}^2 + g_{320}^2) - 4g_{210}g_{220}g_{310}g_{320}}{m^2 g_{210}^2}\right.$$

$$+2\frac{n\beta}{m}\frac{(g_{210}^2 + g_{220}^2)g_{310}g_{320} - g_{210}g_{220}(g_{310}^2 + g_{320}^2)}{m^2 g_{210}^2}\right)\varepsilon^{11/4}$$

$$-\left(\frac{k\beta}{m}\frac{g_{310}(g_{210}^2 + g_{220}^2)(g_{310}^2 + 3g_{320}^2) - 2g_{320}g_{210}g_{220}(3g_{310}^2 + g_{320}^2)}{m^4 g_{210}^2}\right.$$

$$\left.\left.+\frac{n\beta}{m}\frac{g_{320}(g_{210}^2 + g_{220}^2)(3g_{310}^2 + g_{320}^2) - 2g_{310}g_{210}g_{220}(g_{310}^2 + 3g_{320}^2)}{m^4 g_{210}^2}\right)\right]\varepsilon^{15/4}$$

$$\gamma^{(4)} = -\frac{1}{32}\left\{\frac{k\beta}{m}\frac{m^6(1+\mu)^2}{g_{09}g_{210}}\frac{(g_{210}^2 - g_{220}^2)}{g_{210}^2}\varepsilon^{3/4}\right.$$

$$-\frac{\gamma_{14}\gamma_{24}}{g_{09}g_{210}}\frac{g_{210}^2 - g_{220}^2}{g_{210}^2}(1+\mu)^2 m^4\left[\frac{k\beta}{m}g_{310} + \frac{n\beta}{m}g_{320}\right]\varepsilon^{7/4}$$

$$+\frac{k\beta}{m}4m^2 n^4 \beta^4(1+\mu)^2\left(\frac{g_{210}^2 - g_{220}^2}{g_{210}^2}\right)^2$$

$$\left.\times\left[\frac{2(g_{210}^2 - g_{220}^2)(1+2\mu) + g_{210}g_{200}(1+\mu)}{4(g_{210}^2 - g_{220}^2)(1+\mu) - g_{210}g_{200}}\right]^2 \varepsilon^{11/4}\right\} \tag{N.7}$$

in the above equations

$$g_{210} = m^4 + 2\gamma_{212}m^2(k^2\beta^2 + n^2\beta^2) + \gamma_{214}(k^4\beta^4 + 6k^2\beta^2 n^2\beta^2 + n^4\beta^4)$$

$$+ \gamma_{14}\gamma_{24}\frac{e_4\Delta_{01} + e_1\Delta_{03} + e_3\Delta_{02} + e_2\Delta_{04}}{\Delta_{00}}$$

$$g_{220} = 4k\beta n\beta\left[\gamma_{222}m^2 + \gamma_{214}(k^2\beta^2 + n^2\beta^2)\right]$$

$$- \gamma_{14}\gamma_{24}\frac{e_4\Delta_{02} + e_1\Delta_{04} + e_3\Delta_{01} + e_2\Delta_{03}}{\Delta_{00}}$$

$$g_{310} = \left[\gamma_{240}m^4 + 2\gamma_{242}m^2\left(k^2\beta^2 + n^2\beta^2\right) + \gamma_{244}\left(k^4\beta^4 + 6k^2\beta^2 n^2\beta^2 + n^4\beta^4\right) \right]$$
$$- \frac{e_1\Delta_{06} + e_2\Delta_{08} + e_3\Delta_{07} + e_4\Delta_{05}}{\Delta_{00}}$$

$$g_{320} = -4k\beta n\beta\left[\gamma_{242}m^2 + \gamma_{244}\left(k^2\beta^2 + n^2\beta^2\right) \right] - \frac{e_1\Delta_{08} + e_2\Delta_{06} + e_3\Delta_{05} + e_4\Delta_{07}}{\Delta_{00}}$$

$$g_{31} = \left[\gamma_{140}m^4 + 2\gamma_{142}m^2\left(k^2\beta^2 + n^2\beta^2\right) + \gamma_{144}\left(k^4\beta^4 + 6k^2\beta^2 n^2\beta^2 + n^4\beta^4\right) \right]$$
$$- \frac{h_4\Delta_{01} + h_3\Delta_{02} + h_1\Delta_{03} + h_2\Delta_{04}}{\Delta_{00}}$$

$$g_{32} = -4k\beta n\beta\left[\gamma_{142}m^2 + \gamma_{144}\left(k^2\beta^2 + n^2\beta^2\right) \right] - \frac{h_4\Delta_{02} + h_3\Delta_{01} + h_2\Delta_{03} + h_1\Delta_{04}}{\Delta_{00}}$$

$$g_{110} = \left[\gamma_{110}m^4 + 2\gamma_{112}m^2\left(k^2\beta^2 + n^2\beta^2\right) + \gamma_{114}\left(k^4\beta^4 + 6k^2\beta^2 n^2\beta^2 + n^4\beta^4\right) \right]$$
$$- \frac{h_1\Delta_{06} + h_2\Delta_{08} + h_3\Delta_{07} + h_4\Delta_{05}}{\Delta_{00}}$$

$$g_{120} = -4k\beta n\beta\left[\gamma_{112}m^2 + \gamma_{114}\left(k^2\beta^2 + n^2\beta^2\right) \right] - \frac{h_1\Delta_{08} + h_2\Delta_{06} + h_3\Delta_{05} + h_4\Delta_{07}}{\Delta_{00}}$$

$$g_{200} = 16\left[m^4 + 2\gamma_{212}m^2 k^2\beta^2 + \gamma_{214}k^4\beta^4 \right] + \gamma_{14}\gamma_{24}\frac{64m^2\left(\gamma_{220}m^2 + \gamma_{222}k^2\beta^2\right)^2}{\gamma_{31} + \gamma_{320}4\left(m^2 + k^2\beta^2\right)}$$

$$g_{302} = \frac{\gamma_{244}\left(\gamma_{41} + \gamma_{432}4n^2\beta^2\right) + \gamma_{233}\left(\gamma_{41} - \gamma_{413}4n^2\beta^2\right)}{\gamma_{214}\left(\gamma_{41} + \gamma_{432}4n^2\beta^2\right) + \gamma_{14}\gamma_{24}\gamma_{233}^2 4n^2\beta^2}$$

$$g_{303} = \frac{\gamma_{214}\left(\gamma_{41} - \gamma_{413}4n^2\beta^2\right) - \gamma_{14}\gamma_{24}\gamma_{233}\gamma_{244}4n^2\beta^2}{\gamma_{214}\left(\gamma_{41} + \gamma_{432}4n^2\beta^2\right) + \gamma_{14}\gamma_{24}\gamma_{233}^2 4n^2\beta^2}$$

$$g_{09} = \gamma_{114} + \gamma_{133}g_{303} + \gamma_{14}\gamma_{24}\gamma_{144}g_{302}$$

$$g_{13} = \frac{\gamma_{41} + \gamma_{432}4n^2\beta^2}{\gamma_{214}\left(\gamma_{41} + \gamma_{432}4n^2\beta^2\right) + \gamma_{14}\gamma_{24}\gamma_{233}^2 4n^2\beta^2}$$

$$g_{14} = -\frac{\gamma_{144}\left(\gamma_{41} + \gamma_{432}4n^2\beta^2\right) - \gamma_{133}\gamma_{233}4n^2\beta^2}{\gamma_{214}\left(\gamma_{41} + \gamma_{432}4n^2\beta^2\right) + \gamma_{14}\gamma_{24}\gamma_{233}^2 4n^2\beta^2}$$

$$g_{11} = g_{14}\frac{g_{210}^2 - g_{220}^2}{g_{210}^2}(1 + 2\mu) + 2\frac{g_{210}g_{310} - g_{220}g_{320}}{g_{210}^2}$$

$$g_{23} = \left[m^4 + 2\gamma_{212}m^2\left(9n^2\beta^2 + k^2\beta^2\right) + \gamma_{214}\left(k^4\beta^4 + 54k^2\beta^2 n^2\beta^2 + 81n^4\beta^4\right) \right]$$
$$+ \gamma_{14}\gamma_{24}\frac{d_1\Delta_{131} + c_1\Delta_{132} + a_1\Delta_{133} + b_1\Delta_{134}}{\Delta_{130}}$$

$$g_{24} = 12k\beta n\beta\left[\gamma_{212}m^2 + \gamma_{214}\left(k^2\beta^2 + 9n^2\beta^2\right) \right]$$
$$+ \gamma_{14}\gamma_{24}\frac{d_1\Delta_{132} + c_1\Delta_{131} + a_1\Delta_{134} + b_1\Delta_{133}}{\Delta_{130}}$$

$$\eta_1 = g_{210}\left(g_{23}^2 - g_{24}^2\right) - g_{23}\left(g_{210}^2 - g_{220}^2\right)(1 + \mu)$$

$$\eta_2 = 3g_{220}\left(g_{23}^2 - g_{24}^2\right) - g_{24}\left(g_{210}^2 - g_{220}^2\right)(1 + \mu)$$

$$\eta_3 = \left(g_{23}^2 - g_{24}^2\right) + (g_{210}g_{23} - g_{220}g_{24})(1 + \mu)$$

$$\eta_4 = (g_{220}g_{23} - g_{210}g_{24})(1 + \mu)$$

$$R_1 = \frac{\eta_3 \eta_1 + \eta_4 \eta_2}{\eta_1^2 - \eta_2^2}, \quad R_2 = \frac{\eta_3 \eta_2 + \eta_4 \eta_1}{\eta_1^2 - \eta_2^2},$$

$$C_3 = \left[1 - \frac{g_{210}g_{310} - g_{220}g_{320}}{m^2 g_{210}}\varepsilon\right]\cos\left(\frac{k\pi}{2n}\right)$$

$$\Delta_{00} = \begin{vmatrix} r_{11} & r_{12} & r_{13} & r_{14} \\ r_{21} & r_{22} & r_{23} & r_{24} \\ r_{31} & r_{32} & r_{33} & r_{34} \\ r_{41} & r_{42} & r_{43} & r_{44} \end{vmatrix}, \quad \Delta_{01} = \begin{vmatrix} e_1 & r_{12} & r_{13} & r_{14} \\ e_2 & r_{22} & r_{23} & r_{24} \\ e_3 & r_{32} & r_{33} & r_{34} \\ e_4 & r_{42} & r_{43} & r_{44} \end{vmatrix}, \quad \Delta_{02} = \begin{vmatrix} r_{11} & e_1 & r_{13} & r_{14} \\ r_{21} & e_2 & r_{23} & r_{24} \\ r_{31} & e_3 & r_{33} & r_{34} \\ r_{41} & e_4 & r_{43} & r_{44} \end{vmatrix}$$

$$\Delta_{03} = \begin{vmatrix} r_{11} & r_{12} & e_1 & r_{14} \\ r_{21} & r_{22} & e_2 & r_{24} \\ r_{31} & r_{32} & e_3 & r_{34} \\ r_{41} & r_{42} & e_4 & r_{44} \end{vmatrix}, \quad \Delta_{04} = \begin{vmatrix} r_{11} & r_{12} & r_{13} & e_1 \\ r_{21} & r_{22} & r_{23} & e_2 \\ r_{31} & r_{32} & r_{33} & e_3 \\ r_{41} & r_{42} & r_{43} & e_4 \end{vmatrix}, \quad \Delta_{05} = \begin{vmatrix} f_1 & r_{12} & r_{13} & r_{14} \\ f_2 & r_{22} & r_{23} & r_{24} \\ f_3 & r_{32} & r_{33} & r_{34} \\ f_4 & r_{42} & r_{43} & r_{44} \end{vmatrix}$$

$$\Delta_{06} = \begin{vmatrix} r_{11} & r_{12} & f_1 & r_{14} \\ r_{21} & r_{22} & f_2 & r_{24} \\ r_{31} & r_{32} & f_3 & r_{34} \\ r_{41} & r_{42} & f_4 & r_{44} \end{vmatrix}, \quad \Delta_{07} = \begin{vmatrix} f_2 & r_{12} & r_{13} & r_{14} \\ f_1 & r_{22} & r_{23} & r_{24} \\ f_4 & r_{32} & r_{33} & r_{34} \\ f_3 & r_{42} & r_{43} & r_{44} \end{vmatrix}, \quad \Delta_{08} = \begin{vmatrix} r_{11} & r_{12} & f_2 & r_{14} \\ r_{21} & r_{22} & f_1 & r_{24} \\ r_{31} & r_{32} & f_4 & r_{34} \\ r_{41} & r_{42} & f_3 & r_{44} \end{vmatrix}$$

$$\Delta_{130} = \begin{vmatrix} s_{11} & s_{12} & s_{13} & s_{14} \\ s_{21} & s_{22} & s_{23} & s_{24} \\ s_{31} & s_{32} & s_{33} & s_{34} \\ s_{41} & s_{42} & s_{43} & s_{44} \end{vmatrix}, \quad \Delta_{131} = \begin{vmatrix} a_1 & s_{12} & s_{13} & s_{14} \\ b_1 & s_{22} & s_{23} & s_{24} \\ c_1 & s_{32} & s_{33} & s_{34} \\ d_1 & s_{42} & s_{43} & s_{44} \end{vmatrix}, \quad \Delta_{132} = \begin{vmatrix} b_1 & s_{12} & s_{13} & s_{14} \\ a_1 & s_{22} & s_{23} & s_{24} \\ d_1 & s_{32} & s_{33} & s_{34} \\ c_1 & s_{42} & s_{43} & s_{44} \end{vmatrix}$$

$$\Delta_{133} = \begin{vmatrix} s_{11} & s_{12} & a_1 & s_{14} \\ s_{21} & s_{22} & b_1 & s_{24} \\ s_{31} & s_{32} & c_1 & s_{34} \\ s_{41} & s_{42} & d_1 & s_{44} \end{vmatrix}, \quad \Delta_{134} = \begin{vmatrix} s_{11} & s_{12} & b_1 & s_{14} \\ s_{21} & s_{22} & a_1 & s_{24} \\ s_{31} & s_{32} & d_1 & s_{34} \\ s_{41} & s_{42} & c_1 & s_{44} \end{vmatrix}$$

$$r_{11} = r_{22} = r_{34} = r_{43} = mn\beta\gamma_{331}, \quad r_{12} = r_{21} = r_{33} = r_{44} = -mk\beta\gamma_{331}$$

$$r_{13} = r_{24} = \gamma_{41} + \gamma_{430}m^2 + \gamma_{432}\left(k^2\beta^2 + n^2\beta^2\right)$$

$$r_{14} = r_{23} = -2k\beta n\beta\gamma_{432}, \quad r_{31} = r_{42} = -2k\beta n\beta\gamma_{322}$$

$$r_{32} = r_{41} = \gamma_{31} + \gamma_{320}m^2 + \gamma_{322}\left(k^2\beta^2 + n^2\beta^2\right)$$

$$e_1 = n\beta \left[\gamma_{231} m^2 + \gamma_{233} \left(3k^2\beta^2 + n^2\beta^2 \right) \right]$$

$$e_2 = -k\beta \left[\gamma_{231} m^2 + \gamma_{233} \left(k^2\beta^2 + 3n^2\beta^2 \right) \right]$$

$$e_3 = -2mk\beta n\beta \gamma_{222}, \quad e_4 = m \left[\gamma_{220} m^2 + \gamma_{222} \left(k^2\beta^2 + n^2\beta^2 \right) \right],$$

$$f_1 = -n\beta \left[\gamma_{41} - \gamma_{411} m^2 - \gamma_{413} \left(3k^2\beta^2 + n^2\beta^2 \right) \right]$$

$$f_2 = k\beta \left[\gamma_{41} - \gamma_{411} m^2 - \gamma_{413} \left(k^2\beta^2 + 3n^2\beta^2 \right) \right]$$

$$f_3 = -2mk\beta n\beta \gamma_{312}, \quad f_4 = -m \left[\gamma_{31} - \gamma_{310} m^2 - \gamma_{312} \left(k^2\beta^2 + n^2\beta^2 \right) \right]$$

$$h_1 = n\beta \left[\gamma_{131} m^2 + \gamma_{133} \left(3k^2\beta^2 + n^2\beta^2 \right) \right]$$

$$h_2 = -k\beta \left[\gamma_{131} m^2 + \gamma_{133} \left(k^2\beta^2 + 3n^2\beta^2 \right) \right]$$

$$h_3 = -2mk\beta n\beta \gamma_{122}$$

$$h_4 = m \left[\gamma_{120} m^2 + \gamma_{122} \left(k^2\beta^2 + n^2\beta^2 \right) \right]$$

$$s_{11} = s_{22} = s_{34} = s_{43} = 3mn\beta \gamma_{331}, \quad s_{12} = s_{21} = s_{33} = s_{44} = -3mk\beta \gamma_{331}$$

$$s_{13} = s_{24} = \gamma_{41} + \gamma_{430} m^2 + \gamma_{432} \left(k^2\beta^2 + 9n^2\beta^2 \right), \quad s_{14} = s_{23} = -6k\beta n\beta \gamma_{431}$$

$$s_{31} = s_{42} = -6k\beta n\beta \gamma_{322}, \quad s_{32} = s_{41} = \gamma_{31} + \gamma_{320} m^2 + \gamma_{322} \left(k^2\beta^2 + 9n^2\beta^2 \right)$$

$$a_1 = 3n\beta \left[\gamma_{231} m^2 + \gamma_{233} \left(3k^2\beta^2 + 9n^2\beta^2 \right) \right],$$

$$b_1 = k\beta \left[\gamma_{231} m^2 + \gamma_{233} \left(k^2\beta^2 + 27n^2\beta^2 \right) \right]$$

$$c_1 = 6mk\beta n\beta \gamma_{222}, \quad d_1 = m \left[\gamma_{220} m^2 + \gamma_{222} \left(k^2\beta^2 + 9n^2\beta^2 \right) \right]$$

$$a_{01}^{(5/4)} = 1, \quad a_{10}^{(5/4)} = \frac{\vartheta}{\phi} g_{17}$$

$$b_{01}^{(9/4)} = b_{01}^{(3)} = \gamma_{24} g_{19}, \quad b_{10}^{(9/4)} = b_{10}^{(3)} = \gamma_{24} \frac{\vartheta}{\phi} g_{20}$$

$$b_{11} = \frac{1}{b} \left[\left(a_{10}^{(5/4)} \right)^2 \phi^2 b + a_{10}^{(5/4)} 2\vartheta\phi c + \left(2\vartheta^4 - \vartheta^2\phi^2 + \phi^4 \right) \right] \tag{N.8}$$

Appendix O

In Equations 5.137 and 5.138

$$\Theta_7 = \frac{1}{C_3} \left\{ \gamma_{14}\gamma_{24} \frac{m^4(1+\mu)}{16n^2\beta^2 g_{09}g_{06}} \varepsilon^{-1} - \gamma_{24}\gamma_{14} \frac{m^2 g_{11}}{32n^2\beta^2 g_{09}} \right.$$
$$\left. + \frac{1}{8} \frac{\gamma_5}{g_8} \left[m^2(1+2\mu)\varepsilon - 2g_{05}\varepsilon^2 + \frac{g_{05}^2}{m^2} \varepsilon^3 \right] + \frac{\gamma_{24}^2 - \gamma_5^2}{\gamma_{24}} \frac{\gamma_{T2}}{g_T} \lambda_T^{(2)} \right\}$$

$$\Theta_8 = \frac{\gamma_{24}^2 - \gamma_5^2}{\gamma_{24}} \frac{\gamma_{T2}}{g_T} \lambda_T^{(0)} \tag{O.1}$$

and

$$\lambda_T^{(0)} = 2\lambda_p^{(0)}$$

$$\lambda_T^{(2)} = 2\lambda_p^{(2)} - \frac{1}{8} \frac{\gamma_{24}}{g_8} \left\{ m^2(1+2\mu)\varepsilon - 2g_{05}\varepsilon^2 + \frac{g_{05}^2}{m^2} \varepsilon^3 \right\}$$

$$\lambda_T^{(4)} = 2\lambda_p^{(4)} + \frac{1}{64} \frac{\gamma_{24}}{g_8} \left\{ \frac{b_{11}}{32\pi\vartheta} \gamma_{14}^2\gamma_{24}^2 \frac{m^8(1+\mu)^2}{n^4\beta^4 g_{09}^2 g_{06}^2} \varepsilon^{-3/2} \right.$$

$$\left. + m^2 n^4 \beta^4 (1+\mu)^2 \varepsilon^3 \left[\frac{g_{06}(1+2\mu) + 8m^4(1+\mu)g_{10}}{g_{06}(1+\mu) - 4m^4 g_{10}} \right]^2 \right\} \tag{O.2}$$

where $\lambda_p^{(j)}$ ($j = 0, 2, 4$) are defined by Equation L.1. In the above equations (with other symbols are defined by Equations L.2 and L.3)

$$C_{11} = \frac{g_8}{g_T}, \quad g_8 = \gamma_{24}^2 - \frac{2}{\pi} \frac{\gamma_5^2}{\gamma_{24}} \left(\vartheta b_{01}^{(2)} - \phi b_{10}^{(2)} \right) \varepsilon^{1/2}$$

$$g_T = (\gamma_{24}^2 \gamma_{T1} - \gamma_5 \gamma_{T2}) + \frac{2}{\pi} \frac{\gamma_5}{\gamma_{24}} \left(\vartheta b_{01}^{(2)} - \varphi b_{10}^{(2)} \right) (\gamma_{T2} - \gamma_5 \gamma_{T1}) \varepsilon^{1/2} \tag{O.3}$$

Index